FRUCTOSE POLYMERS IN PLANTS AND MICRO-ORGANISMS

FRUCTOSE POLYMERS IN PLANTS AND MICRO-ORGANISMS

edited by

J. F. Farrar and C. J. Pollock

School of Biological Sciences,
University of Wales Bangor,
Bangor, Gwynedd LL57 2UW, UK

and

Institute of Grassland and Environmental Research,
Plas Gogerddan, Aberystwyth,
Dyfed SY23 3EB, UK

Proceedings of the Third International Symposium on Fructan,
Logan, Utah, USA, 21–24 July 1996;
reprinted from
The New Phytologist **136**, 1997

Published for the New Phytologist Trust by
Cambridge University Press

CAMBRIDGE
UNIVERSITY PRESS

Cambridge University Press

The Edinburgh Building, Cambridge CB2 2RU, United Kingdom
40 West 20th Street, New York, NY 10011–4211, USA
10 Stamford Road, Oakleigh, Melbourne, 3166, Australia

Printed in the United Kingdom by the University Press, Cambridge

PREFACE

The papers in this book were submitted at the Third International Fructan Symposium held at the Utah State University campus in Logan, USA, 21–24 July, 1996. As with the previous symposia, all aspects of the physiology, biochemistry, chemistry and molecular biology of fructans in plants and micro-organisms were covered in a series of reviews and research presentations. On this occasion, however, only research papers were considered for publication by *The New Phytologist*, resulting in a volume which gives a comprehensive and detailed account of current research in this area.

The four years which had passed since the second symposium had seen a number of significant achievements in fructan biology, including the cloning and sequencing of a gene from higher plants coding a kestose-sucrose fructosyl transferase and the demonstration of the synthesis *in vitro* by purified enzymes of the full range of fructans observed *in vivo* in tubers of *Helianthus tuberosus*. There was, therefore, considerable discussion of fructan enzymology, of the unusual properties of plant fructosyl transferases and of the additional complexities involved in fructan synthesis in species where more than one homologous series occurs. There was also considerable interest in the properties of transgenic plants containing bacterial fructan genes, where existing patterns of resource allocation were significantly disrupted. Such plants obviously show potential as research tools but can also produce novel carbohydrate structures.

Physiological studies covered a range of issues, including the interactions between fructans and abiotic stresses and the pivotal role of sucrose in regulating the induction of fructan synthesis. An increasing range of plant species is being studied, and useful comparisons at the structural and biochemical level are resulting. There is also an increasing appreciation of the potential utility of fructans as foodstuffs and as raw materials, and a number of presentations dealt with the challenges of commercial fructan production. The overall impression, borne out by the range of papers presented here, is of a rapidly changing area of research where there has been a number of major recent advances and which is contributing significantly to our understanding of plant primary metabolism. The chairman of the organizing committee, Dr N. J. Chatterton, deserves the thanks of all the participants for hosting such an interesting and stimulating conference and for supporting the publication of this book.

Editors C. J. Pollock and J. F. Farrar

New Phytol. (1997), **136**, 3–10

Fructan oligomers in *Poa ampla*

BY N. J. CHATTERTON* AND P. A. HARRISON

USDA-Agricultural Research Service, Forage and Range Research Laboratory, Utah State University, Logan, UT 84322-6300, USA

(*Received 24 July 1996*)

SUMMARY

Ion exchange chromatography of water extracts was used to classify the fructans in leaves of 40 temperate dicots and 110 temperate grass species. Of the 150 species studied, only *Poa ampla* (Merrill) contained a single β-2,6 linked fructan series. All other species grown at 10/5 °C day/night temperatures (16 h day) contained some β-1,2 linked oligofructans, even when the major fructan series was β-2,6 linked. Leaves of *P. ampla* plants grown at 15/10 °C day/night temperatures contained some β-1,2 linked fructans and raffinose in addition to the dominant β-2,6 series, but did not contain significant amounts of 1-kestose or bifurcose (1&6-kestotetraose). Leaves of *Poa arctica* R.Br. and *Poa bulbosa* L. contained significant amounts of raffinose and various β-1,2 linked oligomers but no 1-kestose. *In vivo* synthesis of 6-kestose in leaves of *P. ampla* grown at 10/5 °C proceeds in the absence of significant amounts of either 1-kestose or bifurcose. Thus *P. ampla* probably has a specific 6-SST (6-sucrose-sucrosefructosyl transferase) and might provide a ready source of β-2,6 linked grass fructan.

Key words: Fructan, cold temperature, levan, 6-SST, oligomers.

INTRODUCTION

In many species of Asteraceae, for example in *Helianthus tuberosus* L. (Jerusalem artichoke), *Cichorium intybus* L. (chicory), and *Taraxacum officinale* Weber (dandelion), fructans are mostly a homologous series of linear β-2,1 linked fructose polymer with a terminal sucrosyl moiety (GF_n, $1 < n < 50$) (Bonnett *et al.*, 1994; Koops & Jonker, 1994; Ernst, Chatterton & Harrison, 1995). *Gomphrena macrocephala* St.-Hil (Amaranthaceae) might contain fructans of a single homologous series that co-chromatograph with leven, a β-2,6 linked fructan polymer (Vieira & Figueiredo-Ribeiro, 1993). Most temperate grasses contain some combination of fructan types based on 1-kestotriose (1-kestose), 6-kestotriose (6-kestose), 1&6-kestotetraose (bifurcose) and/or 6_G-kestotriose (neokestose) but many appear to contain a β-2,6 based polymer series (Chatterton *et al.*, 1990; Bancal, Carpita & Guadillere, 1992; Chatterton *et al.*, 1993*a*, *b*: Livingston, Chatterton & Harrison, 1993).

* To whom correspondence should be addressed.
E-mail: njchatt@cc.usu.edu

Cooperative investigations of the USDA Agricultural Research Service and the Utah Agricultural Experiment Station, Logan, Utah 84322. Approved as journal paper no. 4805.

Mention of a trademark, proprietary product or vendor does not constitute a guarantee or warranty of the product by the US Department of Agriculture or Utah State University, and does not imply its approval to the exclusion of other products or vendors that may also be suitable.

The preponderance of β-2,6 linked polymers in many grasses might imply a pivotal role for either 6-kestotriose or 1&6-kestotetraose in biosynthesis of grass fructan. Bancal *et al.* (1992) proposed that bifurcose might be an intermediate in the synthesis of 6-kestose through an elongation-trimming pathway. Simmen *et al.*, (1993) and Penson & Cairns (1994) demonstrated that 6-kestose is synthesized in barley and wheat directly from sucrose without other intermediates or the catabolism of bifurcose. 6-Kestose accumulated early with 1-kestose and bifurcose in cell-free preparations using sucrose as the substrate. By contrast, *in vivo* expression of a 6-fructosyl transferase in wheat occurs later (Bancal *et al.*, 1992).

Variations in the amount of DP (degree of polymerization) three and DP four fructan oligomers in grass leaves indicate the complexity of the enzymes involved in fructan synthesis or the effects of environmental conditions (Cairns & Pollock, 1988; Bancal *et al.*, 1991; Sims, Pollock & Horgan, 1992; Livingston *et al.*, 1993). Some grass species such as orchardgrass and cheatgrass (Chatterton *et al.*, 1993*a*, *b*) have a single dominant fructan series, but also synthesize multiple DP 3 and DP 4 oligomers. By contrast, Jerusalem artichoke synthesizes a simple β-2,1 linked fructan series (inulin) and contains a single DP 3 oligomer (1-kestotriose), so Jerusalem artichoke is a comparatively simple system for studying the enzymes involved in synthesis of β-2,1 linked fructan.

Study of the enzymology of β-2,6 linked fructan synthesis in temperate grasses that synthesize multiple DP 3 and DP 4 fructans and multiple fructan series would be facilitated by identifying grass species that synthesize only one DP 3 (6-kestose) and only one DP 4 (6,6-kestotetraose) fructan. The existence of a plant that synthesizes 6-kestose but lacks the enzymes required to synthesize 1-kestose, neokestose, 1-nystose, and bifurcose would indicate that 6-kestose synthesis *in vivo* is independent of 1-kestose and bifurcose synthesis. Furthermore, such a plant would be a possible source for enzymes involved in the synthesis of β-2,6 linked fructans, a source free from 1-SST and the other enzymes involved in the synthesis of the 1-kestose- and neokestose-based fructans series.

We characterized fructan oligomer types in approx. 150 species of temperate dicots (40) and grasses (110) in an attempt to identify a plant species that contains single DP 3 and DP 4 fructan oligomers and a single series of β-2,6 linked fructans. Fructan-accumulating grasses used in the survey included one or more species from the following genera: *Agropyron, Agrostis, Alopecurus, Avena, Bromus, Calamagrostis, Critesion, Dactylis, Deschampsia, Elymus, Elytrigia, Festuca, Hordeum, Leymus, Lolium, Pascopyrum, Phalaris, Poa, Phleum, Polypogon, Psathyrostachys, Pseudorogneria, Secale, Sitanion, Stipa, Thinopyrum* and *Triticum*. An accession of *Poa ampla* (Merrill), big bluegrass (PI434231) was collected in Wyoming, USA as *Poa arida* [Vasey]). We describe the fructans present in *Poa ampla* and compare its fructan chromatographic profile with those of *Dactylis glomerata* L. (orchardgrass) and two other Poa species – *P. arctica* R. Br. (rhizomatous) and *P. bulbosa* L. (non-rhizomatous).

MATERIALS AND METHODS

Plant materials and extractions

Foliage of vegetative plants was harvested from plants grown in a controlled-environment chamber (10/5 °C day/night or 15/10 °C day/night and a 16 h photoperiod) following a minimum of 28 d regrowth, or from field-grown plants harvested in early spring. Tissues were frozen in liquid nitrogen and lyophilized. Leaves, leaf sheaths, and rudimentary stems (here collectively referred to as leaves) constituted each grass sample. Lyophilized tissues were ground through 0·5 mm screen in a Tecator® cyclone mill and stored in airtight plastic containers at −20 °C.

Carbohydrates were purified by extracting 1 g tissue samples in 50 ml centrifuge tubes. Deionized water (40 ml) was added to the samples which were immediately placed in a water bath at 95 °C for 20 min. Thirty-five ml of water was used for the second and third extractions. After each extraction, samples were centrifuged for 20 min at 15000 ***g***. Supernatants from the three extractions were pooled. Larger samples were extracted for the purification of individual fructans. Six 10 g samples were placed in separate 250 ml centrifuge bottles and extracted three times in *c.* 240 ml boiling water. Bottles containing the boiling water and samples were placed immediately in a water bath and incubated at 90–95 °C for 20 min, then centrifuged for 20 min at 13000 ***g***. Supernatants from the three extractions were pooled. Extracts of both sample sizes were treated similarly after centrifugation except for differences due to sample size. Before concentration by flash evaporation, the pH of each extract was measured and adjusted to neutrality when necessary. If the pH of an extract is acidic, the pH will further decrease with concentration and fructans might be hydrolysed. Phleins might hydrolyse more readily than do inulins.

Removal of protein

Protein in leaf extracts was precipitated by adding 20 μl of 10 % (w/v) lead acetate per ml of extract and heating to 50–60 °C in a water-bath, and removed by centrifugation (20 min at 13000 ***g***). Additional protein was removed by freezing, thawing, and re-centrifugation. Samples used to determine the full range of fructan sizes were refrigerated but not frozen (freezing precipitates large fructans, sometimes irreversibly). Inulins appear to precipitate at cold temperatures more readily than do phleins. Frozen samples were thawed, then warmed (60–70 °C for 15 min) to dissolve larger fructans. Samples were centrifuged as described above to remove any precipitate.

Deionization of extract

Samples were deionized by gradual addition of ion-exchange resins with constant agitation on a magnetic stirrer and continuous monitoring of pH. Anion-exchange resin (Dowex-1, Sigma 1X4-200) was converted to the hydroxyl form by two 30-min washes with 2 M NH_4OH and then rinsed with deionized water until the pH of the eluant was < 7. Cation exchange resin (Dowex-50W) was converted to the chloride form by two 30 min washes with 2 M HCl, followed by rinsing with deionized water until the pH of the rinse water was *c.* 5·5. Dowex-1 was added, while the sample was being stirred, until the pH of the extract was *c.* 9, then Dowex-50W was added until the pH was *c.* 6. This cycle was repeated until the volume of resin (after brief settling) was *c.* 25–30 % of the total extract volume. We carefully monitored the pH of the extract while adding ion-exchange resins, because the addition of too much cation-exchange resin can lower the pH enough to hydrolyse fructans. Carbohydrates did not appear to bind to the ion-exchange resins within 15–20 min. The pH of the extract was brought to neutrality by addition of the appropriate resin; all resin was removed by centrifugation. After decanting, the pellet was washed with deionized water and re-centrifuged to recover additional carbohydrate. The two supernatants were combined and concentrated by flash evaporation.

Solid-phase extraction

Solid-phase extraction of the concentrated extracts was carried out using C_{18} or tC_{18} Sep-Pack Cartridges from Waters Chromatography Division of Millipore Corp. Cartridges were activated with 10 column-volumes of methanol followed by 10 volumes of deionized water. Carbohydrates applied to the cartridges eluted in water without significant retention, and most if not all coloured material remained behind. Some samples required a second cycle of ion-exchange and C_{18} cartridges if they remained strongly coloured following the first cycle. After the carbohydrates were eluted from the Sep-Pak cartridge, samples were neutralized with either NH_4OH or HCl and then concentrated. Extracted carbohydrates were stored at −20 °C until analysed.

Removal of large fructans

Large (DP > 15) fructans were precipitated from oligosaccharides by adding ethanol to the hot water extracts until the ethanol concentration was 60–70 %. The resultant ethanol mixture was stirred with a spatula and centrifuged as above. Supernatants were decanted and flash-evaporated to remove the ethanol and to concentrate. Pellets were re-dissolved in deionized water and brought to their original volume. Both the supernatant (containing fructan up to about DP 20) and the re-dissolved pellet were subjected to a second ethanol precipitation. After flash evaporation or re-dissolving of the pellet, the respective fractions were combined. A second precipitation improved the fractionation of large from small DP fructans. If a complete fructan extract was desired, ethanol precipitation was omitted.

Purification and analysis of individual fructans

Individual fructan oligosaccharides were separated by gel filtration, using a Toyopearl HW-40S TSK gel column (4·8 × 240 cm) (Chatterton *et al.*, 1993*a*, *b*). Carbohydrates eluted from the Toyopearl column were pooled by size to accumulate fructans into predominately single DP fractions. Partially purified fractions were re-chromatographed until each sample contained a single DP as determined by anion exchange chromatography and pulsed amperometric detection (AE-PAD) (Chatterton *et al.*, 1993*a*). Fructans were analysed on a Dionex Carbo Pac 100 column (4 × 250 mm) using a sodium acetate gradient in 150 mM NaOH (1 cm^3 min^{-1}). The elution programme consisted of 25 mM sodium acetate (0–1 min), 25 to 400 mM (1–20 min), 500 mM (20–22 min), and 25 mM (22–30 min).

Fructan structures

Fructan structures (DP 3–DP 8) were deduced from their retention times on anion exchange chromatography by comparison with known standards (Chatterton *et al.*, 1993*a*). Degree of polymerization for each fructan oligomer was established from the glucose to fructose ratio following complete hydrolysis, assuming one glucose residue per molecule. Oligomers were further characterized by determining the structures of their partial hydrolysates (Chatterton *et al.*, 1993*a*). We characterized the fructans present in *P. ampla* and compared their chromatographic profile with those of the other species.

RESULTS

Water-soluble carbohydrates in leaf extracts of all but one species examined contained either β-2,1 linked fructans (dicots) or a combination of β-2,1 and β-2,6 linked fructans (grasses). The exception was *P. ampla* grown at 10/5 °C day/night, which contained a single fructan series and only one oligomer per DP (Fig. 1). Leaf extracts of *P. ampla* (10/5 °C) contained no measurable 1-kestose, neokestose or nystose. Partial hydrolysis of the oligomers DP ⩾ 4 yielded the next smallest fructan in the series plus fructose. Retention times of each oligomer up to DP 12 were compared with standards previously isolated from *Bromus tectorum* and *Dactylis glomerata* (Chatterton *et al.*, 1993*a*, *b*). Fructan oligomers from *P. ampla* co-eluted with the linear β-2,6 linked series found in *D. glomerata* (Fig. 2) as is evident in the chromatogram of a 50–50 mixture of extracts from *P. ampla* and *D. glomerata* (Fig. 3). The peak heights of oligomers common to both species are means of those occurring in the two species when run singly. Oligomers present in only one of the two species are represented by peaks about half their original heights.

All fructans in *P. ampla* leaves grown at 10/5 °C belong to the linear β-2,6 series that includes 6-kestotriose, 6,6-kestotetraose, 6,6,6-kestopentaose and 6,6,6,6-kestohexaose (Chatterton *et al.*, 1993*b*). *P. ampla* (10/5 °C) contained only a small amount of raffinose (Fig. 1). The presence of only one fructan of DP 3 (6-kestose) and one of DP 4 (6,6-kestotetraose) and the absence from *P. ampla* of the fructans of DP 3 (1-kestose and neokestose) and of

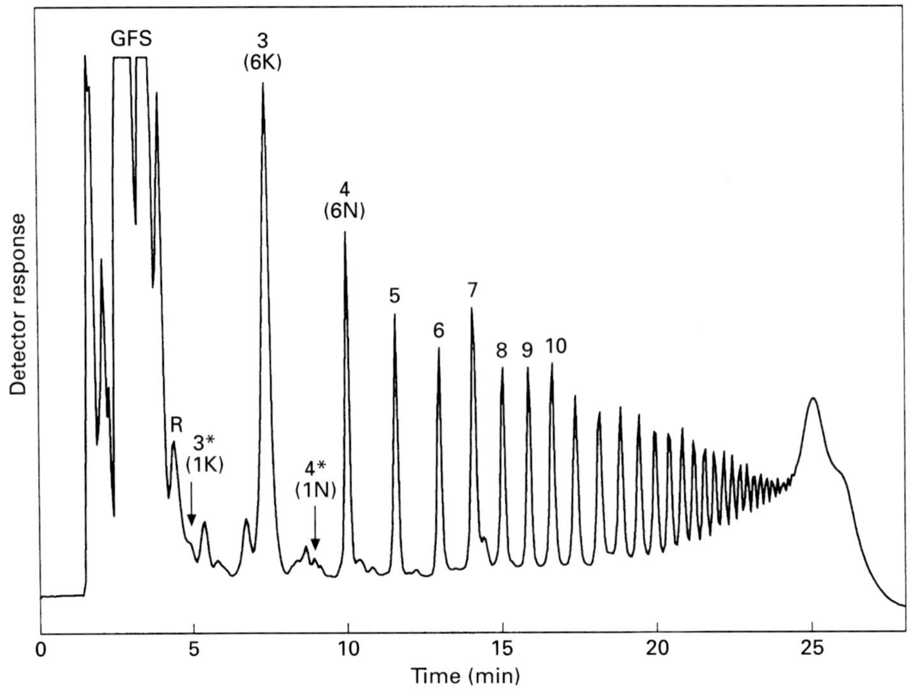

Figure 1. Chromatogram of water-soluble carbohydrates in leaf extracts of *Poa ampla* grown at 10/5 °C day/night temperatures and 16 h photoperiod. G, glucose; F, fructose; S, sucrose; R, raffinose, 3*(1 K) = 1-kestose (β-2,1 linked series absent), 3(6 K) = 6-kestose (β-2,6 series), 4*(1N) = nystose (β2,1 series), 4(6N) = 6,6-kestotetraose (β-2,6 series), 5–10 = DP 5–10 (β-2,6 series).

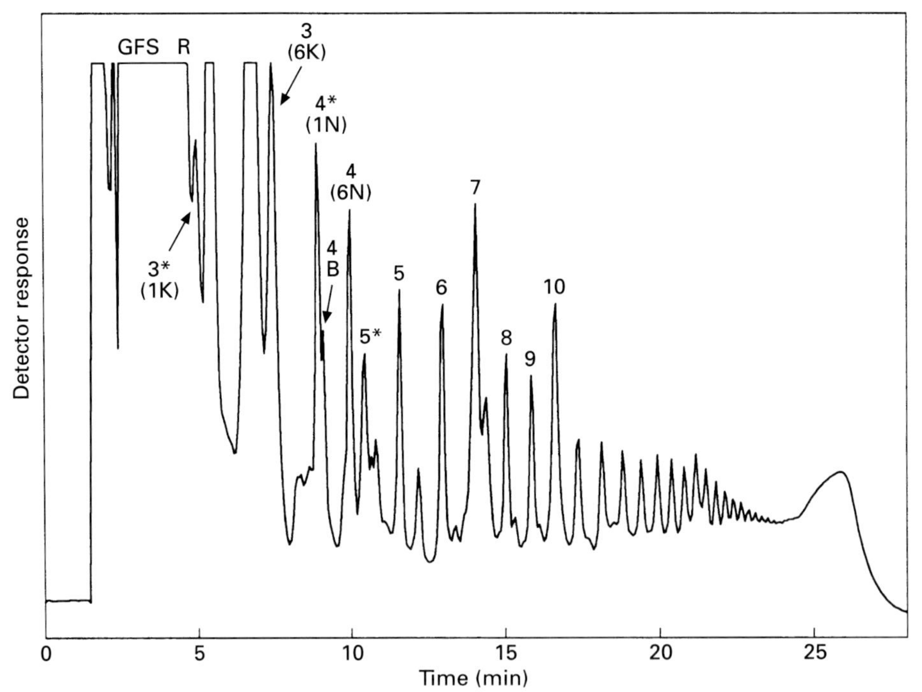

Figure 2. Chromatogram of water-soluble carbohydrates in leaf extracts of *Dactylis glomerata* grown at 10/5 °C day/night temperatures and 16 h photoperiod. G, glucose; F, fructose; S, sucrose; R, raffinose, 3*(1 K) = 1-kestose (β-2,1 linked series), 3(6 K) = 6-kestose (β-2,6 series), 4*(1N) = nystose (β-2,1 series), 4B = bifurcose (1 & 6-kestotetraose), 4(6N) = 6,6-kestotetraose (β-2,6 series), 5* = 1,1,1-kestopentaose (β-2,1 series), 5–10 = DP 5–10 (β-2,6 series).

DP 4 1-nystose and bifurcose (10/5 °C) was unusual. *P. arctica* (Fig. 4) and *P. bulbosa* (Fig. 5) contained varying amounts of 1,1-kestotetraose and 1,1,1-kestopentaose in addition to the linear β-2,6 linked series, but no measurable 1-kestose. *P. arctica* (Fig. 4) and *P. ampla* grown at 15/10 °C (Fig. 6) contained an unknown carbohydrate that eluted just before DP 7 (6,6,6,6,6-kestoheptaose) of the β-2,6 series.

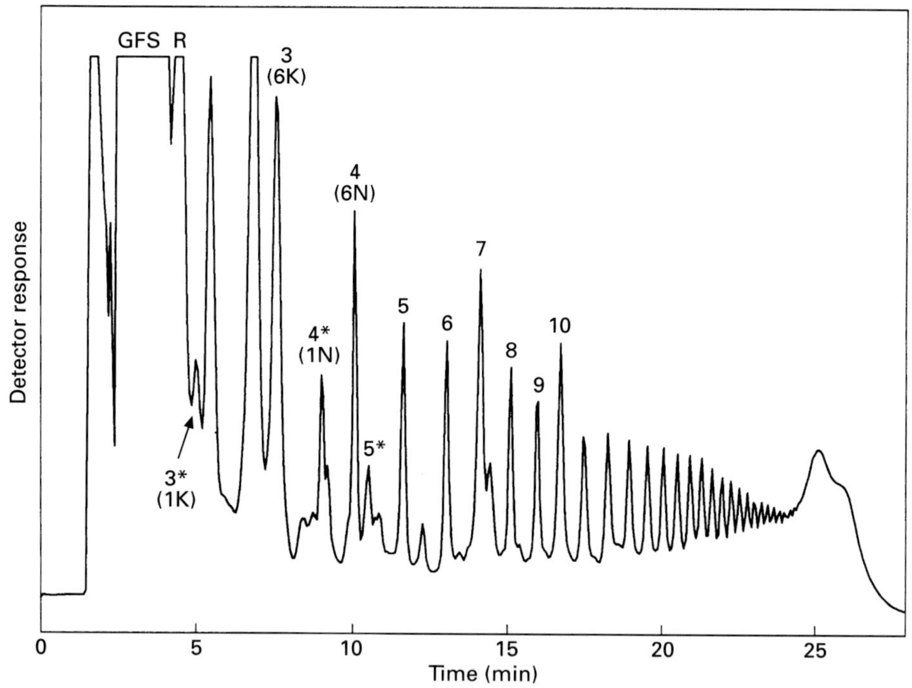

Figure 3. Chromatogram of water-soluble carbohydrates in leaf extracts of *Dactylis glomerata* and *Poa ampla* (50%/50% mix) grown at 10/5 °C day/night temperatures and 16 h photoperiod. G, glucose; F, fructose; S, sucrose; R, raffinose, 3*(1 K) = 1-kestose (β-2,1 linked series), 3(6 K) = 6-kestose (β-2,6 linked series), 4*(1N) = nystose (β-2,1 series), 4(6N) = 6,6-kestotetraose (β-2,6 series), 5* = 1,1,1-kestopentaose 5–10 = DP 5–10 (β-2,6 series).

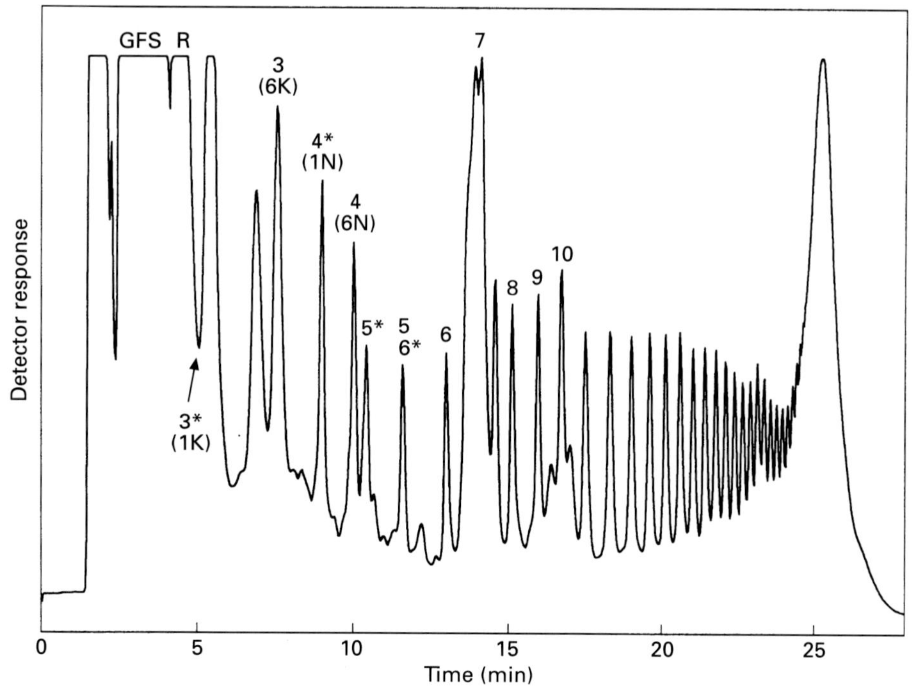

Figure 4. Chromatogram of water-soluble carbohydrates in leaf extracts of *Poa arctica* grown at 10/5 °C day/night temperatures and 16 h photoperiod. G, glucose; F, fructose; S, sucrose; R, raffinose, 3*(1 K) = 1-kestose (β-2,1 series absent?), 3(6 K) = 6-kestose (β-2,6 linked series), 4*(1N) = nystose (β-2,1 series), 4(6N) = 6,6-kestotetraose (β-2,6 series), 5* = 1,1,1-kestopentaose, 6* = 1,1,1,1-kestohexaose, 5–10 = DP 5–10 (β-2,6 series).

P. ampla harvested after growth at 15/10 °C day/night temperatures contained oligomers belonging to multiple fructan series (Fig. 6). The fructan profiles of *P. ampla*, *P. arctica* and *P. bulbosa* were similar when plants were grown at warm temperatures. In addition to the β-2,6 linked series

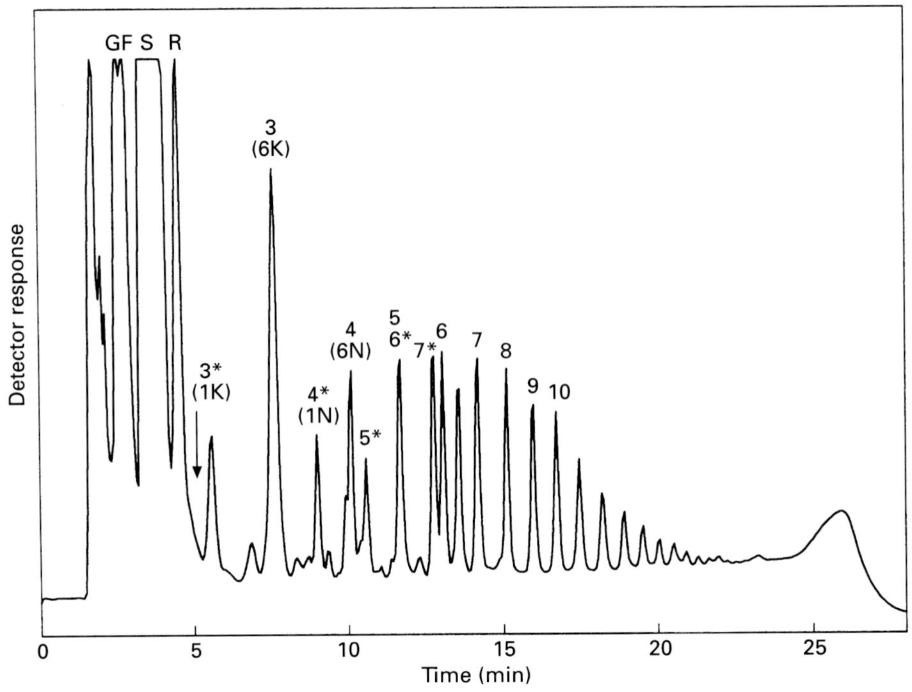

Figure 5. Chromatogram of water-soluble carbohydrates in leaf extracts of *Poa bulbosa* grown at 10/5 °C day/night temperatures and 16 h photoperiod. G, glucose; F, fructose; S, sucrose; R, raffinose, 3*(1 K) = 1-kestose (β-2,1 series-absent), 3(6 K) = 6-kestose (β-2,6 series), 4*(1N) = nystose (β-2,1 series), 4(6N) = 6,6-kestotetraose (β-2,6 series), 5* = 1,1,1-kestopentaose, 6* = 1,1,1,1-kestohexaose, 7* = 1,1,1,1,1-kestoheptaose, 5–10 = DP 5–10 (β-2,6 series).

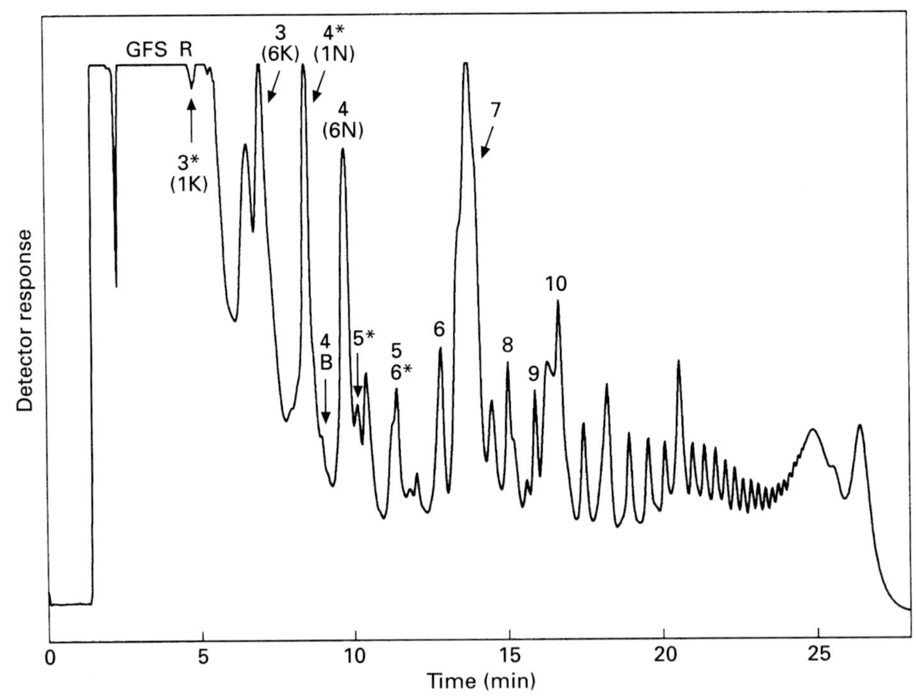

Figure 6. Chromatogram of water-soluble carbohydrates in leaf extracts of *Poa ampla* grown at 15/10 °C day/night temperatures and 16 h photoperiod. G, Glucose; F, fructose; S, sucrose; R, raffinose, 3*(1 K) = 1-kestose (β-2,1 linked series absent), 3(6 K) = 6-kestose (β-2,6 series), 4*(1N) = nystose (β-2,1 series), 4(B) = 1 & 6-kestotetraose (bifurcose), 4(6N) = 6,6-kestotetraose (β-2,6 series), 5* = 1,1,1-kestopentaose, 6* = 1,1,1,1-kestohexaose, 5–10 = DP 5–10 (β-2,6 series).

present in plants grown at 10/5 °C, some β-2,1 linked inulins were also found, including 1-nystose and larger linear inulins. However, no measurable 1-kestose, the logical substrate for nystose synthesis, was found in any of the three species of *Poa* at either growth temperature. It is possible that the presence of very small quantities of 1-kestose might have been masked by large neighbouring peaks (Figs 1, 5).

Although *P. arctica* (rhizomatous), *P. bulbosa* (non-rhizomatous) and *P. ampla* grown at 15/10 °C had β-2,1 linked oligimers, bifurcose, if present, occurred in only trace quantities. By contrast, extracts of *D. glomerata* (Fig. 2), *B. tectorum* (Chatterton *et al.*, 1993*a*) and wheat (Bancal & Gandillere, 1989) contain significant amounts of bifurcose.

The presence of raffinose in the leaf extracts of each species of *Poa* is consistent with earlier observations (Chatterton *et al.*, 1990). Raffinose, like the other non-β-2,6 linked fructans, is present in only minimal amounts in *P. ampla* grown at 10/5 °C (Fig. 1). Other members of the raffinose series (e.g. stachyose, verbascose) were not present. The peak that eluted just after sucrose but before raffinose (Fig. 1) is unidentified. In some other instances, the relative concentrations precluded complete resolution of raffinose from sucrose (Figs 2–6).

DISCUSSION

The presence of a 'clean,' simple, single β-2,6 linked series in cold-grown (10/5 °C day/night) *P. ampla*, is the first example of an extract of a grass containing no inulin (1-kestose and nystose), neokestose or bifurcose-based fructans. *P. ampla* was the only species in the 110 temperate grasses that we evaluated that contained only linear β-2,6 linked oligofructosides. However, smaller amounts of β-2,1 linked fructans (including 1-kestose and nystose) and others of unknown structure did occur in leaves of *P. ampla* grown at warmer (15/10 °C) day/night temperatures. Because no bifurcose was detected (if present it was in very minute quantities) in any of the *Poa* species grown in either warm or cold temperatures, it is unlikely that the elongation-trimming pathway (ETP), a specific β-2,1 hydrolysis of bifurcose, is the mechanism for 6-kestose synthesis in *Poa*, as proposed for wheat (Bancal *et al.*, 1992). Penson & Cairns (1994) could not substantiate the ETP hypothesis in wheat. Instead, they and Simmen *et al.* (1993) believe 6-kestose is synthesized anabolically from sucrose in cell-free enzyme preparations. Our studies (unpublished) of cell-free enzyme preparations from *P. ampla*, using sucrose as a substrate, show an early synthesis of 6-kestose, small but measurable amounts of 1-kestose, and trace amounts of bifurcose.

1-Kestose-synthesizing activity (by invertase?) was present in cell-free extracts of *P. ampla* grown at 10/5 °C (unpublished data). The accumulation of nystose (the synthesis of which is thought to require 1-kestose) was observed in plants grown at 15/10 °C, even though significant levels of 1-kestose were not detected. AE-PAD separation of carbohydrates in a dilution of the *P. ampla* extract (grown at 15/10 °C grown; Fig. 6) demonstrated the absence of 1-kestose (data not shown). Perhaps 1-kestose is synthesized *in vivo* and is so rapidly used as a substrate in nystose synthesis that it never accumulates to measurable quantities. Alternatively, the synthesis of 1-kestose observed in cell-free extracts to which sucrose had been added might have resulted from an ancillary activity of invertase.

Fructan metabolism in leaves of vegetative *P. ampla* was altered by a 5 °C change in the day/night growth temperature. *P. ampla* plants grown in the colder environment contained only β-2,6 linked fructans. Apparently, this seldom occurs in grass plants and could provide researchers with a source of β-2,6 fructan oligomers. Other cool temperatures may result in the accumulation of only β-2,6 fructans in some species; however, we did not test plants at temperatures cooler than 10/5 °C. Raffinose and β-1,2 linked oligomers (except 1-kestose) were present only in *P. ampla* grown at 15/10 °C but occurred in *P. arctica* and *P. bulbosa* grown at both temperatures. We conclude that under closely controlled environmental conditions *Poa ampla* produces a single DP 3 and a single DP 4 fructan and only members of the β-2,6 linked linear fructan series. Our data further support the hypothesis that 6-kestose synthesis occurs in the absence of significant amounts of either 1-kestose or bifurcose, and that a specific 6-SST (sucrose-sucrosefructosyl transferase) is present in *P. ampla*. Additional studies will attempt to purify the enzymes from *P. ampla* involved in β-2,6 linked fructan synthesis.

REFERENCES

Bancal P, Carpita NC, Gaudillère JP. 1992. Differences in fructan accumulated in induced and field-grown wheat plants: an elongation-trimming pathway for their synthesis. *New Phytologist* **120**: 313–321.

Bancal P, Gaudillére JP. 1989. Oligofructan separation and quantification by high performance liquid chromatography. Application to *Asparagus officinalis* and *Triticum aestivum*. *Plant Physiology and Biochemistry* **27**: 745–750.

Bancal P, Henson CA, Gaudillère JP, Carpita NC. 1991. Fructan chemical structure and sensitivity to an exohydrolase. *Carbohydrate Research* **217**: 137–151.

Bonnett GD, Sims IM, John JAS, Simpson RJ. 1994. Purification and characterization of fructans with β-2,1- and β-2, 6-glycosidic linkages suitable for enzyme studies. *New Phytologist* **127**: 261–269.

Cairns AJ, Pollock CJ. 1988. Fructan biosynthesis in excised leaves of *Lolium temulentum* L. I. Chromatographic characterization of oligofructans and their labelling patterns following $^{14}CO_2$ feeding. *New Phytologist* **109**: 399–405.

Carpita NC, Kanabus J, Housley TL. 1989. Linkage structure of fructans and fructan oligomers from *Triticum aestivum* and *Festuca arundinacea* leaves. *Journal of Plant Physiology* **134**: 162–168.

Chatterton NJ, Harrison PA, Thornley WR, Bennett JH. 1993*a*. Structure of fructan oligomers in cheatgrass (*Bromus tectorum* L.). *New Phytologist* **124**: 389–396.

Chatterton NJ, Harrison PA, Thornley WR, Bennett JH. 1993*b*. Structures of fructan oligomers in orchardgrass (*Dactylis glomerata* L.). *Journal of Plant Physiology* **142**: 552–556.

Chatterton NJ, Harrison PA, Thornley WR, Draper EA. 1990. Oligosaccharides in foliage of *Agropyron*, *Bromus*, *Dactylis*, *Festuca*, *Lolium* and *Phleum*. *New Phytologist* **114**: 167–171.

Ernst M, Chatterton NJ, Harrison PA. 1996. Purification and characterization of a new fructan series from species of Asteraceae. *New Phytologist* **132**: 63–66.

Koops AJ, Jonker HH. 1994. Purification and characterization of the enzymes of fructan biosynthesis in tubers of *Helianthus tuberosus* 'Columbia'. *Journal of Experimental Botany* **45**: 1623–1631.

Livingston DP, Chatterton NJ, Harrison PA. 1993. Structure and quantity of fructan oligomers in oats (*Avena* spp). *New Phytologist* **123**: 725–734.

Meier H, Reid JSG. 1982. Reserve polysaccharides other than starch in higher plants. In: Loewus, FA, Tanner, W, eds. *Plant Carbohydrates I, Encyclopedia of Plant Physiology (New Series), Intracellular Carbohydrates.* New York: Springer-Verlag, 418–471.

Penson SP, Cairns AJ. 1994. Fructan biosynthesis in excised leaves of wheat (*Triticum aestivum* L.): a comparison of de novo synthesis *in vivo* and *in vitro*. *New Phytologist* **128**: 395–402.

Simmen U, Obenland D, Boller T, Wiemken A. 1993. Fructan synthesis in excised barley leaves. Identification of two sucrose-sucrose fructosyltransferases induced by light and their separation from constitutive invertases. *Plant Physiology* **101**: 459–468.

Sims IM, Pollock CJ, Horgan R. 1992. Structural analysis of oligomeric fructans from excised leaves of *Lolium temulentum*. *Phytochemistry* **31**: 2989–2992.

Vieira CJ, Figueiredo-Ribeiro RCL. 1993. Fructose-containing carbohydrates in the tuberous root of *Gomphrena macrocephala* St.-Hil. (Amaranthaceae) at different phenological phases. *Plant, Cell and Environment* **16**: 919–928.

New Phytol. (1997), **136**, 11–17

Structural diversity of fructan in relation to the taxonomy of the Poaceae

BY G. D. BONNETT[1]*, I. M. SIMS[2], R. J. SIMPSON[1] AND A. J. CAIRNS[1,2]†

[1] *CSIRO Division of Plant Industry, GPO Box 1600, Canberra, ACT, 2601, Australia*
[2] *Plant Cell Biology Research Centre, School of Botany, The University of Melbourne, Parkville, Victoria, 3052, Australia*

(*Received 24 July 1996*)

SUMMARY

To test the hypothesis that fructan structure can be used as a taxonomic character for the Poaceae we examined the accumulation of the linear, β-2,6-linked fructan series with a terminal glucose residue within the tribes Aveneae and Poeae. Only *Dactylis glomerata* L. (Poeae) has been shown to contain this series, making it unique among fructan structures and a potential taxonomic marker. To this end, 13 members of the tribes Aveneae and Poeae were surveyed for the presence of this fructan series by comparing extracts of water-soluble carbohydrate by TLC with an extract of *D. glomerata*. The fructans of four species were similar to those of *D. glomerata*, as determined by TLC. Further comparison by anion-exchange chromatography and linkage analysis demonstrated that *Phalaris aquatica* L., *Puccinellia stricta* (Hook. f.) C. H. Blom and *Holcus lanatus* L. contained a linear, β-2,6-linked series of fructan with a terminal glucose residue but that *Lagurus ovatus* L., although containing predominantly β-2,6-linked fructan was composed of more than one series of fructan, one with a terminal glucose residue and one with an internal glucose residue. A linear β-2,6-linked series of fructans with a terminal glucose residue was the dominant fructan accumulated by some species of genera belonging to both the Aveneae (*Phalaris* and *Holcus*) and the Poeae (*Dactylis* and *Puccinellia*). As both of these tribes also contain genera that accumulate fructans with an internal glucose residue, e.g. *Avena* and *Lagurus* (Aveneae) and *Lolium* (Poeae), structure of fructan cannot be used to distinguish these tribes. Review of the literature, however, showed that it is possible to separate the supertribes Triticodae and Poodae on the basis of the structure of the fructan that they accumulate.

Key words: Fructan, Poaceae, Aveneae, Poeae.

INTRODUCTION

Fructan, a polymer of fructose based on sucrose, is found in several families of angiosperms (Hendry, 1993). The fructan present in the Asteraceae is a linear series of molecules containing almost exclusively β-(1 → 2)-glycosidic linkages between adjacent fructose residues (Incoll & Bonnett, 1996). In the Poaceae the accumulation of fructan is confined to the subfamily Pooideae, within which a range of fructan structures, including β-2,1- and/or β-2,6-linked fructose residues can be found (e.g. *Bromus tectorum*, Chatterton *et al.*, 1993*a*). The glucose residue can be internal to the molecule (e.g. many of the fructan molecules from *Avena* sp., Livingston, Chatterton & Harrison, 1993) or can be the terminal residue (e.g. fructan from *Triticum aestivum* L., Bancal, Carpita & Gaudillère, 1992). In some species, fructans are branched, and thus possess fructose residues which are linked to three other fructose residues (e.g. *Triticum aestivum*, Bancal *et al.*, 1992).

Although a wide range of fructan sizes and structures is synthesized by grasses, each species appears to accumulate a characteristic suite of fructans. The potential to use presence of fructan and variation in fructan structures as a taxonomic tool for distinguishing members of the Poaceae has been recognized (Smith, 1968; Smouter & Simpson, 1989; Chatterton *et al.*, 1990). However, only the absence of starch in culm bases (indicating storage of fructan or sucrose) and the storage of long or short chain fructans have been used as taxonomic characters (Watson & Dallwitz, 1992). Their division of fructan-accumulating grasses into those which store long or short chain fructans is contentious, as

* To whom correspondence should be addressed at (current address): CSIRO Tropical Agriculture, Davies Laboratory, Private Mail Bag, PO Aitkenvale, Queensland 4814, Australia.

† Permanent address: Environmental Biology Department, Institute of Grassland and Environmental Research, Plas Gogerddan, Aberystwyth, SW23 3EB, UK.

the length of fructan molecules found in a tissue can depend on the stage of development and recent environmental conditions. Fructan structure might be a more reliable taxonomic character by which to distinguish groups within the grass family.

There is some evidence from the literature which indicates that fructan structure might have potential as a taxonomic character; however, the structure of the fructan from only a few species of grass has been elucidated. For instance, species within the supertribe Triticodae characteristically contain fructans with branch points (multiple branching in larger molecules) and a terminal glucose residue, e.g. *Triticum aestivum*, *Secale cereale* L., *Hordeum* sp. (Schlubach and co-workers, summarized in Hirst, 1957 and Meier & Reid, 1982; Bancal *et al.*, 1992), *Elytrigia repens* (L.) Nevski [*Agropyron repens* (L.) P. Beauv.] (Arni & Percival, 1951) and *Bromus tectorum* L. (Chatterton *et al.*, 1993*a*). Much of the fructan accumulated by members of the tribes Aveneae and Poeae that have been studied contain an internal glucose residue (*Avena* sp. Livingston *et al.*, 1993, *Lolium* spp., Sims, Pollock & Horgan, 1992; Bonnett *et al.*, 1994; St. John *et al.*, 1997). The only known exception is *Dactylis glomerata* L. (Poeae), which accumulates a dominant series of linear, exclusively β-2,6-linked fructan with a terminal glucose residue (Chatterton *et al.*, 1993*b*). If the accumulation of this series is limited to a related group of species, this could also be used as a taxonomic marker.

We have screened a number of species from the tribes Aveneae and Poeae for the presence of a fructan series similar to that found in *D. glomerata*. The results are discussed in the context of how fructan structure might be used as a taxonomic character to distinguish species within the Poaceae.

MATERIALS AND METHODS

Plant material

Plants of some of the species within the tribes Aveneae [*Dichelachne crinita* (L. f.) Hook. f., *Holcus lanatus* L., *Lagurus ovatus* L., *Phalaris aquatica* L. cv. Holdfast] and Poeae (*Festuca arundinaceae* Schreb., *Poa ensiformis* Vickery, *Poa labillardieri* Steud., *Poa morrisii* Vickery, *Poa poiformis* (Labill.) Druce, *Poa sieberiana* Sprengel, *Puccinellia stricta* (Hook. f.) C. H. Blom, *Vulpia bromoides* (L.) Gray and *Vulpia myuros* (L.) C. C. Gmelin.) were grown in pots containing a mixture of river loam, peat moss, river sand, vermiculite and perlite that had been composted, limed to pH 6–6·5, fertilized to promote optimum growth, and sterilized. The plants were grown in a glasshouse where the temperature was maintained around 22 °C during the day and 14 °C during the night. Six fully expanded leaf blades of each species were cut from the plants. Three of the leaves were weighed immediately after harvest, frozen in liquid nitrogen and freeze-dried. The other three leaves were placed in individual test tubes with the cut end immersed in water. These leaves were illuminated continuously in a growth cabinet for 23 h at 22 °C to induce fructan synthesis. Induction of fructan synthesis in the system described usually increases the abundance of oligosaccharides. The leaves were then weighed, frozen in liquid nitrogen and freeze-dried.

Screening species of grass for the presence of oligosaccharides similar to those of Dactylis glomerata

The three leaves in each set were pooled and extracted once in 8 ml of 80% (v/v) ethanol at 80 °C, then twice in 8 ml of water at 60 °C. The extracts were pooled and water-soluble carbohydrate (WSC) was measured by the anthrone method (Yemm & Willis, 1954). Aliquots containing 160 μg of WSC were freeze-dried then dissolved in 10 μl water and applied to the origin on a 20×20 cm glass-backed TLC plate precoated with silica gel to a depth of 250 μm (SI250, J. T. Baker, Phillipsburg, NJ, USA). Plates were developed twice in butan-1-ol/propan-2-ol/water 3:12:4 (v/v/v) at room temperature (Kanaya, Chiba & Shimoura, 1978). Compounds were detected by heating to 110 °C after spraying with urea-phosphoric acid (Wise *et al.*, 1955).

Preparative separation of linear, exclusively β-2,1- and exclusively β-2,6-linked, fructan oligosaccharides

β-2,1-Linked oligosaccharides were purified from juice of tubers of *Helianthus tuberosus* L., and β-2,6-linked oligosaccharides were purified from shoots of *D. glomerata*. Plants of *D. glomerata* were grown in gravel and periodically irrigated with a full nutrient solution. Details of the nutrient solution, and of how plants were induced to accumulate fructans by cooling the roots and shoot apex whilst the leaves were at 22 °C and continuously illuminated, have been described previously (Smouter & Simpson, 1991). The shoots were extracted with 80% (v/v) ethanol at 80 °C to obtain an extract enriched in fructan of DP 3–10. Oligosaccharides of different DP (3–10) were separated and purified from the extracts using a combination of gel-filtration columns as previously described (Bonnett *et al.*, 1994). TLC of the individual oligosaccharides was as described above except that the plate used was Kieselgel 60 F_{254}, (Merck, Darmstadt, Germany).

Anion-exchange HPLC

Before anion-exchange HPLC, the fructan extracts were passed through ion-exchange resins [Serdolit CS-2 (H^+) and Serdolit AS-6 ($HCOO^-$), Serva,

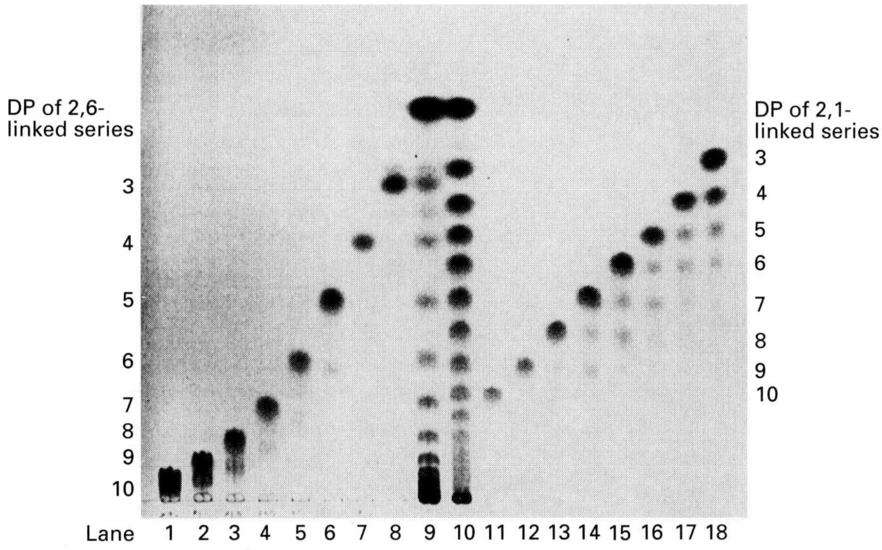

Figure 1. Thin-layer chromatogram of an extract of *Dactylis glomerata* (lane 9), oligosacharides of DP 10–3 from *D. glomerata* purified by gel-filtration (lanes 1–8), extract of *Helianthus tuberosus* (lane 10), oligosaccharides of DP 10–3 from *H. tuberosus* purified by gel-filtration (lanes 11–18).

Heidelberg, Germany] to remove charged compounds, and through C 18 cartridges (Sep-Pak tC18, Millipore, Milford, USA) to remove hydrophobic materials. An extract of *D. glomerata* and oligosaccharides purified from the juice of *H. tuberosus* (by passage through gel filtration, Bio-Gel P4, BioRad, Richmond, USA) were also prepared for anion-exchange HPLC for comparison with the extracts of the grasses. Neutral fructan extracts were analysed on a CarboPac PA-1 column (4 × 250 mm; Dionex, Sunnyvale, CA, USA) using a Dionex BioLC system equipped with a pulsed amperometric detector (Dionex). Fructans were separated in a linear gradient of CH_3COONa in 150 mM NaOH increasing from 30 mM 0–2 min to 1 M after 50 min at a flow rate of 1 ml min^{-1}. The elution times of the trisaccharides were determined from purified standards. To obtain a sample of the fructans in each extract for linkage analysis a bulk sample of all of the fructans was collected after elution from the column and immediately neutralized with acetic acid. Sodium ions were removed on a Dowex 50W-X8, H^+ column at 4 °C (Bio-Rad) and purified carbohydrates were freeze-dried.

Linkage analysis

Samples (50 μg) dissolved in 50 μl dimethylsulphoxide were methylated using NaOH and methyl iodide and converted to partially methylated alditol acetates as described by St. John *et al.* (1997). The derivatives were separated and identified by gas chromatography-mass spectrometry on a fully automated Finnigan MAT 1020B GC-MS (Sunnyvale) equipped with a BPX70 fused silica capillary column (Lau & Bacic, 1993) as described by Bonnett *et al.* (1994).

RESULTS

Analysis of extracts of grasses by TLC and anion-exchange chromatography

Fructans from *D. glomerata*, which form a linear, β-2,6 series of oligosaccharides with a terminal glucose residue, elute with a different pattern from those of *H. tuberosus* which form a linear, β-2,1 series with a terminal glucose residue (Fig. 1). When the two series of oligosaccharides are plotted as log $R_f/(1-R_f)$ against degree of polymerization (DP), as described by French & Wild (1953), two distinct straight lines result (Fig. 2), which intersect at DP 2, sucrose, the compound upon which both series of oligo-

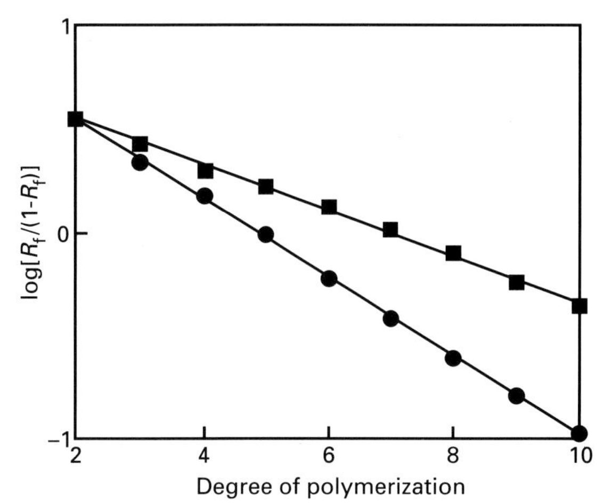

Figure 2. Plot of log $[R_f/(1-R_f)]$ against degree of polymerization for oligosaccharides extracted from *Helianthus tuberosus* (■) and *Dactylis glomerata* (●). R_f values were taken from the positions of each compound after developing in the ascending direction once in butanol:propan-2-ol:water 3:12:4 (v/v/v).

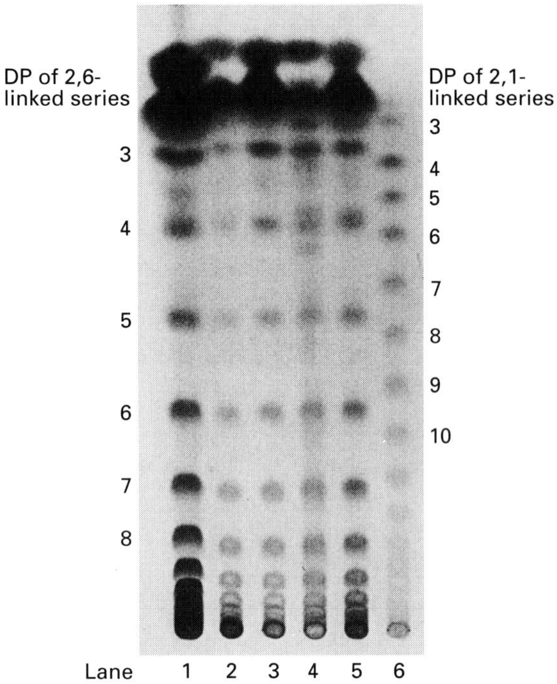

Figure 3. Thin-layer chromatogram of extracts from *Dactylis glomerata* (lane 1), *Phalaris aquatica* (lane 2), *Puccinellia stricta* (lane 3), *Lagurus ovatus* (lane 4), *Holcus lanatus* (lane 5) and *Helianthus tuberosus* (lane 6).

saccharides are synthesized. The distinct pattern of the oligosaccharides from *D. glomerata* on TCL was the basis of the preliminary screen to identify extracts containing the linear, β-2,6-linked series of fructan with a terminal glucose residue.

Of the species examined, extracts from four contained oligosaccharides with mobilities on TLC similar to those in the extract from *D. glomerata*. These were *Phalaris aquatica*, *Holcus lanatus*, *Lagurus ovatus* and *Puccinellia stricta* (Fig. 3). The extracts from *Dichelachne crinita*, *Festuca eliator* (*Festuca arundinacea*), *Vulpia bromoides* and *V. myuros* were more similar to those of *Lolium rigidum* (data not shown), and the extracts from the *Poa* spp. contained high-molecular-mass fructan (present on the origin of the TLC) but did not accumulate enough oligosaccharide to allow comparison with the oligosaccharides from *Dactylis glomerata*.

Further analysis of the extracts that were similar to *D. glomerata* on TLC by anion-exchange HPLC showed that the major trisaccharide present in *D. glomerata*, *Phalaris aquatica*, *Puccinellia stricta* and *Holcus lanatus* was 6-kestose. 1-Kestose was also present though in very small amounts in *H. lanatus* (Fig. 4*b–d*). *Lagurus ovatus* contained 1-kestose, 6-kestose and 6_G-kestose, the latter appearing as a shoulder on the 6-kestose peak (Fig. 4*f*).

The extracts of *D. glomerata*, *H. lanatus* and *L. ovatus* all contained an unidentified peak eluting before 6-kestose at about 9 min (Fig. 4*b*, *e*, *f*). An unidentified compound eluting at the same relative position in an extract from *D. glomerata* was also reported by Chatterton *et al.* (1990). The extract of *H. lanatus* also contained two major unidentified peaks eluting at 5·6 and 6·3 min (Fig. 4*e*).

The elution profiles of *D. glomerata*, *Phalaris aquatica* and *Puccinellia stricta* indicated a similar single series of fructans to be present in all three species. The extract of *H. lanatus* also contained this series but another, minor, series in addition. The elution profile of fructan from *L. ovatus* was more complex than that from *D. glomerata* and the three other species. There appeared to be more than one series of fructan present in *L. ovatus* and the major series did not co-elute with the series from *D. glomerata* although it might have been similar to the minor series present in the extract of *H. lanatus*.

Methylation analysis

When considered together, the anion-exchange elution profiles and the results of the methylation analyses revealed some structural features of the fructans in these extracts. The presence of predominantly terminal glucose, 2,6-linked fructose and terminal fructose in the extracts from *D. glomerata*, *Phalaris aquatica*, *Puccinellia stricta* and *H. lanatus* (Table 1) indicated that these grasses contained predominantly a single, linear series of fructans with a terminal glucose residue and 2,6-linked fructose residues. By contrast, the presence of both terminal glucose and linked (internal) glucose residues in the extract from *L. ovatus* indicated that both fructans with a terminal glucose residue and fructans with an internal glucose residue were present in this species. These two types of glucose residue were probably the basis of the different fructan series shown by anion-exchange HPLC. Both series, however, had predominantly ($6 \rightarrow 2$)-linkages between adjacent fructose residues. The small but significant proportion of linked glucose present in the fructan from *H. lanatus* might have been included in the minor series of fructan present in this species that had similar retention times to the major series found in *L. ovatus*.

DISCUSSION

Extracts from the five species *D. glomerata*, *H. lanatus*, *Puccinellia stricta*, *Phalaris aquatica* and *L. ovatus* had fructans that co-chromatograph on TLC, but the predominant fructan series from *L. ovatus* did not co-elute on anion-exchange HPLC with the fructan series from the other species. These observations are an unequivocal demonstration that caution must be employed when interpreting the DP and structure of particular fructans on the basis of their chromatographic position on TLC alone. It also demonstrates that, at least in some chromatography systems different series of oligosaccharides

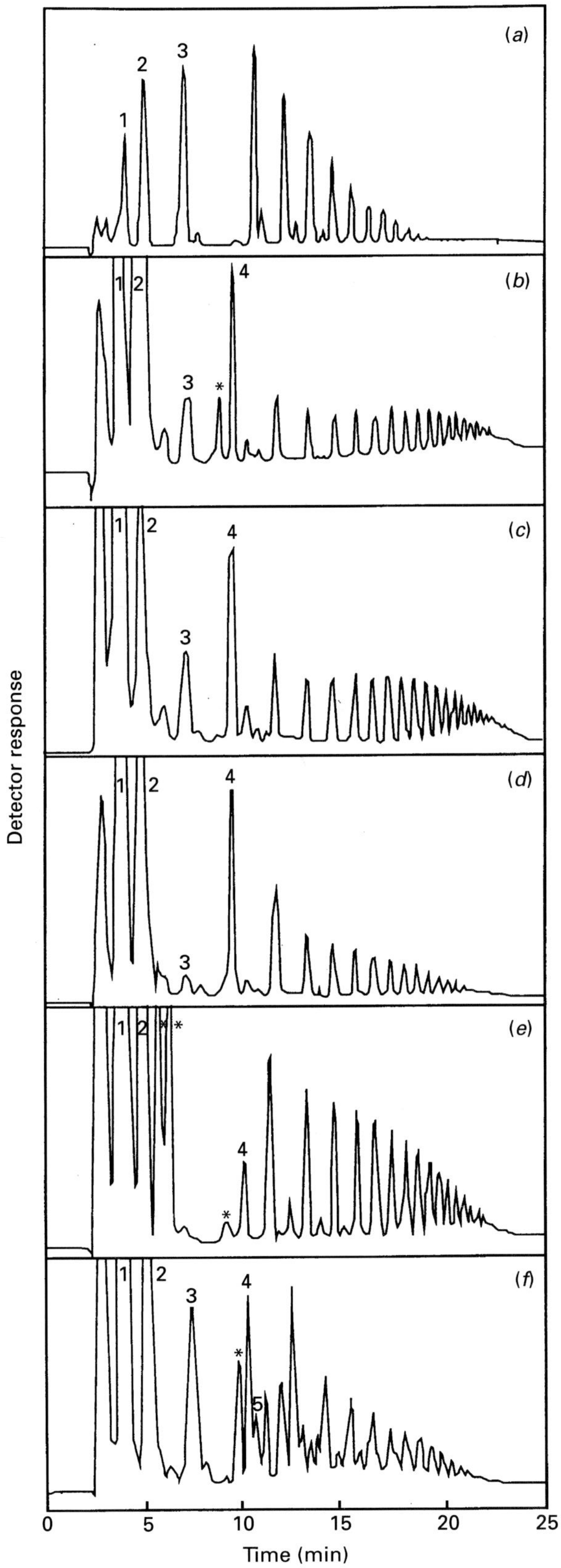

Figure 4. Separation by analytical anion-exchange chromatography of carbohydrates extracted from (*a*) *Helianthus tuberosus*, (*b*) *Dactylis glomerata*, (*c*) *Phalaris aquatica*, (*d*) *Puccinellia stricta*, (*e*) *Holcus lanatus*, (*f*) *Lagurus ovatus*. Carbohydrates were separated on a CarbopacPA-1 column in a gradient of 30–1000 mM CH_3COONa (in 150 mM NaOH) over 48 min. Separated compounds were detected by pulsed amperometric detection; monosaccharides (1), sucrose (2), 1-kestose (3), 6-kestose (4), 6_G-kestose (5) and unidentified (*).

cannot be distinguished by plots of log $R_f/(1-R_f)$ against DP. Anion-exchange HPLC can better resolve compounds of different structure with the same DP than can TLC, and can detect compounds present in amounts low compared with those of the major compounds. This makes anion-exchange chromatography the best method available for separating complex mixtures of fructans, although the possibility remains that some compounds may still co-elute. Analysis of the linkages of extracts of fructan in the present experiments indicates that at least major components were not co-eluting.

The oligosaccharides of *D. glomerata* contain predominantly β-2,6 linkages (Chatterton *et al.*, 1993*b*). On the basis of co-elution with fructans from *D. glomerata* on anion-exchange HPLC and methylation analysis, some species of the tribe Aveneae (*Phalaris aquatica* and *H. lanatus*) and the tribe Poeae (*Pucinellia stricta* and *D. glomerata*) contain a series of linear, β-2,6-linked oligosaccharides with a terminal glucose residue. The very low proportion of 2,1-linked fructose residues present in the methylation analyses was probably due predominantly to 1-kestose, indicating that the fraction of higher molecular mass is solely β-2,6-linked.

L. ovatus also appears to accumulate β-2,6-linked fructan but with about half the molecules containing an internal glucose residue and half a terminal glucose residue. Therefore, in addition to the species that predominantly accumulate a linear β-2,6-linked series of fructan with a terminal glucose residue, some species of the tribes Aveneae (*L. ovatus*, *Avena* spp., Livingston *et al.*, 1993) and Poeae (*Lolium temulentum* L., Sims *et al.*, 1992 and *L. rigidum* Gaudin, Bonnett *et al.*, 1994; St. John *et al.*, 1997) contain many fructan molecules with an internal glucose residue. These data are summarized in Table 2 with the structures of fructan found in other tribes of the subfamily Pooideae.

Evidence from both morphological and phylogenetic studies (Clayton & Renvoize, 1986; Soreng, Davis & Doyle, 1990; Watson & Dallwitz, 1992; Davis & Soreng, 1993; Hsiao *et al.*, 1995) separates the tribes Aveneae and Poeae, with good agreement on which genera should be in which tribe. It is therefore likely that evolution of multiple fructan forms within these tribes has occurred independently. A hypothesis describing convergent evolution of a major biochemical pathway in the Poaceae has been reported previously. Hattersley & Watson (1992) have proposed that C_4 photosynthesis has evolved on more than one occasion in the Poaceae. In addition, it has been suggested that the evolution of fructan metabolism in the plant kingdom has occurred on several occasions (Hendry, 1993).

Although it is not possible to separate the members of the tribes Aveneae and Poeae on the basis of their fructan structure, it does appear to be possible to separate the supertribes Triticodae and Poodae,

Table 1. *Proportion of the different residues in fructan from* Dactylis glomerata, Phalaris aquatica, Puccinellia stricta, Holcus lanatus *and* Lagurus ovatus *deduced from mass spectroscopy after separation of partially methylated alditol acetates by gas chromatography*

Substitution position of derivative	Deduced residue	Proportion of each residue (mol %)				
		Dactylis glomerata	*Phalaris aquatica*	*Puccinellia stricta*	*Holcus lanatus*	*Lagurus ovatus*
2,5-Diacetyl-(2-deuterio)-1,3,4,6-tetramethyl hexitol	Terminal fructose (furanose)	19	21	33	20	36
2,6-Diacetyl-(2-deuterio)-1,3,4,5-tetramethyl hexitol	Terminal fructose (pyranose)	—	—	1	—	—
1,2,5-Triacetyl-(2-deuterio)-3,4,6-trimethyl hexitol	2,1-Linked fructose	3	1	1	—	4
2,5,6-Triacetyl-(2-deuterio)-1,3,4,-trimethyl hexitol	2,6-Linked fructose	65	51	32	60	24
1,2,5,6-Tetraacetyl-(2-deuterio)-4,6-dimethyl hexitol	2,1,6-Linked fructose	—	—	0·5	—	1
1,5-Diacetyl-(1-deuterio)-2,3,4,6-tetramethyl hexitol	Terminal glucose	12	27	31	17	19
1,5,6-Triacetyl-(1-deuterio)-2,3,4-trimethyl hexitol	1,6-Linked glucose	0·5	—	1	3	17

Table 2. *Summary of the characteristics of the fructan present in the tribes of the Pooideae*

Supertribe/Tribe/Genera	Fructan characteristics
Triticodae[1]*	
Triticeae	Mixed 2,1 and 2,6 linkages, branch points and a terminal glucose residue
Triticum[2,3], *Secale*[2], *Hordeum*[2], *Elytrigia*[4]	
Bromeae	Mixed 2,1 and 2,6 linkages, branch points and a terminal glucose residue
Bromus[5]	
Poodae[1]	
Aveneae	Predominantly 2,6 linkages, many molecules with an internal glucose residue
Type 1: *Avena*[6], *Lagurus*[7]	
Type 2: *Phalaris*[7], *Holcus*[7]	Predominantly 2,6 linkages with a terminal glucose residue
Poeae	
Type 1: *Lolium*[8,9]	Predominantly 2,6 linkages, many molecules with an internal glucose residue
Type 2: *Dactylis*[7,10], *Puccinellia*[7]	Predominantly 2,6 linkages with a terminal glucose residue

* **1** There is currently no information available for the tribes Brachypodieae (Triticodeae), Seslerieae and Meliceae (Poodae); **2** Hirst (1957); **3** Bancal *et al.* (1992); **4** Arni & Percival (1951); **5** Chatterton *et al.* (1993*b*); **6** Livingston *et al.* (1993); **7** this study; **8** Sims *et al.* (1992); **9** Bonnett *et al.* (1994); **10** Chatterton *et al.* (1993*a*).

albeit on a small sample of genera (Table 2). Branch points are a characteristic of the fructan from the Triticodae, but are absent from or are very few in fructan from species in the Poodae.

ACKNOWLEDGEMENTS

Phillip Phelps and David Marshall are thanked for growing the plants and preparing the extracts, Dr James St. John for Figure 3, Liza Apps and Dr Rex Oram for providing seed and Dr Mike Lazarides for his advice on grass taxonomy. G.D.B. and R.J.S. received funding from the Australian Wool Research and Promotion Organization. I.M.S. acknowledges the infra-structural support of the Australian Commonwealth Government. A.J.C. acknowledges IGER, CSIRO and The University of Melbourne for support during a period of sabbatical leave.

REFERENCES

Arni PC, Percival EGC. 1951. Studies on fructosans. Part II. Triticin from the rhizomes of couch grass (*Triticum repens* L.) *Journal of the Chemical Society* 1822–1830.

Bancal P, Carpita NC, Gaudillère JP. 1992. Differences in fructan accumulated in induced and field-grown wheat plants: an elongation-trimming pathway for their synthesis. *New Phytologist* **120**: 313–321.

Bonnett GD, Sims IM, St. John JA, Simpson RJ. 1994. Purification and characterization of fructans with β-2,1- and β-2,6-glycosidic linkages suitable for enzyme studies. *New Phytologist* **127**: 261–267.

Chatterton NJ, Harrison PA, Thornley WR, Bennett JH. 1993*a*. Structure of fructan oligomers in cheatgrass (*Bromus tectorum* L.). *New Phytologist* **124**: 389–396.

Chatterton NJ, Harrison PA, Thornley WR, Bennett JH. 1993*b*. Structure of fructan oligomers in orchardgrass (*Dactylis glomerata* L.) *Journal of Plant Physiology* **142**: 552–536.

Chatterton NJ, Harrison PA, Thornley WR, Draper EA. 1990.

Oligosaccharides in foliage of *Agropyron*, *Bromus*, *Dactylis*, *Lolium*, and *Phleum*. *New Phytologist* **114**: 167–171.

Clayton WD, Renvoize SA. 1986. *Genera Graminum Grasses of the World*. London: HMSO.

Davis JI, Soreng RJ. 1993. Phylogenetic structure in the grass family (Poaceae) as inferred from chloroplast DNA restriction site variation. *American Journal of Botany* **80**: 1444–1454.

French D, Wild GM. 1953. Correlation of carbohydrate structure with papergram mobility. *Journal of the American Chemical Society* **75**: 2612–2616.

Hattersley PW, Watson L. 1992. Diversification of photosynthesis. In: Chapman GP, ed. *Grass Evolution and Domestication*. Cambridge: Cambridge University Press, 38–116.

Hendry GAF. 1993. Evolutionary origins and natural functions of fructans – a climatological, biogeographic and mechanistic appraisal. *New Phytologist* **123**: 3–14.

Hirst EL. 1957. Some aspects of the chemistry of fructosans. *Journal of the Chemical Society* 193–204.

Hsiao C, Chatterton NJ, Asay KH, Jensen KB. 1995. Molecular phylogeny of the Pooideae (Poaceae) based on nuclear rDNA (ITS) sequences. *Theoretical and Applied Genetics* **90**: 389–398.

Incoll LD, Bonnett GD. 1996. Fructans in the Compositae – a short review. In Caligari PDS, Hind DJN, *Compositae : Biology and Utilization*. Proceedings of the International Compositae Conference, Kew, 1994. Kew, Royal Botanic Gardens, Vol 2, 401–413.

Kanaya KI, Chiba S, Shimomura T. 1978. Thin-layer chromatography of linear oligosaccharides. *Agricultural and Biological Chemistry* **42**: 1947–1948.

Lau E, Bacic A. 1993. Capillary gas chromatography of partially methylated alditol acetates on a high-polarity, cross-linked, fused silica BPX70 column. *Journal of Chromatography* **637**: 100–103.

Livingston DP, Chatterton NJ, Harrison PA. 1993. Structure and quantity of fructan oligomers in oat (*Avena* spp.). *New Phytologist* **123**: 725–734.

Meier H, Reid JSG. 1982. Reserve polysaccharides other than starch in higher plants. In: Loewus FA, Tanner W, eds. *Encyclopedia of Plant Physiology New Series. Volume 13A, Plant Carbohydrates. I Intracellular Carbohydrates*. Berlin: Springer-Verlag, 418–471.

Sims IM, Pollock CJ, Horgan R. 1992. Structural analysis of oligomeric fructans from excised leaves of *Lolium temulentum*. *Phytochemistry* **31**: 2989–2992.

Smith D. 1968. Classification of several native north American grasses as starch or fructosan accumulators in relation to taxonomy. *Journal of the British Grassland Society* **23**: 306–309.

Smouter H, Simpson RJ. 1989. Occurrence of fructans in the Gramineae (Poaceae). *New Phytologist* **111**: 359–368.

Smouter H, Simpson RJ. 1991. Fructan metabolism in leaves of *Lolium rigidum* Gaudin I. Synthesis of fructan. *New Phytologist* **119**: 509–516.

Soreng RJ, Davis JI, Doyle JJ. 1990. A phylogenetic analysis of chloroplast DNA restriction site variation in Poaceae subfam. Pooideae. *Plant Systematics and Evolution* **172**: 83–97.

St. John JA, Sims IM, Bonnett GD, Simpson RJ. 1997. Identification of products synthesised by a fructan:fructan fructosyltransferase activity from *Lolium rigidum* Gaudin. *New Phytologist* **135**: 249–257.

Watson L, Dallwitz MJ. 1992. *The grass genera of the world*. Wallingford: CAB International.

Wise CS, Dimler RJ, Davis HA, Rist CE. 1955. Determination of easily hydrolysable fructose units in dextran preparations. *Analytical Chemistry* **27**: 33–36.

Yemm EM, Willis AJ. 1954. The estimation of carbohydrates in plant extracts by anthrone. *Biochemistry Journal* **57**: 508–514.

New Phytol. (1997), **136**, 19–28

Cytosolic expression of the *Bacillus amyloliquefaciens SacB* protein inhibits tissue development in transgenic tobacco and potato

By P. G. CAIMI*, L. M. McCOLE, T. M. KLEIN and H. P. HERSHEY

Agricultural Products, E. I. DuPont de Nemours & Co., Wilmington, DE 19880-0402, USA

(*Received 24 July 1996*)

SUMMARY

Fructans are linear or branched polymers containing a single sucrose and repeating fructose residues. An early model for fructan biosynthesis in higher plants suggested that partial synthesis of the polymer occurred in the cell cytosol. The current model suggests that synthesis requires the interaction of two separate fructosyltransferases located in the vacuole. Tobacco lines containing a chemically induced promoter, directing expression of the *Bacillus amyloliquefaciens SacB* gene in the present study, provided an opportunity to regulate and target fructan synthesis to the cytosol of transgenic plants. Induced expression of the gene led to rapid destruction of leaf tissue. Amino acid substitution at a highly conserved site (Arg^{331}) in the *SacB* gene reduced the fructosyltransferase efficiency without reducing the invertase activity of the enzyme. Expression of the mutant gene in transgenic tobacco also resulted in leaf damage. However, the appearance of necrotic tissue was greatly delayed. The results suggest that the phenotype is due to accumulation of fructan in the cytosol. Fructan metabolism in the cytosol of potato tubers was also detrimental to development. Tuber size and starch synthesis was significantly reduced in lines containing the untargeted gene. Transgenic tobacco and potato containing the *SacB* gene offer an opportunity to study the metabolism of fructan and the effect of accumulation on plant cell development.

Key words: Fructan, *SacB*, cytosolic expression, transgenic plants, fructosyltransferase.

INTRODUCTION

Fructans are water-soluble polymers consisting of repeating $\beta(2,1)$ or $\beta(2,6)$ linked fructose residues. Like starch, their function in higher plants is as a transitory or long-term storage reserve (Housley & Daughtry, 1987; Edelman & Jefford, 1968). The advantages of accumulating fructan in place of starch remain obscure. However, physiological roles in addition to that of serving as a carbohydrate reserve have been reported, including involvement in cell expansion, conferring tolerance to cold or drought and the general role as an osmotic regulator (Bieleski, 1993; Pollock, 1986; Points & Del Campillo, 1985; Hendry, 1993).

Fructan metabolism is believed to require the interaction of at least two separate enzymes (Edelman & Jefford, 1968). The initial step in fructan synthesis is catalysed by the enzyme sucrose : sucrose fructosyltransferase (SST), producing a trisaccharide (glucose-fructose-fructose) and glucose from two sucrose residues. The trisaccharide (kestose) produced by SST is thought to be a substrate for the second enzyme fructan : fructan fructosyltransferase (FFT), in the synthesis of larger fructan polymers. FFT acts through the reversible transfer of fructosyl residues between donor and acceptor fructan.

There is strong evidence to suggest that sucrose acts as the sole substrate for SST in all fructan-storing plants (reviewed in Pollock *et al.*, 1996). The two largest pools of sucrose in a plant cell are located in the cytosol, where sucrose is synthesized, and the vacuole, where sucrose might be transiently stored (Heineke *et al.*, 1994). The original model, proposed to explain fructan metabolism in higher plants (Edelman & Jefford, 1968), suggested that the cytosolic sucrose pool was utilized in the synthesis of kestose by the enzyme SST. In this model, transport

* To whom correspondence should be addressed.

of kestose to the vacuole would be necessary for elongation by FFT. More recent studies indicate that the fructan pathway is strictly confined to the vacuole (Darwen & John, 1989; Frehner, Keller & Wiemken, 1984; Carpita *et al.*, 1991), suggesting that transport of sucrose from the cytosol to the vacuole is critical to fructan metabolism. Localization of the fructan biosynthetic pathway to the vacuole in all fructan-storing plant species is, however, based largely on data from a single species, *Helianthus tuberosus* (Housley & Pollock, 1993).

Very little is known about the factors which regulate sucrose partitioning between the cytosol and vacuole. Transgenic plants have been used recently as models to study the role of sucrose in these two subcellular compartments (reviewed in Sonnewald, von Schwaen & Willmitzer, 1993). Cytosolic expression of a yeast-derived invertase was shown to be particularly disruptive to plant development (Sonnewald *et al.*, 1991). Severely stunted plant growth was attributed to interference with sucrose synthesis in the cytosol, and accumulation of photosynthate in the leaves. Transgenic plants containing chimeric bacterial fructosyltransferases have also provided useful information concerning the role of fructan metabolism in plants and the effect of diverting sucrose into this new pathway (van der Meer *et al.*, 1994; Pilon-Smits *et al.*, 1995). Plants containing a fructosyltransferase fused to vacuole-targeting signals accumulate fructan (*c.* 3–8% of tissue d. wt) with no discernible effect on plant development (Ebskamp *et al.*, 1994). However, accumulation of similar levels of fructan in the cytosol of maize endosperm resulted in an extreme detrimental phenotype (Caimi *et al.*, 1996).

The purpose of this study was to extend expression of a fructosyltransferase to additional plant sink and source tissues and to determine whether the cytosolic pool could be effectively diverted into fructan by this 'new' biosynthetic pathway. A mutant *Bacillus amyloliquefaciens SacB* gene together with tissue-specific and chemically-induced promoters were used to determine the effects of accumulation on plant development.

MATERIALS AND METHODS

Construction of chimeric SacB genes

pCIP-Sac. An inducible *SacB* expression cassette was constructed, consisting of a 1·0 Kb promoter region from the maize chemically induced 2-2 gene (Hershey & Stoner, 1991), a 59 bp 5′ untranslated leader from the chorophyll *a/b* binding protein (*Cab*) gene (Dunsmuir, 1985) and the 1·3 Kb mature coding sequence from the *Bacillus amyloliquefaciens SacB* gene, obtained from the plasmid pBE311 (Nagarajan & Borchert, 1991). The cassette also contained a 503 bp 3′ end from the maize chemically induced 2-1 gene (Hershey & Stoner, 1991). The bacterial secretion signal was removed from the *SacB* gene and unique restriction enzyme sites were added by PCR, previously described in Caimi *et al.* (1996). Restriction enzyme sites were also added by linker ligation or PCR to the promoter region, *Cab* gene leader and 3′ end for subcloning purposes. The final construct was designated pCIP-Sac (Fig. 1*a*).

pSSU-Sac and pSSU-Sac(Lys). Substitution of the native Arg^{331} in the *Bacillus subtilis SacB* gene with Lys^{331} was described by Chambert & Petit-Glatron, (1991). Substitution with Lys^{331} (AAG) in the *Bacillus amyloliquefaciens SacB* gene was accomplished by PCR, using the oligonucleotide primer (5′-GAT TCA AAG GGATCC AAA ATG-3′). The primer also creates a unique BamHI site (underlined), but does not alter the amino acid sequence in that region. Correct substitution was confirmed by DNA sequencing (Sanger, Nicklen & Coulson, 1977). The 1·3 Kb mutant *SacB* gene was isolated and ligated into the plasmid pDH51 containing a CaMV 35S transcription termination region (Pietrzak *et al.*, 1986). The small subunit carboxylase promoter region was obtained from the plasmid pNtSS23 (Mazur & Chui, 1985) and ligated to the final construct designated pSSU-Sac(Lys), (Fig. 1*a*). The expression cassette pSSU-Sac (Fig. 1*a*) was constructed by replacing the 1·3 Kb SacB-Lys^{331} gene with the native *SacB* gene (Arg^{331}) gene from the plasmid pBE311 (Nagarajan & Borchert, 1991).

pPat-Sac. The patatin promoter region was isolated by PCR using oligonucleotide primers based on the published sequence by Rocha-Sosa *et al.* (1989). The PCR primers, containing an NcoI site at the initiation codon, produced a 1·0 Kb fragment. The promoter region was ligated to the pCIP-Sac plasmid, replacing the chemically induced promoter. The final expression cassette was designated pPat-Sac (Fig. 1*b*).

Invertase assay of the wild-type and mutant SacB protein

The wild-type *SacB* gene was replaced with the mutant (Lys^{331}) gene in the plasmid pBE311, containing the native bacterial promoter, and transformed into *E. coli* (Sambrook, Fritsch & Maniatis, 1989). Overnight cultures of *E. coli* containing the wild-type or mutant gene were vortexed with glass beads (150–212 μm, Sigma, St Louis, MO) to release protein retained in the periplasmic space. Cell debris was removed by centrifugation, and crude protein extract was loaded on a 12% native acrylamide gel and detected by western analysis using antibodies against the *Bacillus SacB* protein. Protein extracts were adjusted, using a 30 Kd Centricon (Amicon,

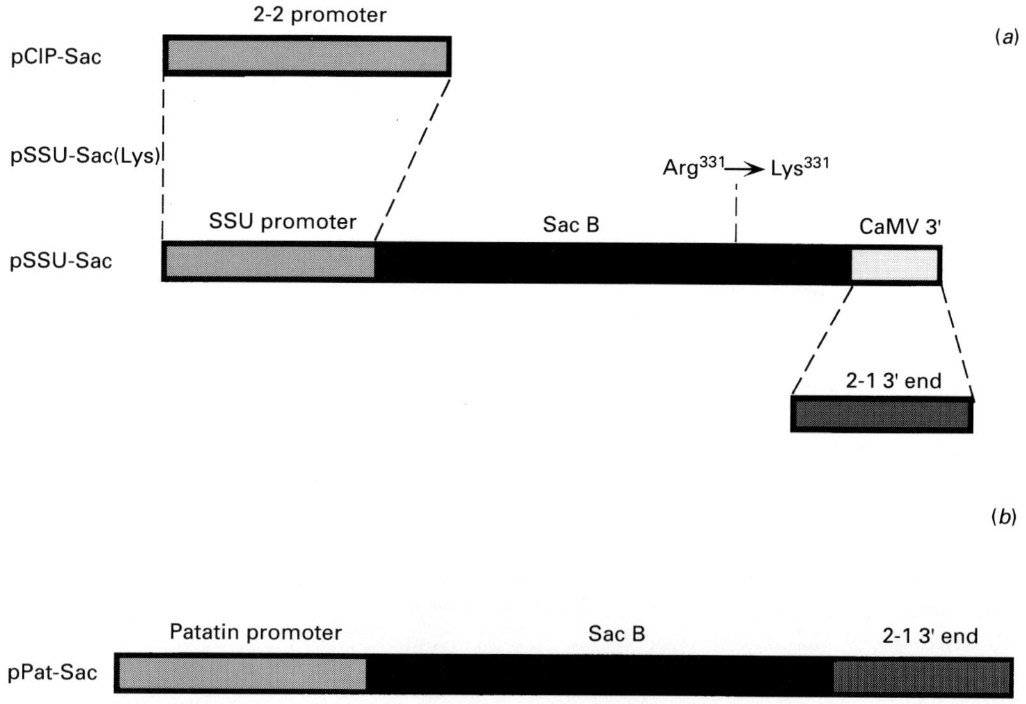

Figure 1. Structure of chimeric *SacB* expression cassettes. (*a*) Expression of the 1·3 kb mature *SacB* gene was directed by a chemically induced, maize 2-1 gene promoter in the cassette pCIP-Sac. The maize 2-1 gene transcription termination region was also contained in this cassette. A tobacco SSU promoter directs expression of *SacB* in leaves containing the pSSU-Sac and pSSU-Sac(Lys) constructs. The *SacB* gene in pSSU-Sac(Lys) contains a Lys^{331} substituted for the native Arg^{331}. (*b*) Tissue and developmental specific expression of the *SacB* gene in potato tubers was directed by a patatin storage protein promoter. The untargeted pPat-Sac construct also contains a transcription termination region from the maize 2-1 gene.

Boston, MA) following western analysis so that equal amounts of mutant and wild-type protein were present in the extracts.

Invertase activity was assayed by incubating the extracts with 100 mM sucrose in 10 mM phosphate buffer, pH 6·0, for 30 min. Invertase activity was defined as the amount of free fructose formed from sucrose. Fructose levels were determined by subtracting the amount of total hexose formed by the amount of free glucose formed in separate enzyme linked assays, described previously (Caimi *et al.*, 1996). Hexose, sucrose and larger fructose oligomers present following incubation with sucrose were also analysed by high-performance anion-exchange chromatography, described below.

Plant material and transformation

Expression cassettes containing a promoter, *SacB* gene and 3′ end were subcloned into the binary vector pZS97K (Bevan, 1984) prior to transformation into *Agrobacterium tumefaciens* LBA4404 (Hoekema *et al.*, 1983). The construct was introduced into *Nicotiana tabacum* L. and *Solenum tuberosum* cv. Desiree by the leaf disk method (Horsch *et al.*, 1985; De Bloch, 1988). Transformed plants were selected for kanamycin resistance, transferred to soil and grown to maturity in the greenhouse. In potato, microtubers were produced by the method of Bourque, Moller & Park (1987).

Chemical induction of tobacco leaves and RNA analysis

Solutions containing 2-chlorobenzenesulfonamide (2-CBSU) were prepared by the methods described in Hershey & Stoner, (1991). Mature tobacco leaves were cut and immediately placed in solutions containing 2-CBSU ($200\ mg\ l^{-1}$) for 12–36 h. Methods for RNA isolation and slot blot analysis are described in Hershey & Stoner, (1991).

Carbohydrate isolation and analysis

Fructan isolation from transgenic tissue and analysis by TLC have been previously described (Caimi *et al.*, 1996). Analysis of soluble carbohydrates was performed by high-performance anion-exchange chromatography using a Dionex chromatograph and pulsed amperometric detector (Sunnyvale, CA). Elution was performed in 150 mM NaOH with a non-linear NaOAc gradient (50 mM–500 mM) by the methods described in Chatterton *et al.* (1993).

RESULTS

Cytosolic expression of the SacB protein

Repeated attempts to recover transgenic plants containing an untargeted *Bacillus amyloliquefaciens*

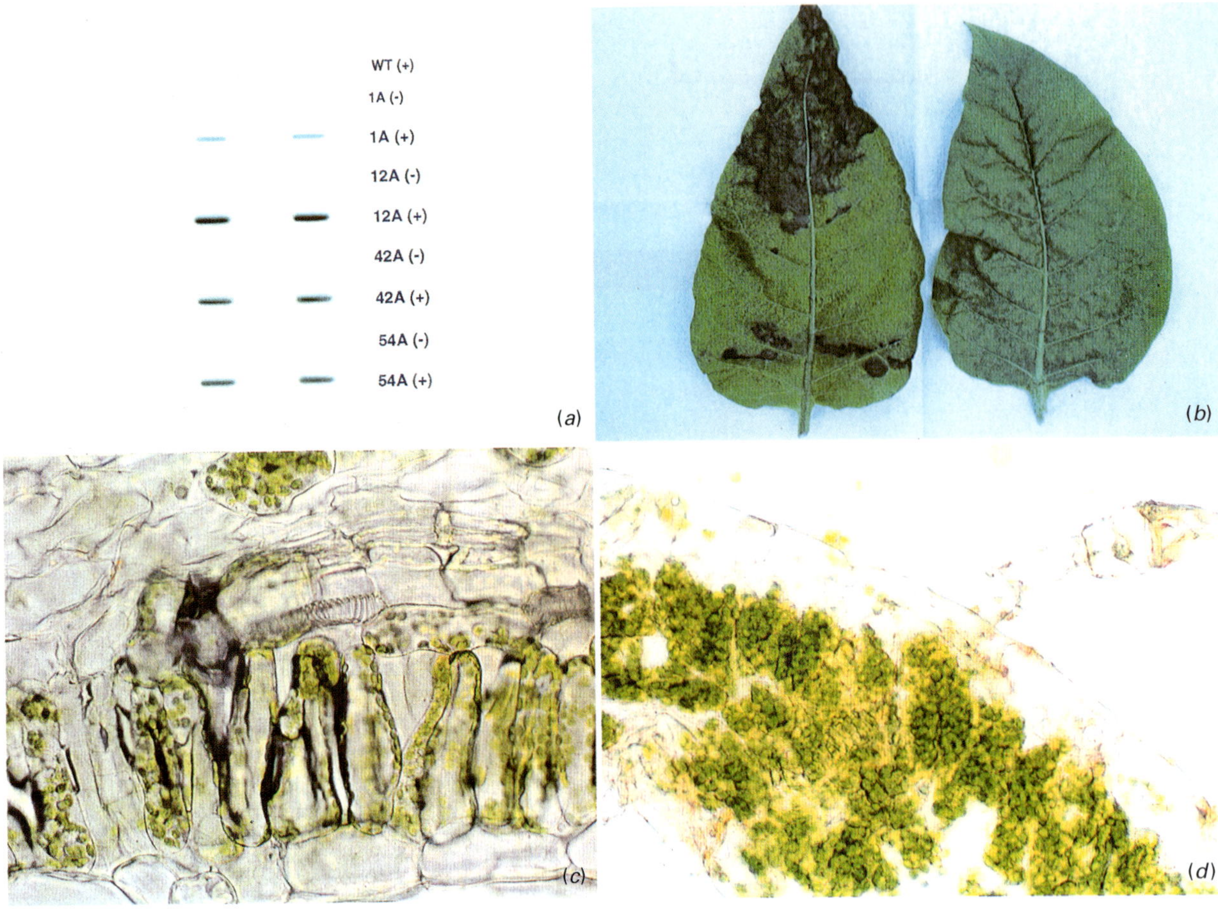

Figure 2. Transgenic tobacco leaves containing the inducible pCIP-Sac cassette (*a*) RNA isolated from a control line (wild-type) and leaves containing the pCIP-Sac construct was characterized by slot blot analysis. Total RNA (10 μg) was applied to a nylon membrane and probed using a labelled 1·3 Kb *SacB* DNA fragment. Transformed leaves were untreated (−) or induced with 2-CBSU (+) for 36 h. (*b*) Tobacco leaves containing the pCIP-Sac construct were induced for 36 h (left). Control leaves were untreated. (*c*) Thin section of an untreated leaf containing the pCIP-Sac cassette. (*d*) Thin section of an induced leaf containing pCIP-Sac.

SacB gene directed by a small subunit carboxylase promoter (SSU) were unsuccessful (pSSU-Sac, Fig. 1*a*). The results were unexpected considering the routinely high recovery rate of transformed tobacco using *Agrobacterium* and the leaf disk transformation method.

Chemically induced expression of the SacB gene

To determine whether expression of the untargeted *SacB* gene prevented regeneration of transformed tobacco, the SSU-promoter was replaced with a maize inducible promoter (Hershey & Stoner, 1991). The maize 2-2 promoter in pCIP-Sac (Fig. 1*a*) is silent in the absence of chemical inducer, 2-chlorobenzenesulfonamide (2-CBSU). Whereas attempts to regenerate transgenic plants containing an actively expressed *SacB* gene were not successful, leaf disk transformation using the pCIP-Sac plasmid, in the absence of chemical inducer, resulted in several kanamycin resistant lines. Northern analysis (Fig. 2*a*) demonstrated that transcripts from the *SacB* gene could not be detected in uninduced tobacco leaves containing pCIP-Sac. However, *SacB* RNA transcripts were found in tobacco leaves induced for 24–36 h. A representative transgenic line (12A) was chosen for additional study.

Table 1. *Soluble carbohydrate analysis of leaves containing the pCIP-Sac cassette*

	Carbohydrate content (mg g^{-1} f. wt)	
	Control	Induced
Sucrose	6·27 ± 0·03	6·67 ± 0·70
Glucose	4·90 ± 0·04	5·67 ± 0·09
Fructose	3·07 ± 0·03	3·26 ± 0·04
Fructan	n.d.	3·98 ± 0·70

Control leaves were untreated. Induced leaves were treated with 2-CBSU for 30 h. n.d., not detected. $n = 3$.

The phenotype of wild-type leaves was unchanged after 36 h of induction (data not shown). However, leaves from line 12A treated with 2-CBSU differed dramatically from uninduced control leaves (Fig. 2*b*). Necrotic tissue appeared at the tip of leaves within 24 h of induction. After 36 h the necrotic

(*a*)

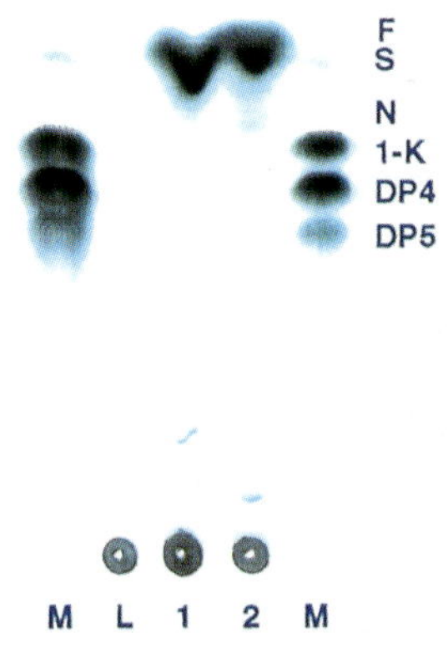

(*b*)

Figure 3. Cytosolic expression of a mutant *SacB* gene in tobacco leaves. (*a*) phenotype of a mature tobacco leaf containing the pSSU-Sac(Lys) cassette. (*b*) TLC analysis of extracts from tobacco leaves containing the pSSU-Sac(Lys) cassette. M, Meioligo standard; L, Bacterial levan standard; 1, soluble extract from a green section of the leaf; 2, soluble extract from a necrotic section of the same leaf. Meioligo standards; F, fructose; S, sucrose; N, Neokestose; 1-K, 1-kestose.

region spread from the margin of the leaves towards the base. Thin sections from uninduced leaves (Fig. 2*c*) show chloroplasts spaced around the perimeter of the cell and the large central region occupied by the vacuole. By contrast, the internal structure of cells induced for 36 h was very different (Fig. 2*d*). The chloroplasts in cells actively synthesizing fructan no longer appeared around the perimeter, but were collapsed into the centre of the cell.

Soluble carbohydrate concentration was determined in leaves induced for 36 h and compared with uninduced controls. The results show that there was very little difference between the sucrose and fructose levels of induced and uninduced leaves (Table 1). Glucose levels were slightly higher than the controls and fructan was detected only in the induced leaves (Table 1).

Cytosolic expression of a mutant SacB protein

A single amino acid substitution in the *Bacillus subtilis SacB* gene was shown significantly to reduce the enzyme's polymerase activity, but did not lower invertase activity (Chambert & Petit-Glatron, 1991). The mutant *SacB* gene containing Lys^{331} substituted for the native Arg^{331} synthesized only the trisaccharide kestose. In this study, the *Bacillus amyloliquefaciens SacB* gene was modified by PCR to determine whether expression of the mutant *SacB* protein in the cytosol of a cell would result in accumulation of kestose, and to determine the effect of accumulation on leaf development.

The SacB-Lys^{331} gene, directed by the native bacterial promoter, was first characterized in *E. coli*. Invertase activity was found to be slightly higher in protein extracts from *E. coli* containing the mutant gene than in those containing the wild-type (Arg^{331}) gene. Activity in the extracts released glucose and fructose, but did not produce high degree of polymerization (DP) fructan after incubation with sucrose (data not shown). The wild-type *SacB* gene was then replaced by the mutant gene in the cassette pSSU-Sac(Lys), (Fig. 1*a*) for expression in tobacco leaves. Except for the single amino acid substitution, pSSU-Sac(Lys) is identical to the pSSU-Sac cassette which could not be recovered in transformed plants. Nevertheless, transformation using the mutant gene resulted in recovery of several transgenic lines.

Plants containing the cytosolically expressed mutant gene could not be distinguished from wild-type controls at early stages of development. However, necrotic regions appeared at the tip of leaves containing the pSSU-Sac(Lys) cassette and progressed towards the base as the plant matured (Fig. 3*a*). Necrotic regions appeared only on older leaves of the plant. The concentration of glucose, an end-

Table 2. *Carbohydrate analysis of leaves containing the pSSU-Sac(Lys) cassette*

	Carbohydrate content (mg g^{-1} f. wt)	
	Green	Necrotic
Glucose	2·58 ± 0·60	3·70 ± 0·78
Fructan	4·27 ± 2·20	10·56 ± 1·99

Tissue was collected from undamaged, green regions of a mature leaf or from an area adjacent to necrotic tissue on the same leaf. $n = 5$.

product of both the polymerase and invertase activity of *SacB*, was determined for normal and damaged sections of leaves containing the mutant gene. Glucose levels were shown to be slightly lower in green sections than in necrotic sections of the same leaf (Table 2). Fructan was shown to accumulate throughout the leaf, but the concentration in necrotic regions was much higher than in the green areas of the leaf (Table 2).

TLC analysis of soluble extracts from necrotic and green regions of leaves containing the pSSU-Sac(Lys) cassette are shown in Figure 3*b*. Marker lanes on the TLC plate demonstrate that high-DP fructan standards (Sigma, St Louis, MO) did not move from the origin after the plate was developed twice. Lanes containing Meioligo® (Meiji Seika Kaisha Ltd., Kawasaki, Japan) standards, consisting of low-DP fructan, sucrose and monosaccharides produced positive signals that ran at or near the solvent front. Soluble extracts from leaves containing the mutant gene also produced a positive signal on TLC plates. The signals from both the green and necrotic sections of the leaf did not move from the plate origin.

Tissue-specific expression of SacB protein in potato tubers

To determine whether the detrimental effect of cytosolic expression of *SacB* in tobacco would be extended to additional plant species and tissue types, the pPat-Sac (Fig. 1*b*) cassette was transformed into potato leaf disks. Tuber-specific expression of the gene was accomplished by replacing the promoter in pSSU-Sac with a 1·0 Kb patatin promoter region (Rocha-Sosa *et al.*, 1989). Transformed lines were selected by growth on kanamycin media and confirmed by PCR using *SacB* coding sequence primers

Figure 4. Tuber-specific expression of an untargeted *SacB* gene. (a) Aboveground tubers in potato line 1A containing high levels of fructan. (*b*) Discolouration of tubers containing the pPat-Sac cassette (right) compared with control tubers. (*c*) Control tuber thin section viewed under polarized light. (*d*) Transformed tuber, accumulating fructan in the cytosol of the cell, viewed under polarized light.

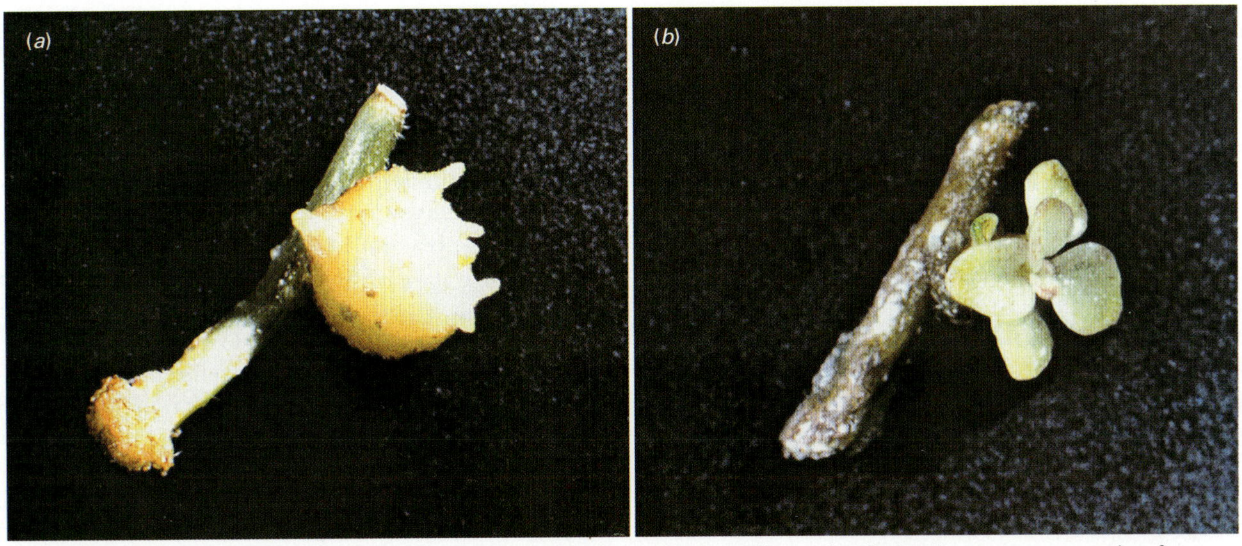

Figure 5. Induction of microtubers from potato stem sections. (*a*) Microtuber induction on a stem section from untransformed potato (wild-type). (*b*) Floret induced on stem section of a potato line transformed with the pPat-Sac cassette.

(data not shown). Independent transformed lines were transferred to soil and grown to maturity in the glasshouse. Transgenic lines could not be distinguished phenotypically from untransformed control plants at early stages of growth (before tuber initiation). However, as the plants matured, growth of transgenic lines lagged behind the controls (transformed with only a binary vector). Tuber size varied between the transgenic lines, the most severely stunted plants producing numerous aboveground, small tubers (Fig. 4*a*). Whether the tubers containing the cytosolically expressed *SacB* protein were large or small, the interior was always darkly stained, compared with controls (Fig. 4*b*). Thin sections of control tubers viewed under polarized light demonstrate the presence of multiple starch granules (Fig. 4*c*). By contrast, starch granules were not found in the smallest transgenic tubers (Fig. 4*d*).

Induction of microtubers

Stem sections from control plants and from the most severely stunted transgenic lines were placed on microtuber-induction medium to study the effect of cytosolic fructan accumulation on tuber initiation. Stem sections from wild-type plants produced microtubers within 5 wk of being placed on induction medium (Fig. 5*a*). However, after 6–8 wk, stem sections from transgenic plants containing the pPat-Sac cassette produced a small floret in place of a microtuber (Fig. 5*b*). The floret elongated with time, but did not develop into a microtuber during the length of the experiment (10 wk).

Analysis of fructan in transgenic tubers

The presence of fructan in mature transgenic tubers was detected by TLC analysis. Positive signals on TLC plates did not migrate from the plate origin, suggesting that the polymers are of high DP (data not shown). Fructan concentration was determined by resorcinol analysis (Caimi *et al.*, 1996). The level of fructan in mature tubers containing the untargeted *SacB* gene was shown to differ between independent lines (Table 3). The d. wt of tubers containing fructan was also shown to be less than that of control lines (Table 3).

Table 3. *Fructan concentration in mature potato tubers containing an untargeted* SacB *gene* (*pPat-Sac*).

Gene/line	Fructan ($mg\ g^{-1}$ d. wt)	D. wt (g)
pPat-Sac		
31B	5·23	1·73
7A	14·31	0·57
1A	48·97	0·15
Wild-type	n.d.	3·23

n.d., none detected.

DISCUSSION

This study demonstrates that expression of the *Bacillus amyloliquefaciens SacB* protein in the cytosol of tobacco leaves and potato tubers results in diversion of sucrose into high-DP fructan. Expression of the *SacB* gene was also shown to have a severe detrimental effect on plant development. Tissue destruction associated with fructan accumulation in the cell cytosol illustrates the significance of confining fructan metabolism to the vacuolar compartment in fructan-storing plants.

Induced expression in tobacco leaves

Recovery of transgenic tobacco containing a constitutively expressed *SacB* gene was not successful.

Transgenic lines were recovered when the constitutive promoter was replaced with a chemically induced promoter from the maize 2-2 gene (Hershey & Stoner, 1991), in the absence of inducer. Tissue destruction rapidly followed induction of the gene with 2-CBSU (Fig. 2*b*).

Sucrose is the sole substrate for the *SacB* enzyme and damage to the leaves might have been due to interference with sucrose metabolism. Disrupting sucrose metabolism by expression of a chimeric yeast invertase in the cytosol of tobacco resulted in severe stunting of plant growth and the appearance of light green or bleached areas on mature leaves (Sonnewald *et al.*, 1991). The severe phenotype in plants containing a yeast-derived invertase was attributed to interference with sucrose transport from the leaf. Hydrolysis of sucrose led to accumulation of hexose and, curiously, an increase in sucrose concentration. The proposed cause of leaf damage in transgenic lines containing the yeast invertase gene was the dedifferentiation of cells due to accumulation of photoassimilate (Heineke *et al.*, 1994).

The invertase activity of *SacB* has been well documented (Chambert & Petit-Glatron, 1991). Expression of an untargeted *SacB* gene in transgenic tissue would have access to sucrose synthesized in the cytosol. However, soluble carbohydrate analysis of leaves containing the pCIP-Sac cassette demonstrated that when leaves are induced for 30 h hexose does not accumulate any more than in uninduced controls (Table 1). In addition, fructan was shown to accumulate only in an induced leaf (Table 1) and necrotic tissue developed only on leaves containing fructan (Fig. 2*b*). The results suggest that invertase activity *per se* is not the case of tissue damage in transgenic lines containing the *SacB* gene.

Expression of a SacB -Lys^{331} gene in tobacco leaves

The cause of tissue damage in transgenic plants might also be related to the size of the bacterial polymer. Fructan synthesized in transgenic maize (Caimi, unpublished) or tobacco (Ebskamp *et al.*, 1994) expressing the *SacB* gene was shown to consist of thousands of individual fructose residues, whereas plant fructans contain an average of only 20–100 residues. The original model for fructan metabolism (Edelman & Jefford, 1968) suggested that the trisaccharide kestose was synthesized in the cytosol by SST and could serve as a donor or acceptor in the synthesis of higher DP fructan, mediated by FFT in the vacuole. Thus, kestose would either be transported immediately to the vacuole, or accumulation, to some degree, would be tolerated in the cytosol. To test whether synthesis of kestose would be tolerated in the cytosol, a mutant *SacB* gene, (SacB–Lys^{331}) reported to produce only kestose (Chambert & Petit-Glatron, 1991), was constructed and transformed into tobacco. Analysis of transgenic tobacco lines containing the mutant *SacB* gene in this study demonstrated that fructan was synthesized in the leaves (Fig. 3*b*). However, it was not kestose, but high-DP fructan that accumulated. Furthermore, the altered polymerase activity of the mutant gene resulted in delayed synthesis, requiring several weeks for high DP fructan to accumulate, compared with rapid synthesis (24–36 h) by the native *SacB* gene (pCIP-Sac).

The presence of high-DP fructan demonstrates that the enzyme has a strong preference for adding fructosyl residues to an existing polymer, rather than creating new 'primer' trisaccharides. The difference in fructan size reported in this study and in the original description of the mutant is likely to be due to the substantially longer incubation time (several weeks) of the enzyme with sucrose when expressed in tobacco leaves compared with a shorter, several minute incubation in the original study (Chambert & Petit-Glatron, 1991). The mutant gene produced only high-DP fructan, therefore we could not test the hypothesis that smaller polymers (i.e. kestose) could accumulate in the cytosol without causing extensive tissue damage. However, the data do support the suggestion that invertase activity is not associated with the development of necrotic tissue or with the inability to recover transgenic plants expressing the wild-type gene. The invertase activity of the mutant was shown to be slightly higher than that of the wild-type gene (Chambert & Petit-Glatron, 1991). The wild-type gene could not be recovered in transgenic plants, and chemical induction of the gene in the pCIP-Sac expression cassette resulted in rapid tissue destruction (Fig. 2*b*). If invertase activity were involved in tissue damage, expression of the mutant gene would have been expected to produce the same or a more severe phenotype.

Transgenic lines containing the mutant gene were recovered and leaves were shown to develop necrotic regions, but only later, as the plant matured (Fig. 3*a*). Glucose levels were slightly lower in green sections of a leaf containing the mutant gene than in damaged regions of the same leaf (Table 2). The concentration of fructan, however, was approximately two to threefold higher in necrotic regions than in green sections of the leaf (Table 2). The results also demonstrate that the phenotype of tobacco leaves expressing the mutant *SacB* gene is likely not to be caused by inhibited carbohydrate transport from the leaf, but to accumulation of fructan in the cytosol.

Accumulation of fructan in the cytosol may result in altered osmotic potential, eventually leading to cell damage. High-molecular-weight polymers would not ordinarily be expected to contribute significantly to the osmotic value of cell sap; however, the data in Table 1 demonstrate that the soluble carbohydrate concentration in an induced leaf is approximately the same as in a control. Therefore, accumulation of fructan, which cannot be degraded or transported, adds to the osmotic value contributed by the carbohydrates present and must be compensated for by some means. Alternatively, cell damage might be related to direct physical pressure exerted by accumulation of a very large polymer in a relatively small cellular compartment. Thin sections from leaves induced with 2-CBSU (Fig. 2*d*) indicate that internal cellular organization is rapidly disrupted, within 30 h of treatment. This observation is consistent with both hypotheses. Future work, using the mutant *SacB* gene and a chemically induced promoter, might help to establish whether damage is due to direct physical contact or to altered osmotic potential within the cell.

Cytosolic expression of SacB in potato

Variation in soluble carbohydrate content among independent lines and between tubers of the same line makes it difficult to determine the precise mechanism of inhibited development (data not shown). However, the results clearly demonstrate that tuber development is severely inhibited in transgenic lines containing an untargeted *SacB* gene (Fig. 4*a*). The reduced size and d. wt of tubers appear to be associated with reduced starch synthesis. The phenotype might be associated with diversion of sucrose away from the starch biosynthetic pathway, resulting in inhibited sink strength and, therefore, reduced dry matter accumulation.

Expression of the fructosyltransferase was also shown to be a powerful inhibitor of tuber initiation (Fig. 5*b*). Controlled expression of the *SacB* gene in transgenic potato might become a useful model system for studies of the role of sink strength in tuber yield and initiation events.

ACKNOWLEDGEMENTS

The authors wish to thank Susan Baker and Elmer Heppard for their excellent technical assistance. We also thank Enno Krebbers and Oremia Penalver for critical review of the manuscript.

REFERENCES

Bevan M. 1984. Binary *Agrobacterium* vectors for plant transformation. *Nucleic Acids Research* **12**: 8711–8720.

Bieleski RL. 1993. Fructan hydrolysis drives petal expansion in the ephemeral daylily flower. *Plant Physiology* **103**: 213–219.

Bourque JE, Moller JC, Park WD. 1987. Use of an *in vitro* tuberization system to study tuber protein gene expression. *In Vitro* **23**: 381–386.

Caimi PG, McCole LM, Klein TM, Kerr, PS. 1996. Fructan accumulation and sucrose metabolism in transgenic maize endosperm expressing a *Bacillus amyloliquefaciens SacB* gene. *Plant Physiology* **110**: 355–363.

Carpita NC, Keller F, Gibeaut DM, Housley TL, Matile P. 1991. Synthesis of inulin oligomers in tissue slices, protoplasts and intact vacuoles of Jerusalem artichoke. *Journal of Plant Physiology* **138**: 204–210.

Chambert R, Petit-Glatron MF. 1991. Polymerase and hydrolase activities of *Bacillus subtilis* leavansucrase can be separately modulated by site-directed mutagenesis. *Biochemical Journal* **279**: 35–41.

Chatterton NJ, Harrison PA, Thornley WR, Bennett JH. 1993. Separation and quantification of fructan (inulin) oligomers by anion exchange chromatography. In: Fuchs A, ed. *Inulin and Inulin-containing Crops*. Amsterdam: Elsevier, 93–99.

Darwen CWE, John P. 1989. Localization of the enzymes of fructan metabolism in vacuoles isolated by a mechanical method from tubers of Jerusalem artichoke. *Plant Physiology* **89**: 658–663.

De Block M. 1988. Genotype-independent leaf disc transformation of potato (*Solanum tuberosum*) using *Agrobacterium tumefaciens*. *Theoretical and Applied Genetics* **76**: 767–774.

Dunsmuir P. 1985. The petunia chlorophyll *a*/*b* binding protein genes: a comparison of *Cab* genes from different gene families. *Nucleic Acids Research* **13**: 2503–2518.

Ebskamp MJM, van der Meer IM, Spronk BA, Weisbeek PJ, Smeekens SCM. 1994. Accumulation of fructose polymers in transgenic tobacco. *Biotechnology* **12**: 272–275.

Edelman J, Jefford TG. 1968. The mechanism of fructosan metabolism in higher plants as exemplified in *Helianthus tuberosus*. *New Phytologist* **67**: 517–531.

Frehner M, Keller F, Wiemken A. 1984. Localization of fructan metabolism in the vacuoles isolated from protoplasts of Jerusalem artichoke tubers (*Helianthus tuberosus L.*). *Journal of Plant Physiology* **16**: 197–208.

Heineke D, Wildenberger K, Sonnewald U, Willmitzer L, Heldt HW. 1994. Accumulation of hexoses in leaf vacuoles: studies with transgenic tobacco plants expressing yeast-derived invertase in the cytosol, vacuole or apoplasm. *Planta* **194**: 29–33.

Hendry GAF. 1993. Evolutionary origins and natural functions of fructans – a climatological, biogeographical and mechanistic appraisal. *New Phytologist* **123**: 3–14.

Hershey HP, Stoner TD. 1991. Isolation and characterization of cDNA clones for RNA species induced by substituted benzenesulfonamides in corn. *Plant Molecular Biology* **17**: 679–690.

Hoekema A, Hirsch P, Hooykaas PJ, Schilperoot R. 1983. A binary plant vector strategy based on separation of vir- and T-region of the *Agrobacterium tumefaciens* Ti-plasmid. *Nature* **303**: 200–204.

Horsch RB, Fry JE, Hoffman NL, Eichholtz D, Rogers SG, Frayley RT. 1985. A simple and general method for transferring genes into plants. *Science* **227**: 1229–1232.

Housley TL, Daughtry CST. 1987. Fructan content and fructosyltransferase activity during wheat seed growth. *Plant Physiology* **83**: 4–7.

Housley TL, Pollock CJ. 1993. The metabolism of fructan in higher plants. In: Suzuki M, Chatterton NJ, eds. *Science and Technology of Fructans*. Boca Raton, FL, USA: CRC Press, 191–225.

Mazur BJ, Chui CF. 1985. Sequence of a genomic DNA clone for the small subunit of ribulose bis-phosphate carboxylase-oxygenase from tobacco. *Nucleic Acids Research* **13**: 2373–2386.

Nagarajan V, Borchert T. 1991. Levansucrase: a tool to study protein secretion in *Bacillus subtilis*. *Research Microbiology* **142**: 787–792.

Pietrzak M, Shillito R, Hohn T, Potrykus I. 1986. Expression in plants of two bacterial antibiotic resistance genes after protoplast transformation with a new plant expression vector. *Nucleic Acids Research* **14**: 5857–5868.

Pilon-Smits EAH, Ebskamp MJM, Paul MJ, Jeuken MJW, Weisbeek PJ, Smeekens SCM. 1995. Improved performance of transgenic fructan-accumulating tobacco under drought stress. *Plant Physiology* **107**: 125–130.

Pollock CJ. 1986. Fructan and the metabolism of sucrose in vascular plants. *New Phytologist* **104**: 1–24.

Pollock CJ, Cairns AJ, Sims IM, Housley TL. 1996. Fructans as reserve carbohydrates in crop plants. In: Zamski E, Schaffer

AA, eds. *Photassimilate Distribution in Plants and Crops*. New York: Marcel Dekker, 97–113.

Pontis HG, Del Campillo E. 1985. Fructans. In: Dey PM, Dixon RA, eds. *Biochemistry of Storage Carbohydrates in Green Plants*. London: Academic Press, 205–227.

Rocha-Sosa M, Sonnewald U, Frommer W, Stratmann M, Schell J, Willmitzer L. 1989. Both developmental and metabolic signals activate the promoter of a class I patatin gene. *EMBO Journal* **8**: 23–31.

Sanger F, Nicklen S, Coulson AR. 1977. DNA sequencing with chain-terminating inhibitors. *Proceedings of the National Academy of Sciences, USA* **74**: 5463–5467.

Sambrook J, Fritsch EF, Maniatis T. 1989. In: Ford N, Nolan C, Ferguson M, eds. *Molecular Cloning: a Laboratory Manual*. Cold Spring Harbor, USA: Cold Spring Harbor Press, 74–84.

Sonnewald U, Brauer A, von Schaewen A, Stitt M, Willmitzer L. 1991. Transgenic tobacco plants expressing yeast-derived invertase either in the cytosol, vacuole or apoplast; a powerful tool for studying sucrose metabolism and sink/source interactions. *The Plant Journal* **1**: 95–106.

Sonnewald U, von Schaewen A, Willmitzer L. 1993. Subcellular manipulation of sucrose metabolism in transgenic plants. *Journal of Experimental Botany* **44**: 293–296.

van der Meer IM, Ebskamp MJM, Visser RG, Weisbeek PJ, Smeekens SCM. 1994. Fructan as a new carbohydrate sink in transgenic potato plants. *The Plant Cell* **6**: 561–570.

New Phytol. (1997), **136**, 29–38

The vacuolar sorting domain of sporamin transports GUS, but not levansucrase, to the plant vacuole

By STEFAN C. H. J. TURK[1]*, KOEN DE ROOS[1], PIER A. SCOTTI[1], KEES VAN DUN[2], PETER WEISBEEK[1] AND SJEF C. M. SMEEKENS[1]

[1] *Department of Molecular Cell Biology, Padualaan 8, 3584 CH Utrecht, The Netherlands*

[2] *D. J. van der Have B.V., P.O. Box 1, 4410 AA Rilland, The Netherlands*

(Received 24 July 1996)

SUMMARY

Fructans (polyfructosylsucrose) are synthesized by a number of plants and micro-organisms. Plant fructans are localized in the vacuole and have a low degree of polymerization (DP), whereas the fructans synthesized by micro-organisms are usually much bigger. There is an increasing interest in fructans for both food and non-food applications. In order to accumulate fructans of high DP in the plant vacuole, the levansucrase protein of *Bacillus subtilis* was fused to the vacuolar targeting sequence of sporamin and expressed in plants. Transgenic tobacco plants in which this fusion gene is expressed accumulate fructans to levels up to 21 % of the d. wt. They showed a reduced translocation of carbohydrates, bleaching of the leaves, stunted growth and increased levels of hexoses and starch. The levansucrase protein was not translocated to the plant vacuole, but retained in the endomembrane system, even though the same targeting signal was able to translocate the *E. coli* GUS protein to the plant vacuole.

Key words: Vacuolar protein sorting, GUS, fructans, levansucrase, sporamin.

INTRODUCTION

Plants use photosynthetically fixed carbon for both growth and storage; two important products of photosynthesis are starch and sucrose. Starch synthesized in the leaf chloroplast plays a role in short-term storage of carbohydrates, whereas starch synthesized in other organs such as seeds and tubers can be stored for longer periods in amyloplasts. Sucrose is also used as a storage carbohydrate in several plants, but its major function in most plants is sugar transport.

Fructan (polyfructosylsucrose) is another storage carbohydrate, found in 15 % of flowering plant species (Hendry & Wallace, 1993), including crops like wheat and barley, and bulb-forming plants such as tulip and onion. Fructans are synthesized from sucrose in the plant vacuole by the action of at least two enzymes (Wagner, Keller & Wiemken, 1983; Pontis & del Campillo, 1985; Pollock & Cairns, 1991) and they seldom exceed a DP of 100 fructose units.

Many micro-organisms can also synthesize fructans (Hendry & Wallace, 1993). Their molecular weight exceeds that of plant fructans by at least two orders of magnitude and the degree of polymerization (DP) can exceed 100000 fructose units.

There is increasing interest in the use of fructans in food and non-food applications (Fuchs, 1991, 1993). Two crops, Jerusalem artichoke and chicory, are currently used for large-scale production of fructans (Fuchs, 1991; Fontana, Herman & Guiraud, 1993). Their fructans are of low DP and, because of the presence of fructan degrading enzymes, the fructan concentration and chain-length can vary considerably (Edelman & Jefford, 1964). One possible way to increase the degree of polymerization of fructans harvested from plants is the introduction of microbial fructosyltransferases into plants that do not make fructans and therefore do not contain the degrading enzymes. In order to test this possibility, tobacco plants were transformed with a construct encoding a levansucrase protein from *Bacillus subtilis* (Ebskamp *et al.*, 1994). Since plant fructans are synthesized and stored in the vacuole, the bacterial levansucrase enzyme was fused to the carboxy peptidase Y (CPY) vacuolar targeting signal from yeast. It was found that the transgenic tobacco plants accumulated fructans to levels of *c.* 8 % of the d. wt

* To whom correspondence should be addressed.
E-mail: s.c.h.j.turk@biol.ruu.nl

and that these fructans were not degraded during plant senescence.

Recent investigations have shown that the yeast CPY is not an efficient vacuolar targeting signal in plants (S. Turk, unpublished). To achieve more efficient vacuolar targeting of the levansucrase protein, the plant vacuolar targeting sequence of the sporamin protein from sweet potato (*Ipomoea batatas* L.) was fused to the levansucrase protein. Sporamin is a vacuolar storage protein and is synthesized as a preproprotein. The N-terminal region contains a signal sequence for translocation across the ER membrane, followed by a prosequence of 16 amino acids, essential for targeting to the vacuole (Hattori, Ichihara & Nakamura, 1987). Deletion of this region leads to secretion of the sporamin protein (Matsuoka & Nakamura, 1991). When the sweet potato sporamin gene is expressed in tobacco protoplasts, the sporamin protein is correctly translocated to the vacuole, showing that the targeting signal is functional in heterologous plant species (Matsuoka *et al.*, 1990; Schroeder *et al.*, 1993). Upon entering the ER, proteins can become glycosylated at conserved sequences, which can have an effect on the enzymatic activity. For the *E. coli* reporter enzyme β-glucuronidase (GUS), it was shown that glycosylation has a negative effect on the enzymatic activity of the protein, and inhibition of glycosylation by the addition of tunicamycin increased the GUS activity more than 100-fold (Farrell & Beachy, 1990). Levansucrase activity, however, is not affected by glycosylation. Scotti *et al.* (1996) used endo H to deglycosylate yeast-derived α-amylase-levansucrase and found no difference in activity between the glycosylated and non-glycosylated sporamin-levansucrase protein. Inhibiting glycosylation, by the addition of tunicamycin, also had no effect on the activity of the levansucrase in a tobacco protoplast transformation assay (M. Ebskamp, unpublished). In this study, we tested whether the vacuolar targeting signal of sporamin was capable of translocating heterologous proteins to the plant vacuole. The N-terminal 111 amino acids were fused to both the GUS N^{358} protein, a mutant protein which no longer can be glycosylated, and to the extended levansucrase protein. The latter has a 20 amino acid C-terminal extension and synthesizes fructan polymers of higher molecular weight than the fructans synthesized by a levansucrase which lacks this extension (Chambert, Rain-Guion & Petit-Glatron, 1992).

MATERIALS AND METHODS

Plasmid construction

Plasmid pSTU8 was constructed by cloning a 1·3 kb *Eco*R I–*Nde* I fragment from plasmid pGMK52 (Chambert *et al.*, 1992), encoding for the extended

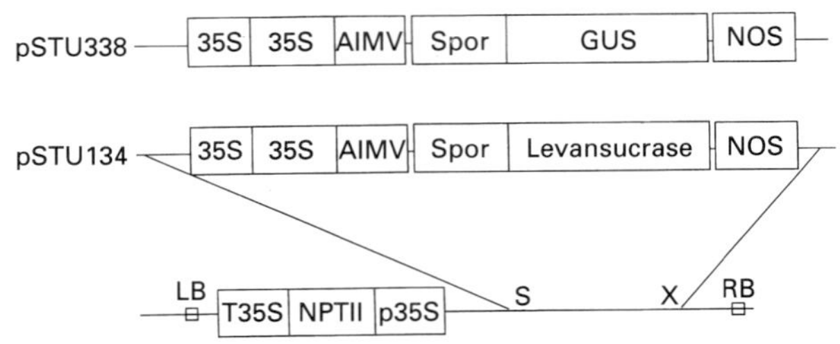

Figure 1. Schematic representation of the constructs used. The *Sma* I (S) and *Xho* I (X) restriction sites, used to clone the constructs between the borders of the binary vector are indicated. Abbreviations: S, *Sma* I; X, *Xho* I; LB, left border; RB, right border; p35S, CaMV 35S promoter; T35S, CaMV 35S terminator.

levansucrase enzyme of *Bacillus subtilis*, into *Eco*R I–*Nde* I digested plasmid pMTL22 (Chambers *et al.*, 1988). In order to introduce a *Nco* I restriction site near the mature processing site of the levansucrase gene a 120 bp *Nco* I–*Eco*R I fragment from plasmid pKP (Ebskamp *et al.*, 1994) was cloned in the *Nco* I–*Eco*R I digested plasmid pSTU8 resulting in plasmid pSTU22. Subsequently, the 1·6 kb *Nco* I–*Bam*H I fragment from pSTU22 was cloned in the *Nco* I–*Bam*H I digested expression vector pPA2, of which the sequence encoding β-glucuronidase was deleted, to yield plasmid pSTU42. Plasmid pPA2 was constructed by cloning the *Eco*R I–*Hind* III fragment of pMOG18 (Sijmonds *et al.*, 1990) in the polylinker of pBluescript II KS (Stratagene cloning systems, La Jolla, CA, USA).

In order to construct the sporamin–levansucrase hybrid construct, a 333 bp DNA fragment, encoding the N-terminal 111 amino acids of the sporamin protein from *Ipomoea batatas*, was amplified by PCR from plasmid gSPO-A1 (Hattori & Nakamura, 1988) using the primers PRST1 5′-TCTCATCTGCCATCATGAAAGCCCTCAC 3′ and PRST2 5′AAAGTCTGGTACGTCGACGT-CATGA 3′. This fragment was digested with *Bsp*H I and cloned into *Nco* I digested plasmid pSTU42 to yield pSTU60.

Subsequently, the 3·0 kb *Sma* I–*Xho* I fragment of pSTU60, containing the sporamin–levansucrase hybrid construct under the control of the 35S promoter and the NOS terminator, was cloned in a pBin19-derived binary vector to yield pSTU134 (Bevan, 1984; Frisch *et al.*, 1995; Fig. 1).

In order to construct the sporamin-β-glucuronidase hybrid construct, a 333 bp PCR fragment encoding the N-terminal portion of the sporamin protein from *I. batatas* was digested with *Bsp*H I and cloned into the *Nco* I site of pPA2, which encodes a glycosylation deficient β-glucuronidase protein, yielding pSTU338 (Fig. 1).

Plant transformation

Plasmid pSTU134 was transformed into *Agrobacterium tumefaciens* strain LBA4404 using electroporation (Mattanovich *et al.*, 1989). The construct was introduced into *Nicotiana tabacum* var. Samson NN using the leaf disk transformation method (Horsch *et al.*, 1985). Regenerated plants, named SP–LS plants, were selected for kanamycin resistance and grown on MS medium (Murashige & Skoog, 1962) containing 3% glucose.

Extraction and analysis of carbohydrates

Tobacco plants were grown for 11 wk in a glasshouse, then *c.* 250 mg of fresh leaf material was harvested at the end of the light period, frozen in liquid nitrogen and pulverized with a mortar and pestle. This material was homogenized in 300 μl of sterile water, incubated for 5 min at 90 °C and centrifuged at 12000 ***g*** for 5 min. This extraction method was repeated once with 300 μl of sterile water at 90 °C, once with 300 μl of 80% ethanol at room temperature and once with 100 μl of sterile water at room temperature, and the supernatants were combined. The pellet was used for starch determination as described by Van der Meer *et al.* (1994). Fructan content of the leaf material was determined by spotting 1 μl of the supernatant on silica gel TLC foils (Schleicher & Schuell) as described by Cairns & Pollock (1988). The TLC was developed three times in 85:15 acetone:water and then treated with urea-phosphoric-spray as described by Wise *et al.* (1995). Quantification of the fructan was determined by scanning the urea-spray-stained TLC foil (Gelscan, Pharmacia-LKB), on which standard concentrations of commercially available fructans (levan, Sigma) were also spotted. Sucrose, fructose and glucose levels were quantified by HPLC as described by Van der Meer *et al.* (1994). For each transformed line at least 20 plants were used.

Protoplast isolation and transformation

Tobacco protoplasts were isolated as described by Goodall *et al.* (1990). Briefly, leaves were collected from 8–14-wk-old tobacco plants grown under sterile conditions. After removal of the midrib the leaves were transferred to a Petri dish containing 11 ml of K_3S (Goodall, Wiebauer & Filipowicz, 1990) supplemented with 0·4% macerozyme and 1·2% cellulase (Yakult Honsha Co. Ltd, Tokyo, Japan), cut into pieces of 1 mm^2 and incubated overnight at 27 °C in the dark. Subsequently the suspension was sieved through a 230 mesh filter and centrifuged at 80 ***g*** for 10 min. Intact protoplasts, floating on top of the suspension, were collected, washed with 10 ml of W5-buffer (154 mM NaCl, 125 mM $CaCl_2.2H_2O$, 5 mM KCl, 5 mM glucose, pH 5·8) and pelleted by centrifugation (5 min at 80 ***g***). The protoplasts were resuspended in W5 buffer, counted using a haemocytometer slide and left at room temperature for at least 30 min. Finally, they were pelleted by centrifugation (5 min at 80 ***g***) and resuspended in $MgCl_2$ solution (15 mM $MgCl_2$, 0·4 M mannitol, 0·1% MES pH 5·6) at a density of 3×10^6 protoplasts ml^{-1}. Next, 0·3 ml of this suspension was transformed by adding 30–40 μg of plasmid DNA, followed by the addition of 0·3 ml of PEG-solution (40% PEG 6000, 0·4 M mannitol, 0·1 M $Ca(NO_3)_2.4H_2O$, pH 8·0) and immediate mixing. After 5–10 min, 1 ml, 2 ml and 7 ml of W5-buffer respectively were slowly added at intervals of 3 min. After centrifugation (5 min at 80 ***g***) the protoplasts were resuspended in 1 ml of K_3G (Goodall *et al.*, 1990) and incubated for 18 h at 27 °C in the dark.

Vacuole isolation

Vacuoles were isolated by resuspending 5×10^6 protoplasts in 35 ml of lysis buffer (0·33 M sorbitol, 2 mM EDTA, 1 mM $MnCl_2$, 1 mM $MgCl_2$, 0·1% fatty-acid-free BSA, 1% Percoll®, 10 μg ml^{-1} neutral red, 50 mM HEPES/KOH, pH 7·3), which was prewarmed at 45 °C. The lysed protoplasts were left for 5–10 min at room temperature after which the lysate was centrifuged at 325 ***g*** for 10 min at 4 °C. Vacuoles were isolated from the top, and the purity of the isolated vacuolar fraction was determined by measuring both the activity of the vacuolar marker enzyme α-mannosidase (Boller & Kende, 1979) and the cytosolic marker enzyme glucose 6-phosphate dehydrogenase (Cairns, 1987).

β-Glucuronidase activity assay

The β-glucuronidase activity was determined by incubating the protein extract in 500 μl of GUS buffer (50 mM sodium phosphate pH 7·4, 10 mM EDTA, 0·1% Triton X-100®, 0·1% sarcosyl, 10 mM β-mercaptoethanol, 0·5 mM leupeptin, 0·5 mM antipain, 1 mM 4-methylumbelliferyl-β-D-glucuronide) at 37 °C. The reaction was stopped by adding 100 μl of the reaction mixture to 2 ml Na_2CO_3, and the formation of the reaction product, 4-methylumbelliferone, was quantified by measuring fluorescence at 455 nm after excitation at 365 nm in a Perkin Elmer fluorimeter.

Levansucrase activity assay

Levansucrase activity was determined by incubating the enzyme overnight at 30 °C in 50 μl of reaction-buffer (0·1 M sucrose, 14·8 kBq [^{14}C]sucrose, 0·1% Triton X-100, 0·5 mM leupeptin, 0·5 mM antipain, 50 mM phosphate buffer pH 6·0). Following incubation, the fructan content was determined by spotting 2 μl on silica-gel TLC foils (Schleicher & Schuell) as described by Cairns & Pollock (1988). The TLC was developed three times in 85:15

acetone:water then exposed to X-ray film for 5 d. The fructan spots at the origin were quantified using the Personal Densitometer SI (Molecular Dynamics) and a concentration range of increasing amounts of fructan to ensure that the signal was in the linear response range.

Binding of levansucrase to concanavalin A

In order to test the binding of the sporamin–levansucrase hybrid protein to concanavalin A (Con A), protoplasts were transformed as described above. After the overnight incubation at 27 °C the protoplasts were pelleted in W5 medium, lysed in 300 μl of protein extract buffer (Con A buffer (20 mM Tris-HCl pH 7·4, 0·5 M NaCl, 1 mM $MnCl_2$, 1 mM $CaCl_2$, 0·5 % NaN_3) supplemented with 1 mM PMSF, 1 $\mu g\ ml^{-1}$ leupeptin, 1 $\mu g\ ml^{-1}$ antipain) and centrifuged (5 min at 9000 ***g***) to remove cell debris. A column containing 300 μl of Con A sepharose (Pharmacia) was equilibrated with 10 ml of Con A buffer, and 275 μl of the protein extract was applied to the equilibrated column after which the column was sealed. After incubation for 2 h at room temperature, the column was allowed to empty by gravity flow, then washed with 20 ml of Con A buffer. Bound proteins were eluted after overnight incubation at 4 °C with 300 μl of elution buffer (50 mM MES pH 6·0, 0·5 M NaCl, 1 mM $MnCl_2$, 1 mM $CaCl_2$, 0·5 % NaN_3, 1 M α-methyl-D-mannoside). Levansucrase activity in the different samples was determined as described above.

RESULTS

The amino terminal part of sporamin can direct GUS to the plant vacuole

Previous reports showed that the N-terminus of sporamin is essential for translocation to the vacuole and that the sporamin protein from sweet potato is correctly targeted to the vacuole, when expressed in tobacco protoplasts (Matsuoka *et al.*, 1990; Schroeder *et al.*, 1993). To test whether or not the sporamin-targeting signal is also capable of translocating heterologous proteins to the vacuole, the N-terminal 111 amino acids of the sporamin protein were fused to a GUS N^{358} protein from which the glycosylation site had been removed (Farrell & Beachy, 1990). This construct was placed under the control of the CaMV 35S promoter, yielding pSTU338 (Fig. 1), and used to transform wild-type tobacco protoplasts, then the GUS activity was determined following overnight incubation. Addition of protease inhibitors to the GUS assay resulted in a fivefold higher GUS activity, indicating that the GUS protein was sensitive to proteases (results not shown). Subsequently, vacuoles were isolated from the transformed protoplasts and the GUS activity determined, together with the activity of the vacuolar marker α-mannosidase and the cytoplasmic marker glucose-6-phosphate dehydrogenase (Table 1). For co-localization experiments, the GUS activity was expressed per unit α-mannosidase or glucose-6-phosphate dehydrogenase. The GUS activity of the protoplasts and vacuoles was 7660 and 8880 pmol MU per unit α-mannosidase and 782 and 2192 pmol MU per unit glucose-6-phosphate dehydrogenase, respectively. These results and similar results obtained in other independent experiments show that the sporamin-targeted GUS protein co-localizes with the vacuolar marker enzyme α-mannosidase and not with the cytoplasmic marker glucose-6-phosphate dehydrogenase. We conclude, therefore, that the targeting sequence of sporamin is indeed capable of targeting a heterologous protein to the vacuole.

The sporamin–levansucrase protein enters the ER

The same sporamin-targeting sequence was fused to the levansucrase gene (Fig. 1) and expressed in protoplasts. In order to determine whether the sporamin–levansucrase hybrid protein is correctly translocated into the endomembrane system, the glycosylation status of the protein was tested by

Table 1. *Vacuolar localization of the sporamin–GUS protein*

	Protoplasts	Vacuoles
GUS activity (pmol $min^{-1}\ \mu l^{-1}$)	4·52	1·83
α-mannosidase activity (ΔOD405 $\mu l^{-1}\ min^{-1}$)	$5{\cdot}90 \times 10^{-4}$	$2{\cdot}06 \times 10^{-4}$
Gluc-6P-dehydr. activity (ΔOD595 $\mu l^{-1}\ min^{-1}$	$57{\cdot}8 \times 10^{-4}$	$8{\cdot}35 \times 10^{-4}$
GUS/α-mannosidase	7660	8880
GUS/Gluc-6P-dehydrogenase	782	2192

Protoplasts from wild-type tobacco plants were transformed with DNA, encoding the sporamin–GUS protein. GUS activity was measured in protoplast and isolated vacuoles together with the vacuolar marker enzyme α-mannosidase and the cytosolic marker enzyme glucose 6-phosphate dehydrogenase.

determining the affinity for concanavalin A (Con A), which binds sugar residues present on glycosylated proteins. The levansucrase protein contains six potential glycosylation sites and it was shown by Scotti *et al.* (1996) that it was glycosylated when targeted to the ER of yeast by an α-amylase signal sequence. The non-glycosylated bacterial levansucrase protein isolated from *B. subtilis*, and the glycosylated α-amylase targeted levansucrase protein, isolated from yeast (Scotti *et al.* 1996), were used as controls. Con A affinity chromatography revealed that the yeast-derived glycosylated levansucrase protein was retained by the column while the non-glycosylated protein did not bind (Table 2). By this procedure, only 30 % of the non-glycosylated protein was detected in the unbound material. It was found that, using a *B. subtilis* levansucrase-specific antibody in a dot blot analysis (Scotti *et al.*, 1996), the remaining 69 % of the levansucrase protein could be detected in the wash solution, indicating that the levansucrase protein exhibits some non-specific affinity for the Con A column (results not shown). Extracts were prepared from tobacco protoplasts transformed with the sporamin–levansucrase construct. Con A chromatography showed that 47 % of the levansucrase activity could be eluted from the column (Table 2). This result, and others obtained from separate experiments, indicate that the sporamin–levansucrase hybrid protein was glycosylated and therefore translocated into the endoplasmic reticulum.

The sporamin–levansucrase protein does not reach the vacuole

Transgenic tobacco plants were produced harbouring plasmid pSTU134 (Fig. 1). Vacuoles were isolated from SP–LS 'line 40' plants (see below) and the vacuolar marker enzyme α-mannosidase was used to isolate the fructans contained in equal amounts of protoplasts and vacuoles. These samples were separated on TLC and fructans visualized by spraying the TLC with a fructose-specific reagent. Fructans larger than 20 fructose units stay at the origin and give the characteristic blue colour. At the origin in the vacuole lane only a brownish precipitate is visible, derived from the neutral red solution which forms a precipitate in the vacuole isolation buffer. This analysis shows that fructans present in the protoplasts do not co-localize with the vacuolar fraction (Fig. 2*a*). Thus, most of the fructan is located outside the vacuole. This finding suggests that either the sporamin–levansucrase protein is not correctly targeted to the vacuole or that the vacuoles of young leaves do not contain sufficient sucrose for fructan synthesis. In order to distinguish between these two possibilities, protoplasts were isolated from wild-type tobacco plants and SP–LS line 40 plants and used for vacuole isolation. Protein extracts, from protoplasts and isolated vacuoles, containing equal amounts of the vacuolar marker enzyme α-mannosidase, were incubated overnight in the presence of 100 mM [^{14}C]sucrose. The fructans made during the overnight incubation were separated on TLC and exposed to an X-ray film. In extracts of wild-type protoplasts and vacuoles no fructans were synthesized, although in the vacuolar extract, probably because of a slightly lower invertase activity, some residual sucrose could be detected. In the vacuolar extract, of SP–LS line 40 plants, 20 times less fructan was synthesized than in the protoplast extract (Fig. 2*b*), providing additional evidence that the sporamin–levansucrase protein was not correctly targeted to the vacuole.

Glycosylation is known to affect protein folding and might prevent proper targeting of the levansucrase to the vacuole. This hypothesis was tested by incubating wild-type tobacco protoplasts, transformed with the sporamin–levansucrase construct, in the presence of 10 μg ml^{-1} tunicamycin. Following incubation, protoplast and vacuolar extracts were incubated in the presence of 100 mM [^{14}C]sucrose and analysed on TLC. This analysis revealed that 14 times more fructan was synthesized with protoplast extract than in the vacuolar extract (results not shown). This suggests that the non-glycosylated levansucrase protein was also not targeted to the plant vacuole. Proteins devoid of sorting information are targeted to the apoplast (Dorel *et al.*, 1989; Denecke, Botterman & Deblaere, 1990). The sporamin–levansucrase protein was translocated into the endomembrane system but did not reach the vacuole and might therefore use the default pathway to the apoplast. In order to test this hypothesis, proteins were isolated from intact leaf and isolated protoplasts of SP–LS line 40. The levansucrase activity was determined in protein extracts containing equal amounts of chlorophyll. Quantitation of the fructan spots revealed that 96 % of the enzymatic activity present in total leaf extract localized inside the protoplasts. Thus we suggest that the sporamin–levansucrase protein is not targeted to the apoplast but is retained in the endomembrane system.

Fructan content of sporamin–levansucrase plants

To determine the fructan content of the sporamin–levansucrase plants, 19 independent primary transformants were grown in soil and tested for fructan accumulation. These primary transformants accumulated fructans in the range from 0·2 to 3·8 % of d. wt. Three transgenic plants (26, 53 and 40) were selected which accumulated 0·6, 1·3 and 3·8 % of the d. wt as fructans respectively. Twenty plants of the S1 generation from each of these three lines were grown in soil in the glasshouse for 11 wk, and leaves of each plant were harvested for fructan analysis. The average fructan content of SP–LS line

Table 2. *Glycosylation status of the sporamin–levansucrase protein*

Sample	Extract	Flow through	Eluted from column
LS (Gly$^-$, *B. subtilis*)	100	30	1
LS (Gly$^+$, Yeast)	100	1·6	10
Spor–LS	100	0	47

All values are percentages.

Non-glycosylated levansucrase from *Bacillus subtilis*, glycosylated amy-levansucrase isolated from yeast and the sporamin–levansucrase protein isolated from transformed tobacco protoplasts were applied to a 0·5 ml Con A column (total extract). After a 2-h incubation period, the column was rinsed with 20 ml of Con A buffer. Bound proteins were eluted from the column with 1 M α-methyl mannoside and levansucrase activity determined.

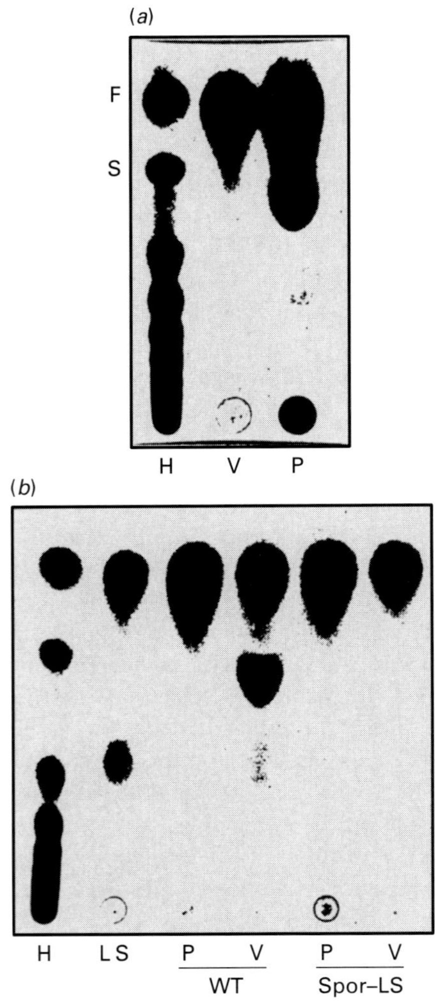

Figure 2. Protoplasts (P) and vacuoles (V) were isolated from a transgenic SP–LS line-40 plant. Fructans were isolated from both protoplasts and vacuolar extracts each containing equal amounts of the vacuolar marker α-mannosidase, and analysed on TLC (*a*). Proteins were extracted from protoplasts (P) and isolated vacuoles (V), from both a wild-type (WT) and a transgenic SP–LS line-40 (Spor–LS) plant. Protein extracts, containing equal amounts of the vacuolar marker α-mannosidase, were incubated in the presence of [^{14}C]sucrose. The products produced were separated on TLC which was exposed to X-ray film (*b*). Fructans larger than 20 fructose units reside at the origin. H, Fructans isolated from *Helianthus tuberosus*; F, fructose; S, sucrose.

Table 3. *Fructan content of the three different SP–LS lines*

SP–LS line	Fructan/ d. wt	Fructan/ f. wt	Range/ d. wt
26	2·3 ± 0·6	0·23 ± 0·07	1·6–3·7
40	6 ± 4	0·8 ± 0·7	1·2–14·2
53	7 ± 5	0·6 ± 0·4	3·8–21·2

All values are percentages.

Fructans were isolated from lower leaves after a light period of 8 h, separated on TLC and quantified. The average fructan content, as well as the fructan range (range) were determined and expressed as mg 100 mg^{-1} d. wt or mg 100 mg^{-1} f. wt.

26, 40 and 53 was 2·3, 6 and 7 %, of d. wt, respectively, with the highest accumulators of each line containing up to 3·7, 14·2 and 21·2 % of the d. wt as fructans. The average f. wt fructan contents of these lines were 0·23, 0·8 and 0·6 %, respectively (Table 3). These data show that, with the sporamin–levansucrase construct, tobacco plants can accumulate fructans up to 21 % of d. wt. Moreover, it is clear that the endomembrane system of tobacco plants contains sufficient amounts of sucrose for fructan synthesis.

Sugar analysis of fructan-accumulating tobacco plants

In order to determine whether fructan accumulation has an effect on the levels of soluble sugars and starch, the average glucose, fructose, sucrose and starch levels of the SP–LS line-40 plants were determined. In wild-type plants, levels of glucose, fructose and sucrose were low. In fructan-accumulating transgenic plants, however, the levels of glucose, fructose and sucrose increased, to up to 10 times the levels found in wild-type plants. Plotting the hexose content of the individual plants against the fructan content shows a correlation between the increase in glucose, fructose and sucrose and the

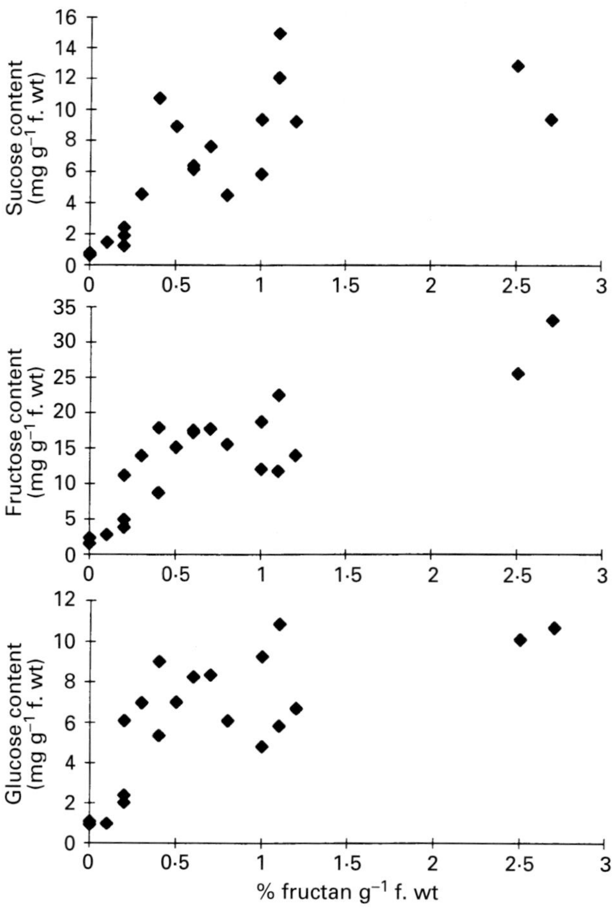

Figure 3. Determination of the glucose, fructose and sucrose content in 11-wk-old SP–LS line-40 plants after a light period of 8 h. Values are expressed as mg g^{-1} fresh leaf tissue.

Table 4. *Effect of fructan accumulation on plant height*

	Average plant height (cm)	Range (cm)
WT	79±12	62–100
SP–LS		
Line 26	74±10	53–90
Line 40	52±11	28–74
Line 53	66±12	50–80

Plant height was determined for wild-type plants and the three transgenic SP–LS lines after 11 wk in the glasshouse. For each line, the average plant height and the lowest and highest plant (range) are shown. WT, wild type.

Figure 4. Effect of fructan accumulation on root growth. Roots of wild-type plants (right) and SP–LS line-40 (left), grown for 11 wk in a glasshouse.

amount of fructan (Fig. 3). To determine whether or not the starch levels were also altered in the leaves of the transgenic plants, starch was isolated from mature leaves of wild-type plants and plants of SP–LS line 40 which did not show any bleaching (see below) even though they contained 10% of d. wt as fructans. Analysis of the starch contents revealed that leaves of the SP–LS plants contained three times more starch than the leaves of the corresponding wild-type plants (30 mg g^{-1} f. wt vs. 10 mg g^{-1} f. wt). These results suggest that, in the SP–LS plants which accumulate high levels of fructan, the translocation of carbohydrates from source to sink cells is impaired.

Phenotypic effects of fructan accumulation

It is known that tobacco plants with a reduced capacity to translocate carbohydrates show stunted growth, bleaching of leaves, reduced root growth and increased levels of hexoses and starch (Sonnewald *et al.*, 1991; Reismeier, Willmitzer & Frommer, 1994). Fructan-accumulating SP–LS plants also had increased levels of hexoses (Fig. 3) and showed stunted growth, as indicated by the average plant height (Table 4). The reduction in plant height was most pronounced in the individual plants with the highest fructan content (data not shown) and was not significantly different from wild-type for the low accumulator SP–LS line 26. The root size of SP–LS line-40 plants, with a high fructan content, was reduced compared with that of the wild type (Fig. 4). The lower stem region of these transgenics was brownish, whereas that of the wild type was green. These observations indicate that, in the transgenic plants, the translocation of carbohydrates is reduced. Plants of all three lines which accumulate high levels of fructan showed bleaching of the leaves, whilst this was not observed in non-transformed plants and low fructan accumulators. The bleaching of the leaves of line-40 plants was most severe. In all three lines, leaves started to bleach at the rim and between the veins and bleaching was only observed in mature leaves.

DISCUSSION

The production of fructans in plants which normally do not accumulate fructans can be advantageous, especially in high-yielding crops such as potato or sugar beet, from which fructan-degrading enzymes appear to be absent. Moreover, novel fructan types can be generated by the introduction of novel enzymatic activities The vacuole is probably the most compatible compartment for high-level fructan accumulation, and constructs were assembled with the intention of stimulating fructan synthesis in the vacuole of transgenic plants. Most soluble proteins destined for the vacuole are first translocated across the ER membrane, then traverse the ER–Golgi system. Such proteins must contain sorting information for vacuolar targeting, which selects these proteins for vacuolar transport probably in a trans-Golgi compartment. Such information can be present in different regions of the protein, and deletion of these sequences leads to missorting (for a review see Nakamura & Matsuoka, 1993). Such proteins follow the 'default' pathway and are usually secreted into the extracellular environment. We previously used the yeast CPY targeting sequence to translocate the bacterial levansucrase protein to plant vacuoles (Ebskamp *et al.*, 1994). Transgenic tobacco plants expressing this hybrid gene accumulated up to 8 % of d. wt as fructans. Subsequent analysis of these plants showed that CPY-targeted levansucrase did not lead to efficient vacuolar fructan accumulation. Recently it was shown that for proper targeting of the CPY protein to the yeast vacuole, the VPS10 receptor protein is required (Marcusson *et al.*, 1994). Such a VPS10 receptor-like molecule might not be present in plants. Therefore, new constructs were created, using the vacuolar targeting sequence of the plant vacuolar protein sporamin. In this study we have demonstrated that a protein fragment containing the N-terminal-targeting sequence and 74 amino acids of the sporamin protein is able to translocate the *E. coli* GUS protein to the plant vacuole. Use of this targeting sequence for vacuolar transport of the *B. subtilis* levansucrase resulted in proper translocation into the endoplasmic reticulum as shown by glycosylation of the protein. However, levansucrase was not properly translocated to the vacuole (Fig. 2). Moreover, all the enzymatic activity is retained inside the protoplast. It is possible that the sporamin–levansucrase protein is already committed to a (pre)vacuolar compartment where it becomes trapped, and this might explain its retention inside the cell. Recently it was shown that plant cells contain functionally distinct vacuolar compartments (Paris *et al.*, 1996). The sporamin–GUS hybrid protein co-localizes with the vacuolar marker enzyme α-mannosidase (Table 1) and sequences in the levansucrase protein might target the sporamin–levansucrase hybrid protein to another 'vacuolar' compartment, or prevent entry into the vacuole. Alternatively, the sporamin–levansucrase protein could be retained in the secretory pathway. In yeast, the signal sequence of α-amylase is able to translocate the levansucrase protein across the endoplasmic reticulum, but the levansucrase activity is retained within the cell (Scotti *et al.*, 1996). A retention signal present in the levansucrase protein might, therefore, interfere with protein secretion both in plant cells and in yeast. Different retention signals are known for both the ER and the Golgi network. Signals such as the C-terminal tetrapeptide KDEL (Munro & Pelham, 1987) retrieve soluble proteins from the Golgi complex back to the ER lumen. Also the tetrapeptide YQRL is sufficient for trans-Golgi localization of proteins (Humphrey *et al.*, 1993). Testing the levansucrase sequence for the presence of known retention signals did not reveal the presence of known ER or Golgi retention signals. More research is necessary to localize the protein in the endomembrane system and to determine which sequences, if any, are responsible for the retention of the sporamin–levansucrase hybrid protein in the secretory pathway.

Glycosylation of the levansucrase is known to affect protein folding and an altered protein structure might affect the transit of the levansucrase to the vacuole. In order to determine whether glycosylation affects vacuolar targeting, vacuoles isolated from protoplast were transformed in the presence of the glycosylation inhibitor tunicamycin. In this vacuolar extract, little levansucrase activity could be detected, indicating that the non-glycosylated sporamin–levansucrase protein is not targeted to the vacuole. This suggests that glycosylation of the levansucrase protein is not causing its retention in the endomembrane system. The levansucrase gene used in this study is a modified *sacB* gene in which a stop codon is changed into a glutamine (pGMK52), encoding a levansucrase with a 20-amino-acid C-terminal extension (Chambert *et al.*, 1992). This extension is rich in hydrophobic amino acids which might slow down transit of proteins from the ER (Okita & Rogers, 1996). The presence of these hydrophobic amino acids delays transit of the levansucrase protein from the ER, thereby preventing proper vacuolar targeting.

Transformation of tobacco with the sporamin–levansucrase hybrid gene resulted in 19 independent transformants which accumulated fructans up to 3·8 % of the d. wt. The amount of fructan present in the S1 offspring is, however, much higher, which can be explained by the fact that the primary transformants are heterozygous, whereas some of the S1 plants are homozygous, for the levansucrase gene. However, the environmental conditions which control growth and the rate of photosynthesis (e.g. temperature and light intensity) might also play a major role in the accumulation of fructan. Analysis of the offspring of three transformants shows that the highest expresser (SP–LS line 40) shows bleaching of the leaves, stunted growth, reduced root growth together with an accumulation of glucose, fructose, sucrose and starch, while the phenotypes of the other two lines were much less severe. These phenotypes are similar to those found in plants with reduced sucrose transporter activity (Riesmeier, Willmitzer & Frommer, 1994). Thus it appears likely that the translocation of carbohydrates is also disturbed in the sporamin–levansucrase plants. The sporamin–levansucrase construct is under the control of the CaMV 35S promoter, which is active in the phloem tissue and the companion cells. Possibly fructans accumulated in these tissues block the transport of carbohydrates to the sink tissue. The amount of fructan accumulated in these plants is not significantly different from the amount which accumulates in several plants of the middle expresser (SP–LS line 53; Table 3). Thus fructan alone might not be the only determinant of the observed phenotype. The levansucrase enzyme from *B. subtilis* can transfer fructosyl units from sucrose to the growing fructan chain, thereby liberating glucose, while the enzyme also exhibits some invertase activity (Dedonder, 1966). These monosaccharides, when translocated from the endomembrane system to the cytoplasm are phosphorylated by hexokinase, resulting in a decrease in the levels of inorganic phosphate. Low cytosolic levels of inorganic phosphate will lead to an accumulation of starch in the source leaves and, possibly, to reduced growth. Hexokinase has also been proposed to act as a key sensor and signal transmitter of sugar-mediated repression in plants (Jang & Sheen, 1994). The addition of glucose, fructose or other substrates of hexokinase represses photosynthetic gene transcription. This is not caused by a depletion of cytosolic phosphate or ATP but is probably mediated by the active hexokinase. Mannoheptulose, a specific inhibitor of hexokinase, blocks the repression caused by glucose, whereas the phosphorylated glucose itself does not act as a repression signal (Jang & Sheen, 1994). Recently it was reported that the accumulation of hexoses results in the induction of systemic acquired resistance and a sugar-induced hypersensitive response (HR) (Herbers *et al.*, 1996). Whether in our transgenic plants the accumulation of hexoses also results in an induction of the HR response remains to be determined.

ACKNOWLEDGEMENTS

We would like to thank Irma Vijn for her assistance with the Con A experiments and Nathalie Gerrits and Michel Ebskamp for critical reading of the manuscript. This work was funded in part by the European Communities 'BIOTECH' programme, as part of the Project of Technological Priority 1993–1996 (S.T. and P.S.).

REFERENCES

Bevan M. 1984. Binary *Agrobacterium* vectors for plant transformation. *Nucleic Acids Research* **12**: 8711–8721.

Boller T, Kende H. 1979. Hydrolytic enzymes in the central vacuole of plant cells. *Plant Physiology* **63**: 1123–1132.

Cairns AJ. 1987. Colorimetric microtiter plate assay of glucose and fructose by enzyme linked formazan production: applicability to the measurement of fructosyl transferase activity in higher plants. *Analytical Biochemistry* **167**: 270–278.

Cairns AJ, Pollock CJ. 1988. Fructan biosynthesis in excised leaves of *Lolium temulentum* L. I. Chromatographic characterization of oligofructans and their labelling patterns following $^{14}CO_2$ feeding. *New Phytologist* **109**: 399–405.

Chambers SP, Prior SE, Barstow DA, Minton NP. 1988. The PMTL *nic*$^-$ cloning vectors. I. Improved pC polylinker regions to facilitate the use of sonicated DNA for nucleotide sequencing. *Gene* **68**: 139–149.

Chambert R, Rain-Guion MC, Petit-Glatron MF. 1992. Readthrough of the *Bacillus subtilis* levansucrase stop codon produces an extended enzyme displaying a higher polymerase activity. *Biochimica et Biophysica Acta* **1132**: 145–153.

Dedonder R. 1966. Levansucrase from *Bacillus subtilis*. *Methods in Enzymology* **8**: 500–505.

Denecke J, Botterman J, Deblaere R. 1990. Protein secretion in plant cells can occur *via* a default pathway. *The Plant Cell* **2**: 51–59.

Dorel C, Voelker TA, Herman EM, Chrispeels MJ. 1989. Transport of proteins to the plant vacuole is not by bulk flow through the secretory system, and requires positive sorting information. *Journal of Cell Biology* **108**: 327–337.

Ebskamp MJM, van der Meer IM, Spronk BA, Weisbeek PJ, Smeekens SCM. 1994. Accumulation of fructose polymers in transgenic tobacco. *Biotechnology* **12**: 272–274.

Edelman J, Jefford TG. 1964. The metabolism of fructose polymers in plants. *Biochemical Journal* **93**: 148–161.

Farrell LB, Beachy RN. 1990. Manipulation of β-glucuronidase for use as a reporter in vacuolar targeting studies. *Plant Molecular Biology* **15**: 821–825.

Fontana A, Herman B, Guiraud J. 1993. Production of high-fructose-containing syrups from Jerusalem artichoke extracts with fructose enrichment through fermentation. In: Fuchs A, ed. *Inulin and Inulin-Containing Crops*. Amsterdam: Elsevier Science Publishers, 251–258.

Frisch DA, Harris-Haller LW, Yokubaitis NT, Thomas TL, Hardin SH, Hall TC. 1995. Complete sequence of the binary vector Bin19. *Plant Molecular Biology* **27**: 405–409.

Fuchs A. 1991. Current and potential food and non-food applications of fructans. *Biochemical Society Transactions* **19**: 555–560.

Fuchs A. (ed.) 1993. *Inulin and inulin-containing crops*. Amsterdam: Elsevier Science Publishers.

Goodall GJ, Wiebauer K, Filipowicz W. 1990. Analysis of pre mRNA processing in transfected plant protoplasts. *Methods in Enzymology* **181**: 148–161.

Hattori T, Ichihara S, Nakamura K. 1987. Processing of a plant vacuolar protein precursor *in vitro*. *European Journal of Biochemistry* **166**: 533–538.

Hattori T, Nakamura K. 1988. Genes coding for the major tuberous root protein of sweet potato: identification of putative regulatory sequence in the 5′ upstream region. *Plant Molecular Biology* **11**: 417–425.

Hendry GAF, Wallace RK. 1993. The origin, distribution, and evolutionary significance of fructans. In: Suzuki M, Chatterton NJ, eds. *Science and Technology of Fructans*. Boca Raton, FL, USA: CRC Press, 119–139.

Herbers K, Meuwly P, Frommer WB, Métraux J, Sonnewald U. 1996. Systemic acquired resistance mediated by the ectopic expression of invertase: possible hexose sensing in the secretory pathway. *The Plant Cell* **8**: 793–803.

Horsch RB, Fry JE, Hoffman NL, Eichholtz D, Rogers SG, Fraley RT. 1985. A simple and general method for transferring genes into plants. *Science* **227**: 1229–1232.

Humphrey JS, Peters PJ, Yuan LC, Bonifacino JS. 1993. Localization of TGN38 to the trans-Golgi network: involvement of a cytoplasmic tyrosine-containing sequence. *Journal of Cell Biology* **120**: 1123–1135.

Jang JC, Sheen J. 1994. Sugar sensing in higher plants. *The Plant Cell* **6**: 1665–1679.

Marcusson EG, Horazdovsky BF, Cereghino JL, Gharakhanian E, Emr SD. 1994. The sorting receptor for yeast vacuolar carboxypeptidase Y is encoded by the *vps 10* gene. *Cell* **77**: 579–586.

Matsuoka K, Matsumoto S, Hattori T, Machida Y,

Nakamura K. 1990. Vacuolar targeting and posttranslational processing of the precursor to the sweet potato tuberous root storage protein in heterologous plant cells. *Journal of Biological Chemistry* **265**: 19750–19757.

Matsuoka K, Nakamura K. 1991. Propeptide of a precursor to a plant vacuolar protein required for vacuolar targeting. *Proceedings of the National Academy of Sciences, USA* **88**: 834–838.

Mattanovich D, Rüker F, da Camara Machado A, Laimer M, Reguer F, Steinkellner H, Himmler G, Katinger H. 1989. Efficient transformation of *Agrobacterium* spp. by electroporation. *Nucleic Acids Research* **17**: 6747.

Munro S, Pelham HRB. 1987. A C-terminal signal prevents secretion of luminal ER proteins. *Cell* **48**: 899–907.

Murashige T, Skoog F. 1962. A revised medium for rapid growth and bioassay with tobacco tissue culture. *Physiologia Plantarum* **15**: 473–497.

Nakamura K, Matsuoka K. 1993. Protein targeting to the vacuole in plant cells. *Plant Physiology* **101**: 1–5.

Okita TW, Rogers JC. 1996. Compartmentation of proteins in the endomembrane system of plant-cells. *Annual Review of Plant Physiology and Plant Molecular Biology* **47**: 327–350.

Paris N, Stanley CM, Jones RL, Rogers JC. 1996. Plant cells contain two functionally distinct vacuolar compartments. *Cell* **85**: 563–572.

Pollock CJ, Cairns AJ. 1991. Fructan metabolism in grasses and cereals. *Annual Review of Plant Physiology and Plant Molecular Biology* **42**: 77–101.

Pontis HG, del Campillo E. 1985. Fructans. In: Dey PM, Dixon RA, eds. *Biochemistry of Storage Carbohydrates in Green Plants*. London: Academic Press, 205–227.

Riesmeier JW, Willmitzer L, Frommer WB. 1994. Evidence for an essential role of the sucrose transporter in phloem loading and assimilate partitioning. *EMBO Journal* **13**: 1–7.

Schroeder MR, Borkhsenious ON, Matsuoka K, Nakamura K, Raikhel NV. 1993. Colocalization of barley lectin and sporamin in vacuoles of transgenic tobacco plants. *Plant Physiology* **101**: 451–458.

Scotti PA, Præstegaard M, Chambert R, Petit-Glatron M. 1996. The targeting of *Bacillus subtilis* levansucrase in yeast is correlated to both the hydrophobicity of the signal peptide and the net charge of the N-terminus mature part. *Yeast* **12**: 953–963.

Sijmonds PC, Dekker BMM, Schrammeijer B, Verwoerd TC, van den Elzen PJM, Hoekema A. 1990. Production of correctly processed human albumin in transgenic plants. *Biotechnology* **8**: 217–221.

Sonnewald U, Brauer M, von Schaewen A, Stitt M, Willmitzer L. 1991. Transgenic tobacco plants expressing yeast-derived invertase in either the cytosol, vacuole or apoplast: a powerful tool for studying sucrose metabolism and sink/source interactions. *The Plant Journal* **1**: 95–106.

Van der Meer IM, Ebskamp MJM, Visser RGF, Weisbeek PJ, Smeekens SCM. 1994. Fructan as a new carbohydrate sink in transgenic potato plants. *The Plant Cell* **6**: 561–570.

Wagner W, Keller F, Wiemken A. 1983. Fructan metabolism in cereals: induction in leaves and compartmentation in protoplasts and vacuoles. *Zeitschrift für Pflanzenphysiologie* **112**: 359–372.

Wise CS, Dimler RJ, Davis HA, Rist CE. 1955. Determination of easily hydrolysable fructose units in dextran preparations. *Analytical Biochemistry* **27**: 33–36.

New Phytol. (1997), **136**, 39–51

Synthesis of fructans by partially purified fructosyltransferase activities from *Lolium rigidum*

By J. A. ST. JOHN[1], I. M. SIMS[2], G. D. BONNETT[3]† and R. J. SIMPSON[3]*

[1] *Department of Agriculture and Resource Management, The University of Melbourne, Parkville, Victoria, 3052, Australia*
[2] *Plant Cell Biology Research Centre, Department of Botany, The University of Melbourne, Parkville, Victoria, 3052, Australia*
[3] *CSIRO Division of Plant Industry, GPO Box 1600, Canberra, ACT, 2601, Australia*

(*Received 24 July 1996*)

SUMMARY

Sucrose:sucrose fructosyltransferase (SST) activity was partially purified from whole shoots of *Lolium rigidum* by a combination of affinity chromatography, gel filtration and anion-exchange chromatography. The SST activity co-eluted with some fructan:fructan fructosyltransferase (FFT) and invertase activities and consequently the partially purified preparation was termed the fructosyltransferase (FT) preparation.

The SST-like activity in the FT preparation was purified 214-fold and had an apparent molecular mass of 84000. The FT preparation contained several peptides with an apparent pI of 4·6–4·7. When assayed with sucrose concentrations up to 600 mM, the FT preparation synthesized 1-kestose at all concentrations, and synthesized 6-kestose at concentrations of 150 mM and greater. The K_m of 1-kestose production was 0·2 M. When the FT preparation was assayed at a concentration of activity approximately half that measured in fresh tissue with 100 mM sucrose, 1-kestose, or 6_G-kestose as substrates, fructans of degree of polymerization (DP) $\leqslant 5$ were synthesized.

A partially purified FFT activity, free of SST and invertase activities, which synthesized β-2,1-glycosidic linked oligofructans of DP $\leqslant 6$, was combined *in vitro* with the FT preparation (FFT-FT preparation) to give a ratio of SST:FFT activities similar to that measured in crude enzyme extracts from *L. rigidum*. The FFT-FT preparation synthesized oligofructans when assayed with 100 mM concentrations of sucrose, 1-kestose or 6_G-kestose, but was not able to synthesize fructans of DP $\geqslant 6$ even after extended assays of up to 10 h. The FFT-FT preparation was also assayed with 100 mM sucrose with small amounts of concentrated sucrose added periodically during the assay to maintain the substrate concentration. In this assay, the FFT-FT preparation synthesized fructans up to an apparent DP of 17 or greater. The fructans of DP $\geqslant 6$ synthesized in the assay appeared to form two molecular series containing both β-2,1- and β-2,6-glycosidic linked fructosyl residues with terminal or internal glucosyl residues. The apparent rate of SST activity in the assay of the FFT-FT preparation was greater than that measured in a similar assay of the FT preparation alone which did not result in fructans with DP $\geqslant 6$. It was concluded that the FFT-FT preparation, when assayed with a continual supply of sucrose, contained a factor which promoted synthesis of fructans of DP $\geqslant 6$ and synthesis of β-2,6-glycosidic linkages between fructosyl residues.

Key words: *Lolium rigidum*, sucrose:sucrose fructosyltransferase (SST), fructan:fructan fructosyltransferase (FFT), fructan.

INTRODUCTION

The fructans accumulated by grasses are either linear or branched and contain β-2,1- and/or β-2,6-glycosidic linkages with either a terminal or internal glucosyl residue (Meier & Reid, 1982). Consequently, the path of fructan synthesis in grasses is more complicated than that in *Helianthus tuberosus* L. which is the model most commonly used to describe synthesis of fructans in plants (Edelman & Jefford, 1968). In this model, fructans are synthesized from sucrose by the concerted action of two enzymes, sucrose:sucrose fructosyltransferase (SST) and fructan:fructan fructosyltransferase

* To whom correspondence should be addressed.
† Present address: CSIRO Tropical Agriculture, Davies Laboratory, Private Mail Bag, PO Aitkenvale, Queensland, 4814, Australia.

(FFT). SST activity (St. John, Bonnett & Simpson, 1993) and FFT activity (St. John *et al.*, 1997*a*) have been partially purified from *Lolium rigidum* Gaudin. However, analysis of the path of synthesis of oligofructans and higher molecular mass fructans in *Lolium* spp. is still incomplete and further enzyme purification and characterization is required.

Evidence for the existence of SST in grasses is increasing (e.g. St. John *et al.*, 1993; Simmen *et al.*, 1993), but is still equivocal because attempts to purify SST have not succeeded in completely separating invertase activity from SST activity. When assayed with high concentrations of sucrose (e.g. > 500 mM), invertase is known to synthesize all three fructan trisaccharides (Pollock & Cairns, 1991). There is some evidence that 1-kestose can be synthesized by a SST-like enzyme in grasses, and in some species 1-kestose might be a key intermediate in the synthesis of oligofructans and fructans of higher molecular mass (Wagner & Wiemken, 1987; St. John *et al.*, 1993). In species such as *Lolium* which accumulate 6_G-kestose in addition to 1-kestose, there is no direct evidence of a SST-like enzyme capable of synthesizing 6_G-kestose. However, in barley, which accumulates 1-kestose and 6-kestose, a partially purified enzyme preparation with both invertase and SST-like activity has been shown to synthesize 6-kestose directly from sucrose when assayed at comparatively low sucrose concentrations (e.g. < 100 mM) (Simmen *et al.*, 1993). This activity thus provides an additional route by which carbohydrate can enter the fructan pool from sucrose (Duchateau *et al.*, 1995).

Lolium spp., including *L. rigidum*, accumulate high-molecular-mass fructans which contain mainly β-2,6-glycosidic linked fructosyl residues with a mean degree of polymerization (DP) of 25–30 (Tomasic, Jennings & Glaudemans, 1978; Sims, Pollock & Horgan, 1992; Bonnett *et al.*, 1994). However, a FFT activity partially purified from *L. rigidum* synthesized only β-2,1-glycosidic linked oligofructans when assayed with 1-kestose and did not synthesize fructans of DP > 6 (St. John *et al.*, 1997*b*). 6-Kestose is not normally accumulated in *L. rigidum* (Smouter & Simpson, 1991*a*), and oligosaccharides with low DP (< 6) contain predominantly β-2,1-glycosidic linkages, with β-2,6-glycosidic linked oligosaccharides being present in smaller proportions (St. John *et al.*, 1997*b*). It is not known by which path the β-2,6-glycosidic linked oligosaccharides and high-molecular-mass fructans are synthesized.

This paper reports the synthesis of a range of oligofructans and fructans of higher molecular mass from sucrose by the partially purified FFT activity when assayed in combination with a partially purified fructosyltransferase preparation containing SST activity from *L. rigidum*.

MATERIALS AND METHODS

Plant material

Plants of *Lolium rigidum* cv. Wimmera were grown in a glasshouse at Melbourne, Australia (lat. 37° 49′ S, long. 144° 45′ E) June 1993–January 1994 and irrigated with nutrient solution as described by St. John *et al.* (1997*a*).

Synthesis of fructans was induced in the plants when they had four fully expanded leaves using the method described previously (St. John *et al.*, 1993). Whole shoots were harvested 96 h after the induction of fructan synthesis was commenced, were frozen immediately in liquid nitrogen and stored at −20 °C.

Partial purification of the fructosyltransferase activities

During purification, enzyme activities were assayed using sucrose (SST and invertase activities) and 1-kestose (FFT activity) as substrates. In addition, 6_G-kestose (FFT activity) was used as a substrate for activities in fractions eluting from the gel filtration column. 6-Kestose was not used as substrate because we had previously been unable to detect enzyme activity in crude extracts using this substrate.

The enzyme preparation was extracted from whole shoots (1·5 kg) of *L. rigidum*, and proteins in the crude extract were precipitated with ammonium sulphate as described by St. John *et al.* (1997*a*). Precipitated proteins were redissolved in 50 mM phosphate–citrate buffer (pH 5·5) containing 1 mM $CaCl_2$ and 1 mM $MnCl_2$ and loaded on an affinity chromatography column (Concanavalin A Sepharose, 145 × 23 mm diameter, Pharmacia, Uppsala, Sweden) in the same buffer at a flow rate of 1 ml min^{-1}. Bound proteins were eluted from the column with 250 mM methyl mannopyranoside dissolved in the same buffer. The proteins were concentrated and the buffer changed to 20 mM histidine buffer (pH 5·5) by ultrafiltration through a size-selective membrane (Centricon 30, Amicon, Beverley, USA). The proteins were then loaded on a gel filtration column (Sephacryl 200 HR, 600 × 16 mm diameter, Pharmacia) and eluted in the histidine buffer at a flow rate of 0·5 ml min^{-1}. The gel filtration column was calibrated for molecular-mass determination with proteins of known molecular mass: catalase (240000), aldolase (158000), bovine serum albumin (68000), hen egg albumin (45000), chymotrypsinogen A (25000); (Pharmacia). One-ml fractions were collected after the void volume and each fraction was assayed for SST, FFT and invertase activities.

Fractions eluting from the gel filtration column which contained relatively high SST and FFT activities and relatively low invertase activity were combined and loaded on an anion-exchange column

(DEAE-MemSep, Millipore, Milford, USA) with 20 mM histidine buffer (pH 5·5). Proteins bound to the anion-exchange column were eluted at a rate of 2 ml min^{-1} in a gradient of 0–500 mM NaCl containing 20 mM histidine (pH 5·5) formed over 60 ml; 1-ml fractions were collected. Those fractions (see Fig. 2) which contained relatively high SST activity were combined and characterized. This enzyme preparation contained SST and FFT activities and was, therefore, designated the fructosyltransferase (FT) preparation. The FT preparation also retained some invertase activity.

In a separate, but otherwise similar, protein purification experiment using a subsample from the same batch of plant material, fractions eluting from the anion-exchange column which had high FFT(1-kestose) activity were combined and subjected to chromatofocusing (St. John *et al.*, 1997*a*). The FFT activity eluting after chromatofocusing was used in this study; it was free of both invertase and SST activities and its characteristics are described by St. John *et al.* (1997*a*).

Six attempts were also made to use chromatofocusing to separate the SST and FFT activities present in the FT preparation. Aliquots of the FT preparation were concentrated (Centricon 30, Amicon, USA), the buffer changed to 25 mM methylpiperazine buffer (pH 5·7) and aliquots loaded on the chromatofocusing column (Mono P HR 200 × 5 mm diameter, Pharmacia). Proteins bound to the column were eluted at 1 ml min^{-1} in a pH gradient from 5·7 to 4·0 formed over 30 ml (Polybuffer 74, Pharmacia). One-ml fractions were collected. Protein eluting from the column was monitored at 280 nm using a spectrophotometer detector (SPD-2A, Shimadzu, Kyoto, Japan). The SST and FFT activities were separated during one of the attempts, but generally the SST activity was not stable and could not be detected after chromatofocusing.

Storage of enzymes

Between the steps of the purification procedure, the enzyme preparations were frozen in liquid nitrogen and stored at -70 °C. The FT preparation obtained from the anion-exchange column was divided into 50 μl aliquots, frozen in liquid nitrogen and stored at −70 °C.

Protein measurement and electrophoresis

Protein was measured according to the procedure of Bradford (1976) using a commercial Coomassie Blue reagent (Bio-Rad protein dye reagent concentrate, Bio-Rad, Richmond, USA). Bovine serum albumin (Sigma, St. Louis, USA) was used as a standard. Proteins in the FT preparation were separated by two-dimensional gel electrophoresis as described by St. John *et al.* (1997*a*).

Purification of substrates

Fructan substrates are not available commercially and were purified as described by St. John *et al.* (1997*a*). 1-Kestose, (1,1)-kestotetraose and (1,1,1)-kestopentaose were purified from a commercial preparation of oligosaccharides (Neosugar P, Meiji Seika Kaisha Ltd, Kawasaki, Japan) and 6_G-kestose was purified from bulbs of *Allium cepa* L.

Enzyme assays

Assays of enzyme activities consisted of 20 μl enzyme and 10 μl substrate in 50 mM phosphate–citrate buffer (pH 5·5). Assays of SST and invertase activities were conducted by incubating the enzyme preparations with 100 mM sucrose (Aristar, BDH Chemicals, Poole, England) and assays of FFT activities were conducted by incubating the enzyme preparations with 100 mM 1-kestose (FFT(1-kestose)) or 6_G-kestose (FFT(6_G-kestose)). During the purification procedure, assays of crude extract and subsamples of enzyme preparations were performed in triplicate before each separation step and one blank assay was also conducted in which enzyme mixture that had been boiled for 5 min was incubated with the substrate. Assays of enzyme activities were conducted for every second fraction which eluted from gel filtration, anion-exchange and chromatofocusing columns. The assays were incubated at 25 °C for 4 h.

Determination of kinetic characteristics of the SST activity of the FT preparation

The FT preparation was assayed using a range of sucrose concentrations up to 600 mM. The enzyme preparation was diluted 12-fold in 50 mM phosphate–citrate buffer (pH 5·5) to give activities which would not lead to substrate limitation during the assay.

The FT preparation alone (diluted threefold in 50 mM phosphate–citrate buffer, pH 5·5) or in combination with the FFT activity (diluted sixfold) was assayed using 100 mM concentrations of either sucrose, 1-kestose or 6_G-kestose for up to 10 h. Additional assays were conducted in the presence of 100 mM sucrose with 1–2 μl of 300 mM sucrose added to prevent substrate depletion.

Analysis of reaction products

Fructose, sucrose and fructans were quantified by scanning densitometry after separation by ascending thin-layer chromatography as described by St. John, Bonnett & Simpson (1996). Fructose, sucrose and fructans were detected with urea phosphoric acid stain (Wise *et al.*, 1955). Bands migrating in positions

$\geqslant$ an apparent DP of 6 were not quantified separately and are reported as fructans of DP $\geqslant$ 6. Authentic standards for fructans up to DP 5 were used for quantification. However, standard fructans of DP $\geqslant$ 6 were not available and therefore fructans of this size were quantified using fructose as the standard. The results shown in Figure 6 are, therefore, presented in mg ml^{-1} rather than in nkat of enzyme activity.

The products of assays in which 1-kestose or 6_G-kestose were used as substrates were also separated by reversed phase (RP) HPLC (DextroPak column, 100 mm × 8 mm, Millipore). The column was eluted with water at a flow rate of 1·0 ml min^{-1}. The sugars were detected using a differential refractometer (model 410, Millipore) and peak areas were quantified using a chromatography computer package (Baseline, Millipore).

Units of enzyme activity

The sucrose-hydrolysing activity of invertase was measured as nkat of fructose released and SST activity was measured as nkat of trisaccharide synthesized from sucrose. The FFT activity synthesized several products when incubated with oligofructan substrates and the effective rate of FFT activity was therefore calculated as the least number of fructosyl transfers per second necessary to synthesize all the products by the most direct route. When a fructan trisaccharide was used as the substrate for FFT activity, synthesis of any trisaccharide other than the trisaccharide substrate was counted as one fructosyl transfer. Synthesis of tetrasaccharide was also counted as one fructosyl transfer and synthesis of pentasaccharide was counted as two fructosyl transfers. FFT activities were calculated this way because it was not possible to know whether synthesis of products had occurred by less direct routes.

Standards for mass spectroscopy and linkage analysis

A sample of Neosugar P (Meiji Seika Kaisha Ltd) with known proportions of sucrose, 1-kestose, (1,1)-kestotetraose, and (1,1,1)-kestopentaose was used as a standard sample for the GC-MS analysis. β-2,6-glycosidic linked oligofructans were obtained from leaves and stems of *Dactylis glomerata* grown in a glasshouse at Melbourne and harvested 2 wk after anthesis. Fructans were extracted from the material in 80% (v/v) ethanol at 80 °C then by four extractions in water at 60 °C. A fraction containing fructans was obtained by gel filtration (Bio-Gel P4, 750 × 25 mm diameter, Bio-Rad) and charged compounds in the resultant fraction were removed by ion exchange (Serdolit CS-2 (H^+) and Serdolit AS-6 ($HCOO^-$), Serva, Heidelburg, Germany). Lipids and hydrophobic materials were removed by passage through Sep-Pak tC18 cartridges (Millipore). β-2,1-glycosidic linked oligofructans were extracted from tubers of *Helianthus tuberosus* that were obtained from a local market; fructans were extracted by grinding the tubers and were treated similarly to the extract from *D. glomerata*.

Determination of degree of polymerization and linkage analysis

Fructans synthesized by the FT and FFT activities were dissolved in deionized water and separated by high-performance anion-exchange chromatography (HPAEC) on a Dionex HPLC using a CarboPac PA-1 column (250 × 4 mm diameter, Dionex Corp., Sunnyvale, CA, USA) equilibrated in 150 mM NaOH containing 25 mM CH_3COONa at a flow rate of 1 ml min^{-1}. Oligosaccharides were eluted with a gradient of CH_3COONa in 150 mM NaOH as follows: 25–400 mM (1–26 min), 400–500 mM (26–28 min) and 500 mM (28–30 min). Eluted material was monitored by pulsed amperometric detection (Dionex). The column was calibrated with a series of β-2,1-glycosidic linked oligofructans from *H. tuberosus*, with a β-2,6-glycosidic linked series of oligofructans from *D. glomerata*, with Neosugar P which had known proportions of sucrose, 1-kestose, (1,1)-kestotetraose and (1,1,1)-kestopentaose, and with 6_G-kestose from *Allium cepa*.

Fructans of DP $\geqslant$ 5 were separated from material of lower molecular mass on a Dionex CarboPac PA-1 preparative column (250 × 9 mm diameter) at a flow rate of 5 ml min^{-1}. Fructans were injected in 150 mM NaOH containing 25 mM CH_3COONa, and eluted with the same gradient used for analytical separations (see above). Fructans of DP $\geqslant$ 5 were collected and neutralized immediately after collection with acetic acid, sodium ions were removed on a Dowex 50W-X8, H^+ column (Bio-Rad Laboratories) at 4 °C and the purified oligosaccharides concentrated under reduced pressure at 40 °C.

Soluble carbohydrates from shoots of *L. rigidum*, which had been induced to accumulate fructans, were extracted from plant material with 80% ethanol at 80 °C followed by two extractions in water at 60 °C. Extracts were combined, rotary-evaporated, charged, and hydrophobic materials were removed as described previously. Fructans of DP $\geqslant$ 5 from *L. rigidum* were separated by anion-exchange chromatography (preparative CarboPac Dionex column) as described above.

Analysis of the linkages was determined by methylation of the samples and gas chromatography-mass spectroscopy as described by St. John *et al.*, (1997*b*). The average DP of the combined oligosaccharides and fructans of higher molecular mass from the assays and *L. rigidum* was estimated from the ratio of glucosyl and fructosyl residues.

RESULTS

Partial purification of the SST preparation

The shoots of *L. rigidum* in which fructan synthesis had been induced contained 2·0 ± 0·1 nkat g^{-1} f. wt of invertase activity, 1·5 ± 0·1 nkat g^{-1} f. wt of SST activity, 8·7 ± 0·7 nkat g^{-1} f. wt of FFT(1-kestose) activity and 2·2 ± 0·2 nkat g^{-1} f. wt of FFT(6_G-kestose) activity.

The fractions which eluted from the gel-filtration column contained differing amounts of invertase, SST and FFT activities (Fig. 1). Invertase had two peaks of activity at fractions 21 and 29, with apparent molecular masses of 115 000 and 61 000, respectively. The SST activity peaked at fraction 25, with an apparent molecular mass of 84000. FFT(1-kestose) activity eluted over a broad range, with two peaks at fractions 19 and 31 (apparent molecular masses of 134000 and 53000, respectively). FFT(6_G-kestose) activity also eluted over a broad range, across fractions 23–29, (apparent molecular mass over that range of 61000 to 106000). Fractions 23–34, which had relatively high SST and FFT activity, were combined and loaded onto the anion-exchange column. SST activity eluted with a peak at fraction 24 (Fig. 2). Invertase and FFT(1-kestose) activities co-eluted with the SST activity, but eluted over a broader range of fractions with peaks at fractions 24 and 35, respectively.

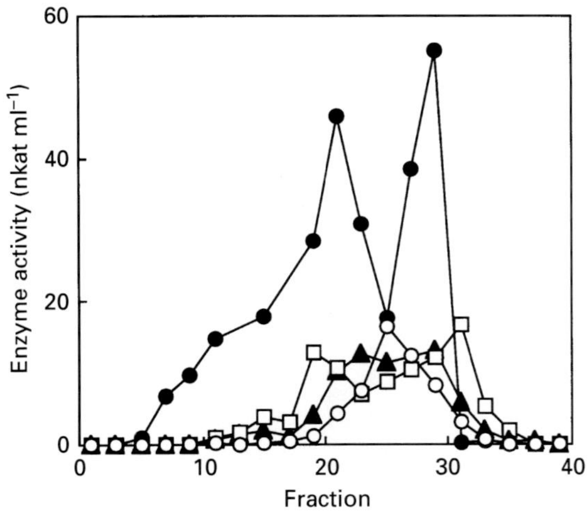

Figure 1. Invertase (●), SST (○), FFT(1-kestose) (□) and FFT(6_G-kestose) (▲) activities from shoots of *Lolium rigidum* in 1 ml fractions eluting from a gel-filtration column (S-200 HR, Pharmacia) eluted at 0·5 ml min^{-1} with 20 mM histidine buffer at pH 5·5. Fractions 23–34 were combined for anion-exchange chromatography.

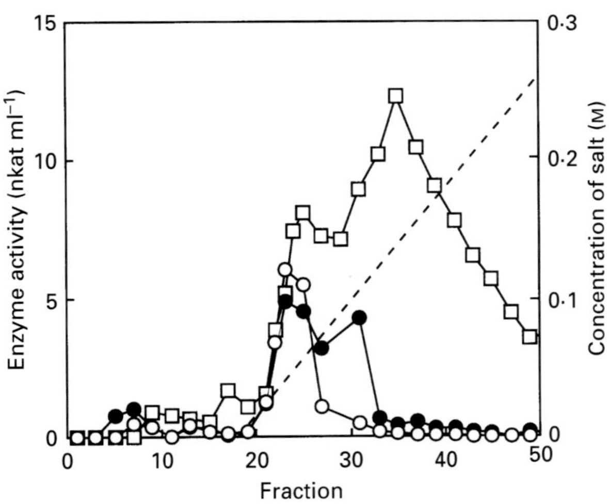

Figure 2. Invertase (●), SST (○) and FFT(1-kestose) (□) activities from shoots of *Lolium rigidum* in 1 ml fractions eluting from an anion-exchange column (MemSep DEAE, Millipore). Proteins were loaded on the column in 20 mM histidine buffer, pH 5·5, and proteins bound to the column were eluted at a rate of 2 ml min^{-1} in a gradient of 0–500 mMNaCl (---) started at fraction 18 and formed over 60 ml. Fractions 21–24 were combined and designated the fructosyltransferase (FT) activity.

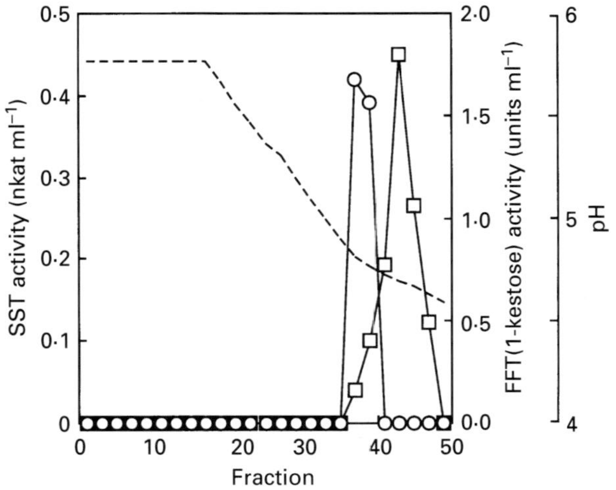

Figure 3. SST (○) and FFT(1-kestose) (□) activities from shoots of *Lolium rigidum* in 1 ml fractions eluting from a chromatofocusing column (Mono P, Pharmacia). Proteins were loaded on the column in 25 mM methylpiperazine buffer (pH 5·7) and were eluted at 1 ml min^{-1} in a gradient (---) of pH 5·7–4·0 formed over 30 ml.

Five of six attempts to improve the separation of SST from FFT and invertase by chromatofocusing were unsuccessful because the SST activity was unstable. Although protein eluted from the column, as indicated by detection at 280 nm (u.v. detector), in almost all cases no SST activity was recovered from the chromatofocusing column. However, at one attempt, SST activity was measurable and a substantial separation of SST from FFT activity was achieved (Fig. 3), but the amount of activity recovered was insufficient to allow characterization of the enzyme. Consequently, characterization of the SST activity was conducted using the FT preparation (Fig. 2, fractions 21–24) obtained from the anion-exchange separation.

Characterization of SST activity

The FT preparation contained SST activity, invertase activity and FFT activities, but the relative

Table 1. *Partial purification of SST activity from shoots of* Lolium rigidum

	Total activity (units)	Specific activity (nkat mg^{-1} protein)	Purification (-fold)	Recovery (%)	Ratios of activity SST:INV:FFT(1-k):FFT(6_G-k)
Crude extract	5030	0·28	1·0	100	1:1·3:5·8:1·5
$(NH_4)_2SO_4$ precipitate	1962	0·53	1·9	39	1:1·4:4·0:1·4
Concanavalin A	246	1·77	6·3	4·9	1:1·6:8·9:1·6
Gel filtration	131	16·5	58·9	2·6	1:2·8:5·7:1·6
Anion exchange	10	60·0	214	0·2	1:0·5:1·2:0·9

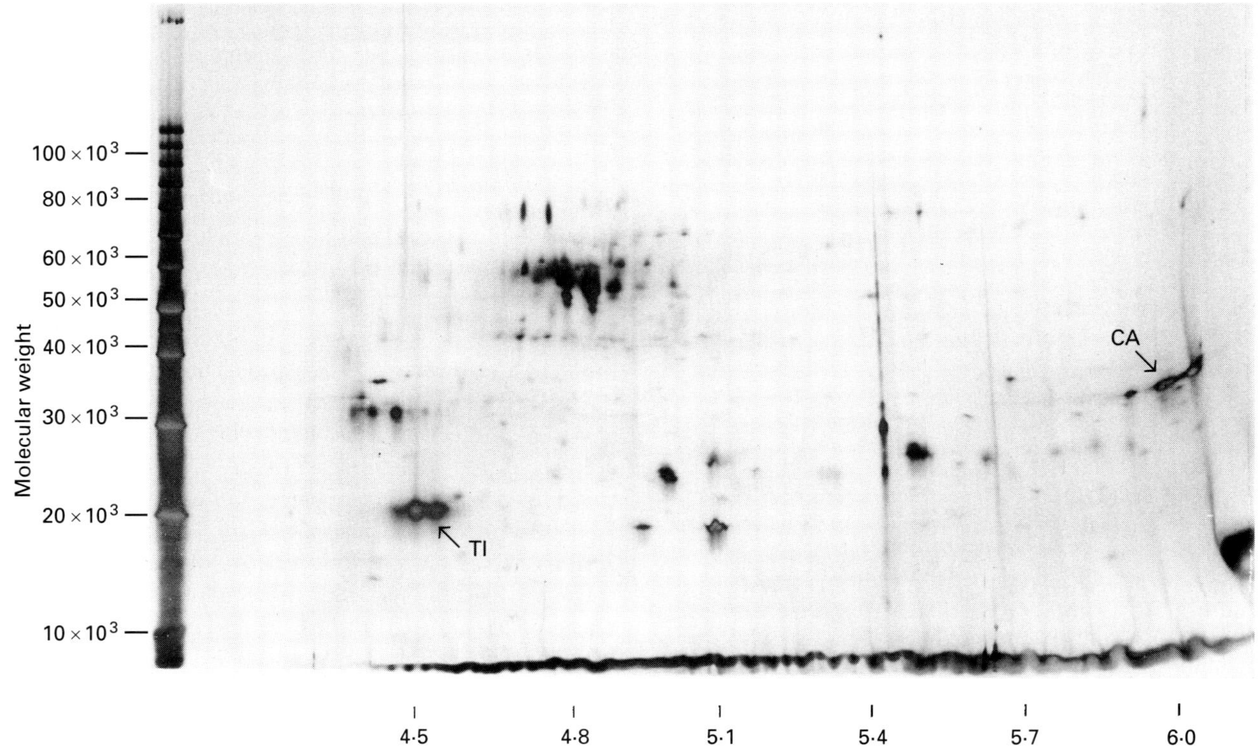

Figure 4. Two-dimensional gel electrophoresis of proteins in the FT preparation which was obtained from fractions 21–24 which eluted from an anion-exchange column. The first dimension was run on a reswollen isoelectric focusing strip (110 mm Immobiline dry strip pH 4–7, Pharmacia) followed by electrophoresis in a precast gradient SDS–PAGE gel (ExcelGel SDS gradient 8–18%, Pharmacia) in the second dimension. Proteins were visualized by silver-staining. The standard proteins trypsin inhibitor (TI) and carbonic anhydrase (CA) are indicated by arrows.

SST activity was higher in this preparation than in the crude extract (Table 1). The FT preparation, when assayed with 100 mM substrate contained 2·4 nkat ml^{-1} of SST activity, 1·2 nkat ml^{-1} of invertase activity, 2·9 nkat ml^{-1} of FFT(1-kestose) activity and 2·1 nkat ml^{-1} of FFT(6_G-kestose) activity. The SST, FFT(1-kestose) and FFT(6_G-kestose) activities in the FT preparation were purified 214-fold, 35-fold and 120-fold, respectively.

Proteins in the FT preparation were separated by two-dimensional gel electrophoresis (Fig. 4). The preparation contained several peptides with a group of strongly staining spots which had pI values between 4·6 and 4·8 and molecular masses between 50000 and 80000.

The FT preparation, when assayed with sucrose, synthesized 1-kestose at all concentrations of sucrose and synthesized 6-kestose when incubated with concentrations of 150 mM or greater (Fig. 5). No 6_G-kestose or tetrasaccharide was synthesized in these assays, which contained 0·2 nkat ml^{-1} of SST activity in the assay mixture when incubated with 100 mM sucrose. The activity was saturated at concentrations of ⩾ 400 mM. The 1-kestose-synthesizing activity in the diluted FT preparation had a K_m of 201 mM and a V_{max} of 0·7 nkat ml^{-1} for 1-kestose production.

Products synthesized by the FFT activity and the FT preparation

The aim of these assays was to determine whether the FT preparation alone, or a mixture of the FT

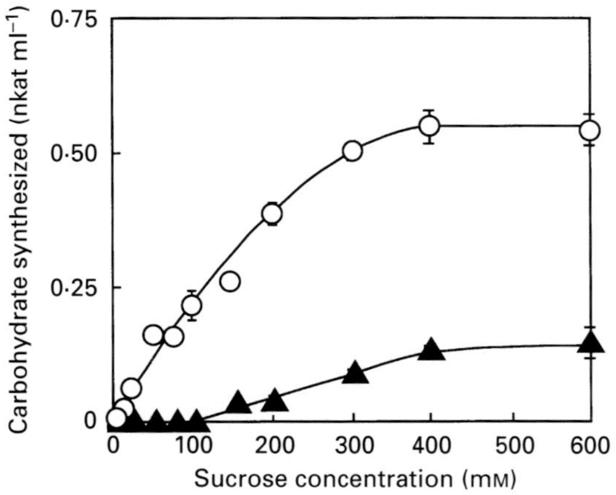

Figure 5. 1-Kestose synthesizing activity (○) and 6-kestose synthesizing activity (▲) in the FT preparation when assayed with a range of concentrations of sucrose. Error bars are 2 × SE; where not shown, the error bar is smaller than the symbol.

preparation and a FFT activity (St. John *et al.*, 1997*a*) (the combined FFT-FT preparation) could synthesize a range of oligosaccharides and fructans of higher molecular mass from sucrose or oligosaccharide substrates.

The FFT activity was diluted to give a concentration of activity, when assayed with 100 mM substrate, of 2·2 nkat ml^{-1} FFT(1-kestose) activity and 0·5 nkat ml^{-1} FFT(6_G-kestose) activity. The FT preparation was diluted to give a concentration of activity, when assayed with 100 mM substrate, of 0·8 nkat ml^{-1} SST activity, 1·0 nkat ml^{-1} FFT(1-kestose) activity and 0·7 nkat ml^{-1} FFT(6_G-kestose) activity.

Assays of the FFT activity with various substrates were reported previously by St. John *et al.*, (1997*a*, *b*). In brief, the FFT activity, when assayed with 100 mM concentrations of 1-kestose, (1,1)-kestotetraose or 6_G-kestose, synthesized fructans up to DP 6 which contained only β-2,1-glycosidic linked fructosyl residues and predominantly had an internal glucosyl residue. The FFT activity was unable to synthesize fructans when sucrose was the sole substrate, but did use sucrose as a fructosyl acceptor when assayed using sucrose and 1-kestose or 6_G-kestose together.

Two assays of the FT preparation were conducted with sucrose (100 mM) as the substrate. In the first assay, the substrate was depleted over the 10 h period and reached a sucrose concentration of 21 mM. In that assay, the FT preparation synthesized, after 2 h, 1-kestose, and released fructose (Fig 6*a*). The rate of SST activity in the assay after 2 h was 0·8 nkat ml^{-1} (Table 2). After 4 h, the amount of 1-kestose did not increase further, but small amounts of 6_G-kestose, tetrasaccharide and pentasaccharide were synthesized by the FT preparation. In the second assay, small volumes of 300 mM sucrose (1–2 μl) were added to the assay after each subsample was removed. This resulted in the sucrose concentration being maintained between 100 mM (0 h) and 80 mM (10 h). The FT preparation maintained its rate of activity and the amount of 1-kestose synthesized increased up to 10 h (Fig. 6*b*). Small amounts of 6_G-kestose, tetrasaccharide and pentasaccharide were synthesized, but fructans of DP > 5 were not detected.

The FT preparation, when incubated with 100 mM 1-kestose, synthesized tetrasaccharide, pentasaccharide and some 6_G-kestose, but not fructans of DP > 5 (Fig. 6*c*). After 2 h, the rate of FFT(1-kestose) activity in the FT preparation was 1·0 nkat ml^{-1} (Table 2).

When incubated with 100 mM 6_G-kestose, the FT preparation synthesized tetrasaccharide and pentasaccharide (Fig. 6*d*). Fructans of DP > 5 were not synthesized in these assays. After 2 h, the rate of FFT(6_G-kestose) activity in the FT preparation was 0·7 nkat ml^{-1} (Table 2).

The combined FFT-FT preparation was reconstituted to have 0·8 nkat ml^{-1} of SST activity, 3·2 nkat ml^{-1} of FFT(1-kestose) activity and 1·2 nkat ml^{-1} of FFT(6_G-kestose) activity to mimic the ratio of these activities measured in the crude extract of enzyme from *L. rigidum*. When the combined FFT-FT preparation was assayed with 100 mM sucrose without periodic additions of substrate, 1-kestose was the major product after 2 h, but 6_G-kestose, tetrasaccharide and some pentasaccharide were also synthesized (Fig. 6*e*). After 4 h, the amount of 1-kestose decreased, while the amounts of the other fructan products increased. Fructans of DP > 5 were not synthesized in this assay. An unexpected observation was that the *apparent* rate of SST activity in the assay after 2 h was 2·8 nkat ml^{-1}, threefold greater than when the FT preparation alone was incubated with 100 mM sucrose. The amount (on a mg basis) of tetrasaccharide and pentasaccharide synthesized after 10 h was 2·6-fold and 6·5-fold higher, respectively, than in the assays of FT preparation alone with 100 mM sucrose.

In assays with 100 mM sucrose supplemented by periodic additions of substrate the concentration of sucrose in the assays oscillated owing to the depletion of substrate followed by the periodic addition. Nevertheless, the concentration of sucrose was maintained between 100 mM (0 h) and 58 mM (after 10 h). The amount of 1-kestose that was synthesized continued to increase up to 10 h and a range of oligosaccharides and fructans of higher molecular mass up to at least an apparent DP of 17 was synthesized (Figs 6*f*, 7). The *apparent* rate of SST activity in the assay after 2 h was also 2·6 nkat ml^{-1} (Table 2). The amount of tetrasaccharide and pentasaccharide synthesized after 2 h was 55 % and 67 % lower, respectively, than in the assays of the combined FFT-FT preparation with sucrose without periodic additions of substrate.

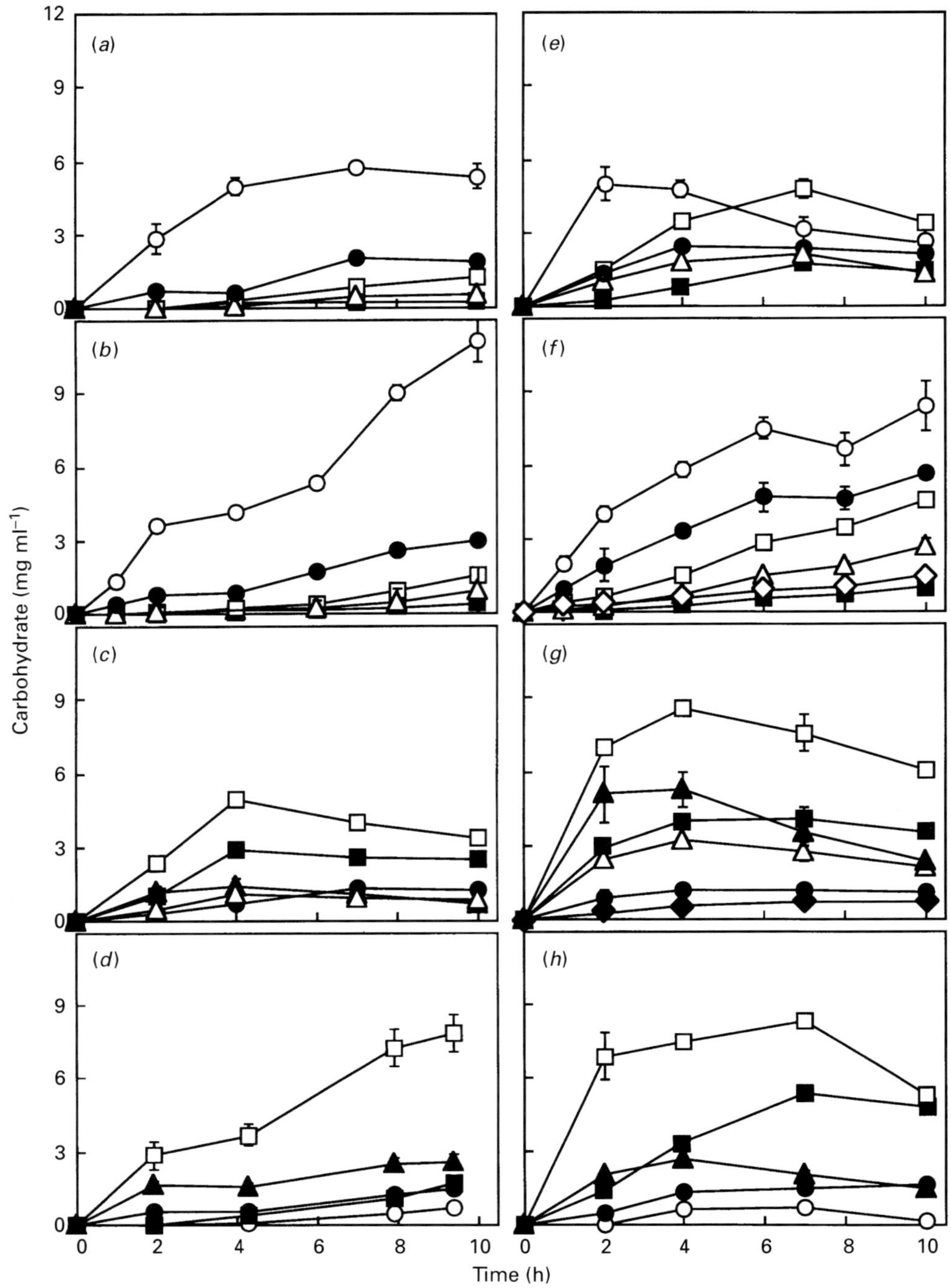

Figure 6. Products synthesized by the combined FT preparation when assayed with (*a*) 100 mM sucrose; (*b*) 100 mM sucrose with periodic additions of small volumes of 300 mM sucrose; (*c*) 100 mM 1-kestose and (*d*) 100 mM 6_G-kestose; and products synthesized by the FFT-FT preparation when assayed with (*e*) 100 mM sucrose; (*f*) 100 mM sucrose with periodic additions of small volumes of 300 mM sucrose; (*g*) 100 mM 1-kestose and (*h*) 100 mM 6_G-kestose. Products are fructose (●), sucrose (▲), 1-kestose (○), 6_G-kestose (△), tetrasaccharide (□), pentasaccharide (■), hexasaccharide (◆) and fructan of DP > 6 (◇).

The combined FFT-FT preparation, when assayed with 100 mM 1-kestose, synthesized tetrasaccharide, 6_G-kestose, pentasaccharide and hexasaccharide, but did not synthesize fructans of DP > 6 even after 10 h incubation (Fig. 6*g*). The amount of tetrasaccharide in the assay reached a maximum at 4 h and declined thereafter, coinciding with increases in the amounts of pentasaccharide and hexasaccharide. The combined FFT-FT preparation, when incubated together with 100 mM 1-kestose, exhibited 3·2 nkat ml^{-1} of FFT activity, which was the equivalent to the sum of the rates of FFT activities exhibited by the two enzyme preparations when assayed separately with 100 mM 1-kestose (Table 2).

When assayed with 100 mM 6_G-kestose, the combined FFT-FT preparation synthesized tetrasaccharide and pentasaccharide, but after 10 h had not synthesized fructans of DP ⩾ 6 (Fig. 6*h*). The amount of tetrasaccharide and pentasaccharide

Table 2. *Rates of activities after 2 h incubation of the FT preparation and the FFT activity extracted from* L. rigidum *(St. John et al., 1997a) when assayed with sucrose, 1-kestose or 6_G-kestose at 100 mM (mean ± SE)*

Assay	Periodic additions of substrate	Apparent SST activity (ml^{-1})	FFT activities (ml^{-1})
FT + sucrose		0·8 ± 0·1	
FT + sucrose	×	1·0 ± 0·1	
FT + 1-kestose			1·0 ± 0·2
FT + 6_G-kestose			0·7 ± 0·1
FFT + sucrose*		0·0	
FFT + 1-kestose*			2·2 ± 0·2
FFT + 6_G-kestose*			0·5 ± 0·1
FT + FFT + sucrose		2·8 ± 0·2	
FT + FFT + sucrose	×	2·6 ± 0·3	
FT + FFT + 1-kestose			3·2 ± 0·1
FT + FFT + 6_G-kestose			1·9 ± 0·3

* Assays of FFT activity shown here for comparison are reported in more detail by St. John *et al.* (1997*a*).

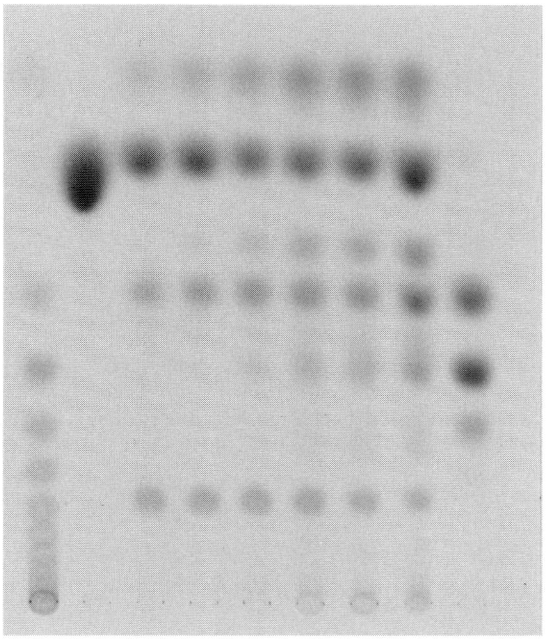

Figure 7. Thin-layer chromatography plate showing the products of the assay of the FFT-FT preparation with 100 mM sucrose as substrate with periodic additions of small volumes of 300 mM sucrose to minimize substrate depletion. Subsamples were taken at 0, 1, 2, 4, 6, 8, and 10 h. An internal standard, stachyose, was added to the samples before thin-layer chromatography. Standards were JA, an extract from tubers of *Helianthus tuberosus*, and N, a sample of NeoSugar P (Meiji Seika Kaisha Ltd).

increased up to 7 h to amounts equivalent to those in the assay of the FFT-FT preparation with 100 mM 1-kestose. However, hexasaccharide was not synthesized. When incubated with 100 mM 6_G-kestose, the combined mixture of the FFT-FT preparation exhibited 1·9 nkat ml^{-1} of FFT(6_G-kestose) activity (Table 2).

HPAEC of fructans synthesized in the assay of the combined FFT-FT preparation with 100 mM sucrose with additions of substrate, showed that compounds eluting with the same retention times as 1-kestose, 6_G-kestose and (1,1)-kestotetraose were present. The elution profile of the fructans was also compared with that of oligofructan standards from *H. tuberosus* and *D. glomerata*. This indicated that fructans of up to at least an apparent DP of 17 and probably higher had been synthesized (Fig. 8*a*). The relative amount of each apparent DP decreased with

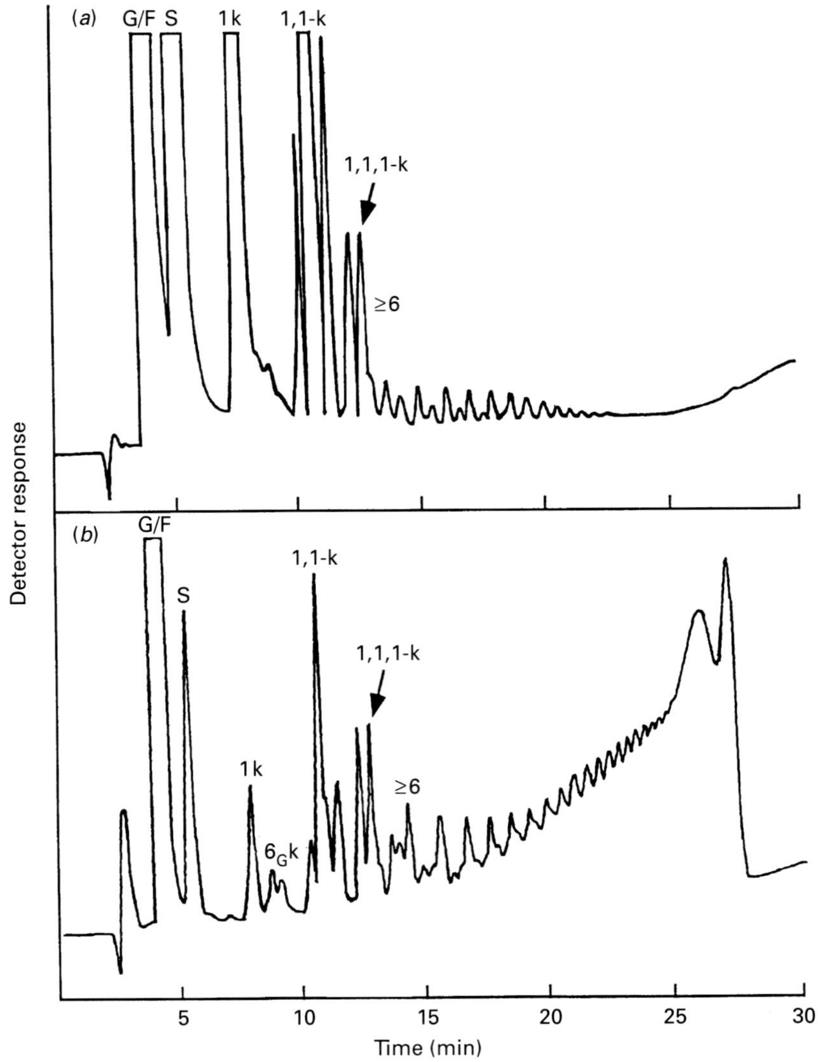

Figure 8. Separation of (*a*) fructan synthesized in assays of the FFT-FT preparation with 100 mM sucrose as substrate with periodic additions of small volumes of 300 mM sucrose and (*b*) fructan extracted from shoots of *Lolium rigidum* by analytical anion-exchange chromatography. Carbohydrates are glucose (G), fructose (F), sucrose (S), 1-kestose (1k), 6_G-kestose (6_Gk), (1,1)-kestotetraose (1,1-k), pentasaccharides (1,1,1-k), and fructans of DP $\geqslant 6$ ($\geqslant$ 6).

Table 3. *Proportion of deduced residues in fructans of DP $\geqslant$ 5 synthesized by the FFT-FT preparation when assayed for 10 h with 100 mM sucrose with periodic additions of sucrose to maintain substrate concentration*

Structure of derivative	Deduced residue	Proportion of each residue (mol %)
2,5 Diacetyl-(2-deuterio)-1,3,4,6-tetramethyl hexitol	Terminal fructose	10
1,2,5-Triacetyl-(2-deuterio)-3,4,6-trimethyl hexitol	2,1-Linked fructose	44
2,5,6-Triacetyl-(2-deuterio)-1,3,4-trimethyl hexitol	2,6-Linked fructose	30
1,2,5,6-Tetraacetyl-(2-deuterio)-3,4-dimethyl hexitol	2,1,6-Linked fructose	—
1,5-Diacetyl-(1-deuterio)-2,3,4,6-tetramethyl hexitol	Terminal glucose	14
1,5,6-Triacetyl-(1-deuterio)-2,3,4-trimethyl hexitol	1,6-Linked glucose	2
Estimation of degree of polymerization		6·25

Derivatives were determined by mass spectroscopy after separation by gas chromatography of partially methylated alditol acetates.

increasing molecular mass. In the extract from leaves of *L. rigidum*, there were at least three molecular species of fructan present at each apparent DP > 5 (Fig. 8*b*). However, only two peaks were apparent in the profile of fructans synthesized by the enzyme preparation. For example, fructans of DP 6 extracted from *L. rigidum* eluted at 13·5, 13·8 and 14·2 min, whereas those synthesized by the enzyme preparation eluted only at 13·6 and 14·2 min. The elution profile of fructans extracted from leaves also showed that there were fructans of higher molecular mass present *in vivo* than those synthesized by the enzyme preparation.

Methylation analysis of the pooled sample fructans (DP > 5) synthesized by the FFT-FT preparation, indicated the presence of fructosyl residues containing both $2 \rightarrow 1$ and $2 \rightarrow 6$ linkages in the ratio of 4·4:3 (Table 3). Fructans containing both terminal glucose residues and internal glucose residues were present. The ratio of terminal to internal glucosyl residues was 7:1. The average DP of the fructans of DP > 5 was 6·25 as determined from the ratio of glucose to fructose obtained from the linkage analysis.

DISCUSSION

Characteristics of the FT preparation

A SST-like activity was partially purified from *L. rigidum*, but complete separation of SST activity from invertase or FFT activities was not achieved. Although, on one occasion, SST activity was separated from FFT activity by chromatofocusing (Fig. 3), in general the SST activity proved unstable during chromatofocusing. The SST activity of the FT preparation had an apparent molecular mass of 84000 and was purified 214-fold, a degree of purity fourfold greater than achieved previously (St. John *et al.*, 1993) and was similar to that reported by others for SST activity from monocotyledonous species (Shiomi, Kido & Kiriyama, 1985; Wagner & Wiemken, 1987). However, Shiomi & Izawa (1980) have reported an 800-fold purification of SST from asparagus.

The FT preparation synthesized predominantly 1-kestose when incubated with sucrose. The 1-kestose-synthesizing activity was saturated at concentrations of sucrose of 400 mM or greater. 6-Kestose, which is not normally accumulated by *L. rigidum* (Smouter & Simpson, 1991*a*), was synthesized at concentrations of sucrose > 150 mM. This might have been the result of the invertase activity present in the FT preparation, since a partially purified invertase activity from *L. rigidum* has been shown to synthesize both 1-kestose and 6-kestose (St. John *et al.*, 1993). The 1-kestose synthesizing activity had an apparent K_m of 201 mM, which is within the range reported for SST activities partially purified from various species (Pollock, 1986; Chatterton *et al.*, 1988; Cairns, Winters & Pollock 1989; Smouter & Simpson, 1991*b*; St. John *et al.*, 1993). However two SST activities partially purified from barley did not saturate at concentrations of up to 600 mM (Simmen *et al.*, 1993).

Synthesis of oligosaccharides and fructans of higher molecular mass by the fructosyltransferase activities

Cairns (1992) found that the synthesis of fructans *in vitro* can be facilitated by assaying with high concentrations of enzyme activity. Therefore, to examine whether fructans of higher molecular mass could be synthesized *in vitro*, we also assayed the frucytosyltransferase activities at approx. half the concentration originally measured per g of fresh mass (assuming 1 g fresh mass contains 1 g of water).

The FT preparation which contained FFT and invertase activities in addition to SST activity did not synthesize fructans of DP ≥ 6 from 100 mM sucrose even after extended assay of up to 10 h (Fig. 6*a*). To compensate for the rapid utilization of the substrate, sucrose was periodically added during some assays. However, maintenance of the concentration of the substrate did not lead to synthesis of fructans of DP ≥ 6 (Fig. 6*b*). The FT preparation was not incubated with concentrations of sucrose of > 100 mM because 6-kestose, the trisaccharide not normally accumulated by *L. rigidum*, would have been synthesized at the higher concentrations. In addition, the maximum concentration of sucrose in vacuoles of cells that are actively photosynthesizing is predicted to be about 100 mM (Farrar & Farrar, 1986; Cairns, *et al.*, 1989). The FT preparation was also unable to synthesize fructans of DP ≥ 6 when incubated with 100 mM concentrations of 1-kestose or 6_G-kestose (Fig. 6*c*, *d*).

In a previous study (St. John *et al.*, 1997*a*, *b*), the partially purified FFT activity from *L. rigidum* was characterized. The FFT activity was free of SST and invertase activities and was unable to utilize sucrose or 6-kestose as sole substrates. When incubated with 100-mM concentrations of 1-kestose, (1,1)-kestotetraose, (1,1,1)-kestopentaose or 6_G-kestose, the FFT activity synthesized oligofructans of DP ≤ 6. However, the FFT activity was unable to synthesize fructans of DP > 6 even after extended assays of up to 10 h. The oligofructans synthesized by the FFT activity contained predominantly β-2,1-glycosidic linked fructosyl residues with an internal glucosyl residue.

As neither the FT preparation alone nor the FFT activity alone could synthesize fructans of DP > 6, the FT preparation and the FFT activity were combined to give a FFT-FT preparation with a ratio of activities (1 SST:1·5 FFT(6_G-kestose):4 FFT (1-kestose)) similar to, but at a concentration of activities approx. half, that measured in the crude extract. The FFT-FT preparation synthesized several oligofructans when assayed with 100 mM sucrose, but fructans of DP > 5 were not synthesized. However, when sucrose was added periodically during the assay to prevent substantial depletion of substrate, fructans of apparent DP of 17 or greater were synthesized by the FFT-FT preparation (Figs

7, 8). HPAEC of the products of the assay revealed that two 'series' of fructans were present at each DP > 5 (Fig. 8*a*). However, HPAEC of fructans extracted from *L. rigidum* indicated that at least three series of fructans were synthesized *in vivo* (Fig. 8*b*), indicating that at least one series of fructan (DP > 5) was not synthesized by the enzyme preparation.

The linkages of the fructans of DP $\geqslant 5$ synthesized by the FFT-FT preparation were analysed as a bulked sample and were found to contain both β-2,1- and β-2,6-glycosidic linked fructosyl residues. The fructans had predominantly a terminal glucosyl residue, although some fructan molecules with an internal glucosyl residue were also present (Table 3). However, it is significant that the FFT component of the FFT-FT preparation was shown previously to synthesize only β-2,1-glycosidic linked oligofructans, with mainly an internal glucosyl residue (St. John *et al.*, 1997*b*). The predominance of terminal glucosyl residues in fructan products of the combined FFT-FT preparation indicated that other fructosyltransferase activities not present in the FFT preparation were probably responsible for the synthesis of the fructans with internal glucosyl residues.

The FFT-FT preparation synthesized only fructans of DP $\leqslant 6$ when assayed with 100 mM concentrations of 1-kestose or 6_G-kestose. Additions of substrate were not made during these assays because it was considered unlikely that substrate limitation could have been the reason why fructans of DP $\geqslant 6$ were not synthesized from trisaccharide substrates. The trisaccharide concentrations used were at least 20-fold higher, and the concentrations of tetrasaccharides and pentasaccharides formed were seven-fold higher after 2 h than the concentrations measured in assays of the FFT-FT preparation with sucrose in which fructans of DP $\geqslant 6$ were synthesized.

The apparent rate of SST activity in the assays of the FFT-FT preparation was calculated to be threefold higher than the rate of SST activity in equivalent assays of the FT preparation alone. The combination of the FFT activity and the FT preparation therefore resulted in an increase in the *apparent* rate of SST activity which might have been essential for the synthesis of fructans of higher molecular mass. The possible reasons for an increase in the *apparent* rate of SST activity include (i) the removal of an inhibitor or addition of an activator of SST activity which resulted in increased rates of synthesis, (ii) the presence of other fructosyltransferase activities that were in the enzyme preparations but which were not apparent when they were assayed in isolation, and/or (iii) the synthesis in the mixture of an intermediate fructan essential for the further synthesis of fructans of DP $\geqslant 6$.

In theory, SST activity, measured in the absence of FFT activity, might be inhibited by high concentrations of its product (e.g. 1-kestose). Thus an apparent increase in SST activity might result from utilization of 1-kestose in the FFT-FT preparation. However, in our assays of the FFT-FT preparation, the concentrations of 1-kestose and other oligofructans were actually greater than in assays of the FT preparation alone, providing little support for this hypothesis.

It is possible that other fructosyltransferases were present in the partially purified preparations, but were rendered inoperative in the conditions used to assay FFT or FT in isolation. Since substrates that were used to detect activities during the purification procedure were limited to sucrose, 1-kestose and 6_G-kestose, activities which required other substrates would therefore not be detected. Such an activity has been reported recently by Duchateau *et al.* (1995) who partially purified a sucrose:fructan fructosyltransferase activity from barley which was capable of transferring a fructosyl residue from sucrose to a fructan (including 1-kestose) to synthesize oligofructans with β-2,6-glycosidic linked fructosyl residues. This activity provided an entry route for carbohydrate into the fructan pool distinct from entry via SST. If a similar activity were present in *L. rigidum* it might account for the increase in *apparent* SST activity observed in the present experiment. However, in assays of the FT preparation alone, sucrose, 1-kestose and other fructan oligosaccharides were present once reactions using sucrose as the primary substrate were under way, and assays of the FFT activity specifically, using both sucrose and 1-kestose (St. John *et al.*, 1997*a*) did not suggest the presence of a sucrose:fructan fructosyltransferase activity with properties similar to that reported by Duchateau *et al.*, (1995).

Finally, in the oligofructans of DP $\leqslant 5$ extracted from shoots of *L. rigidum*, fructosyl residues were predominantly β-2,1-glycosidic linked. The lowest DP fructan which contained β-2,6-glycosidic linked fructosyl residues was a fructan of DP 5 (St. John *et al.*, 1997*b*). The FFT activity used in this study synthesized the β-2,1-glycosidic linked oligomers (DP $\leqslant 5$) but not the β-2,6-glycosidic linked DP 5 (St. John *et al.*, 1997*a*). It is therefore possible that an intermediate fructan essential for the synthesis of fructans of DP > 6 was synthesized by the combined FFT-FT preparation.

CONCLUSION

Considered together, our observations indicate that synthesis of fructans of DP > 6 in *L. rigidum* requires the concerted action of at least two fructosyltransferase activities which were recombined in the FFT-FT preparation. However, this combined enzyme preparation appeared to synthesize from sucrose only two of the three fructan series present in extracts from *L. rigidum*. This indicated that other fructosyltransferase activities are probably also present *in vivo* and are possibly essential for synthesis of the full complement of fructans found in this grass species.

ACKNOWLEDGEMENTS

Neosugar P was a gift from Meiji Seika Kaisha Ltd, Japan. J. A. St. J. held an Australian Wool Research and Promotion Organization Postgraduate Award and an RWS Nicholas Agricultural Science Scholarship. R. J. S. and G. D. B. received funding from the Australian Wool Research and Promotion Organization and I. M. S acknowledges the infrastructural support of a Cooperative Research Centre of the Australian Commonwealth Government. Dr G. Tanner (CSIRO Division of Plant Industry) is thanked for conducting the separation of proteins by two-dimensional gel electrophoresis.

REFERENCES

Bonnett GD, Sims IM, St. John JA, Simpson RJ. 1994. Purification and characterization of fructans with β-2,1- and β-2,6-glycosidic linkages suitable for enzyme studies. *New Phytologist* **127**: 261–269.

Bradford MM. 1976. A rapid and sensitive method for the quantitation of microgram quantities of protein utilizing the principle of protein-dye binding. *Analytical Biochemistry* **72**: 248–254.

Cairns AJ. 1992. Fructan biosynthesis in excised leaves of *Lolium temulentum* L. V. Enzymatic *de novo* synthesis of large fructans from sucrose. *New Phytologist* **122**: 253–259.

Cairns AJ, Winters A, Pollock CJ. 1989. Fructan biosynthesis in excised leaves of *Lolium temulentum* L. III. A comparison of the *in vitro* properties of fructosyl transferase activities with the characteristics of *in vivo* fructan accumulation. *New Phytologist* **112**: 342–352.

Chatterton NJ, Harrison PA, Thornley WR, Bennett JH. 1988. Characterization of sucrose:sucrose fructosyltransferase from crested wheatgrass. *New Phytologist* **109**: 29–33.

Duchateau N, Bortlik K, Simmen U, Wiemken A, Bancal P. 1995. Sucrose:fructan 6-fructosyltransferase, a key enzyme for diverting carbon from sucrose to fructan in barley leaves. *Plant Physiology* **107**: 1249–1255.

Edelman J, Jefford TG. 1968. The mechanism of fructosan metabolism in higher plants as exemplified in *Helianthus tuberosus*. *New Phytologist* **67**: 517–531.

Farrar SC, Farrar JF. 1986. Compartmentation and fluxes of sucrose in intact leaf blades of barley. *New Phytologist* **103**: 645–657.

Meier H, Reid JSG. 1982. Reserve polysaccharides other than starch in higher plants. In: Loewus FA, Tanner W, eds. *Encyclopedia of Plant Physiology, new series, vol 13A*, Berlin: Springer-Verlag, 418–471.

Pollock CJ. 1986. Fructans and the metabolism of sucrose in vascular plants. *New Phytologist* **104**: 1–24.

Pollock CJ, Cairns AJ. 1991. Fructan metabolism in grasses and cereals. *Annual Review of Plant Physiology* **42**: 77–101.

Shiomi N, Izawa M. 1980. Purification and characterisation of sucrose:sucrose 1-fructosyl transferase from the roots of asparagus (*Asparagus officinalis* L.). *Agricultural and Biological Chemistry* **44**: 603–614.

Shiomi N, Kido H, Kiriyama S. 1985. Purification and properties of sucrose:sucrose 1^{F}-β-D-fructosyltransferase in onion seeds. *Phytochemistry* **24**: 695–698.

Simmen U, Obenland D, Boller T, Wiemken A. 1993. Fructan synthesis in excised leaves of barley. Identification of two sucrose-sucrose fructosyltransferases induced by light and their separation from constitutive invertases. *Plant Physiology* **101**: 459–468.

Sims IM, Pollock CJ, Horgan R. 1992. Structural analysis of oligomeric fructans from excised leaves of *Lolium temulentum* L. *Phytochemistry* **31**: 2989–2992.

Smouter H, Simpson RJ. 1991*a*. Fructan metabolism in leaves of *Lolium rigidum* Gaudin. I. Synthesis of fructan. *New Phytologist* **119**: 509–516.

Smouter H, Simpson RJ. 1991*b*. Fructan metabolism in leaves of *Lolium rigidum* Gaudin. II. Fructosyltransferase, invertase and fructan hydrolase activity. *New Phytologist* **119**: 517–526.

St. John JA, Bonnett GD, Simpson RJ. 1993. Fructosyltransferase activities from *Lolium rigidum* Gaudin. *New Phytologist* **123**: 705–715.

St. John JA, Bonnett GD, Simpson RJ. 1996. A method for rapid quantification of sucrose and fructan oligosaccharides suitable for enzyme and physiological studies. *New Phytologist* **134**, 197–203.

St. John JA, Bonnett GD, Simpson RJ, Tanner GA. 1997*a*. A fructan:fructan fructosyltransferase activity from *Lolium rigidum*. *New Phytologist* **135**, 235–247.

St. John JA, Sims IM, Bonnett GD, Simpson RJ. 1997*b*. Identification of products synthesized by a fructan:fructan fructosyltransferase activity from *Lolium rigidum*. *New Phytologist* **135**, 249–257.

Tomasic J, Jennings HJ, Glaudemans PJ. 1978. Evidence for a single type of linkage in a fructofuranan from *Lolium perenne*. *Carbohydrate Research* **62**: 127–133.

Wagner W, Wiemken A. 1987. Enzymology of fructan synthesis in grasses. *Plant Physiology* **85**: 706–710.

Wise CS, Dimler RJ, Davis HA, Rist CE. 1955. Determination of easily hydrolysable fructose units in dextran preparations. *Analytical Chemistry* **27**: 33–36.

New Phytol. (1997), **136**, 53–60

Biosynthesis *in vitro* of high-molecular-mass fructan by cell-free extracts from tuberous roots of *Viguiera discolor* (Asteraceae)

BY NAIR M. ITAYA, MARCOS S. BUCKERIDGE AND RITA C. L. FIGUEIREDO-RIBEIRO*

Seção de Fisiologia e Bioquímica de Plantas, Instituto de Botânica de São Paulo, Caixa Postal 4005, CEP 01061-970 – São Paulo, SP, Brazil

(*Received 24 July 1996*)

SUMMARY

Tuberous roots of *Viguiera discolor* accumulate up to 80 % of their dry mass as fructans. The distribution pattern of oligomers suggests the predominance of anabolic reactions at the beginning of dormancy, when a continuous series of fructans between sucrose and higher polymers is present. This paper describes the synthesis *in vitro* of fructans of high molecular mass by enzyme extracts prepared from growing tuberous roots of *V. discolor* at the beginning of dormancy. Sucrose:sucrose fructosyl transferase activity was characterized by the synthesis of 1-kestose from sucrose and fructan:fructan fructosyl transferase activity by the production of nystose from 1-kestose. The optimal temperature for both activities was 30 °C and the optimal pH range was between 5·5 and 6·0. The apparent K_m for sucrose with respect to 1-kestose synthesis was 173 mol m^{-3}, whereas the K_m for nystose synthesis from 1-kestose was much higher (592 mol m^{-3}). Long incubation periods (up to 264 h) with sucrose, 1-kestose or nystose produced fructans of high molecular mass *in vitro* that resembled those found *in vivo*. Our data show that the higher the molecular mass of the substrate, the higher the average molecular mass of the product that was formed.

Key words: Fructan biosynthesis, fructosyl transferase, *Viguiera discolor*, fructose polymers.

INTRODUCTION

Tuberous roots of *Viguiera discolor* Baker (Asteraceae) accumulate up to 80 % of their dry mass as fructan of the inulin type. The chromatographic profile of oligomers and the chain length of polymers vary with the phenological cycle. Sprouting and flowering phases are characterized by the predominance of oligofructans up to degree of polymerization (DP) 4, whereas dormancy is associated with the accumulation of a continuous series between sucrose and polymers with a mean molecular mass of up to 28 kDa (DP c. 170; Isejima & Figueiredo-Ribeiro, 1993). This suggests that the biosynthesis of fructans occurs principally at the end of the growing period, which occurs naturally during autumn in plants from subtropical areas of cerrado (Figueiredo-Ribeiro *et al.*, 1992; Figueiredo-Ribeiro, 1993). A similar pattern has been described for plants from temperate climates that accumulate fructans, as exemplified by *Helianthus tuberosus* (Edelman & Jefford, 1968). In this species, two enzymes are considered to be responsible for synthesis of fructans: sucrose:sucrose fructosyl transferase (SST, EC 2.4.1.99), which produces 1-kestose from sucrose, and fructan:fructan fructosyl transferase (FFT, EC 2.4.1.100), which catalyses chain elongation. FFT may also redistribute fructosyl units among fructan molecules during tuber dormancy and sprouting (Edelman & Jefford, 1968; Lüscher, Frehner & Nösberger, 1993; Koops & Jonker, 1994).

The established model for fructan synthesis has been questioned recently, especially with respect to grasses (Cairns, 1993). The particular case of *Festuca arundinacea* illustrates the difficulty of describing a general model for fructan biosynthesis in higher plants, since the purified fructosyltransferase activity

* To whom correspondence should be addressed.

had the ability to transfer fructosyl units to water in addition to transferring to sucrose and oligofructan (Lüscher & Nelson, 1995). Similarly, a purified enzymatic preparation from tuberous roots of *Cichorium intybus* (Asteraceae) produced 1-kestose at high sucrose concentrations and fructose and glucose at lower sucrose concentrations (Van den Ende & Van Laere, 1993). Recently, Koops & Jonker (1996) reported that reactions other than the conversion of sucrose into 1-kestose were catalysed by a purified SST from growing tubers of *H. tuberosus*. Furthermore, cell-free synthesis of polymeric fructans in the presence of high concentrations of sucrose has been demonstrated recently using enzymatic preparations from different sources (Penson & Cairns, 1994, and references therein). These studies allowed the assessment of physiologically significant fructan biosynthesis by the selection of appropriate assay systems, even though performed using crude extracts.

In this paper we describe the most favourable conditions for assaying SST and FFT activities, and report the biosynthesis *in vitro* of high-molecular-mass fructans by extracts of tuberous roots of *V. discolor* harvested at the beginning of dormancy.

MATERIALS AND METHODS

Plant material

Tuberous roots of *V. discolor* Baker (Asteraceae) were collected from plants growing naturally in a preserved area of cerrado near Moji-Guaçu, SP, Brazil (22° 15–18′ S, 47° 8–12′ W) harvested at the beginning of dormancy (March 1995), as defined previously (Isejima & Figueiredo-Ribeiro, 1993). Tuberous roots of equal size (0·5–1·0 cm) from 10 plants were excised, washed thoroughly in running tap-water, peeled, cut into small pieces and used for both soluble carbohydrate and enzyme extractions.

Preparation of cell-free enzyme extract

Two samples (15–20 g f. wt each) were homogenized in a chilled mortar and pestle with McIlvaine buffer pH 5·4, containing 2 mol m^{-3} EDTA, 2 mol m^{-3} β-mercaptoethanol, 0·05% sodium azide and 20 mol m^{-3} pyridoxal hydrochloride (to inhibit invertase activity). Two freshly broken Pasteur pipettes were also used during homogenization to form a smooth paste (Cairns, 1992). The slurry was

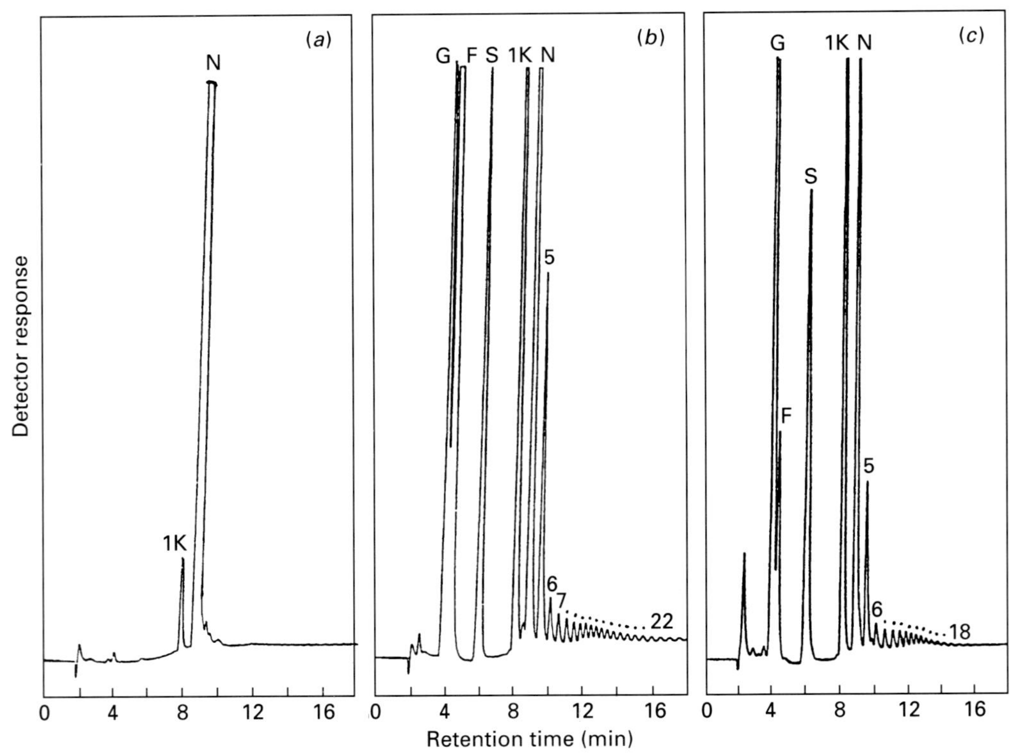

Figure 1. HPAEC/PAD analysis of fructans produced after 264 h incubation at 30 °C of the enzyme preparation from tuberous roots of *Viguiera discolor* with nystose (*b*) compared with soluble neutral carbohydrates (SNC) from the same tissue collected from plants at early dormancy (*c*). Enzyme preparation is equivalent to 108·5 μg protein cm^{-3} and SNC are equivalent to 1 mg fresh mass of tuberous roots. Markers G, F, S, 1-K and N refer to retention times of authentic standards of glucose, fructose, sucrose, 1-kestose and nystose, respectively. The other peaks were identified by comparison with the retention time of oligofructans extracted from tubers of *Helianthus tuberosus*. Control: 0 min incubation (*a*).

filtered through two layers of nylon, then centrifuged for 30 min at 31000 ***g***. The supernatant was dialysed for 24 h against 2 m^3 of buffer (diluted fivefold), with eight changes. The dialysed suspension was centrifuged at 13000 ***g*** for 10 min and the supernatant concentrated threefold using Centriflo filters CF 25 (Amicon). All procedures were carried out at 5 °C. The content of protein in the tuberous roots, on a fresh mass basis, was 214 $\mu g\ g^{-1}$.

Extraction and analysis of soluble carbohydrates from tuberous roots and from enzyme assays

Tuberous roots (1 g f. wt) were killed instantly by chopping into boiling 80% (v/v) ethanol and processed for soluble carbohydrate extraction according to the method described by Pollock & Jones (1979).

Oligomeric carbohydrates in the tuberous root extracts and produced during enzyme reactions were analysed by high-performance anion-exchange chromatography with pulsed amperometric detection (HPAEC/PAD) using a 4 × 250 mm CarboPac PA-1 column on a Dionex System Mod. DX 300 (USA), according to Vieira, Braga & Figueiredo-Ribeiro (1995). Sugars were identified by co-chromatography with authentic standards. Purified samples of nystose and 1-kestose were kindly provided by Dr Norio Shiomi (Rakuno Gakuen University, Japan). The apparent DP of the oligomers was determined by comparison with the chromatographic mobilities of fructans extracted from dormant tubers of *Helianthus tuberosus*. The polymers synthesized by enzyme reactions were analysed on a CarboPac PA-100 column using a linear gradient of 25 mol m^{-3} to 500 mol m^{-3} sodium acetate in 150 mol m^{-3} sodium hydroxide at a flow rate of 1 cm^{-3} min^{-1} (Shiomi, 1993). In some samples, free and combined fructose was measured by a ketose-specific modification of the anthrone reaction (Jermyn, 1956), using inulin from *H. tuberosus* as standard. The mean apparent DP of fructo-polysaccharides synthesized *in vitro* was determined by gel-permeation chromatography on Bio-Gel P-10 according to Carvalho & Dietrich (1993). The separated fructo-oligosaccharides and polysaccharides were further analysed by HPAEC/PAD, using CarboPac PA-1 and CarboPac PA-100 columns respectively.

Measurement of enzyme activity and optimization of fructan synthesis

Samples of enzyme preparations (100 mm^3) were mixed with an equal volume of 800 mol m^{-3} sucrose as substrate for SST activity and 800 mol m^{-3} 1-kestose as substrate for FFT activity, both in McIlvaine buffer pH 5·4. The mixture was incubated at 30 °C for 4 h (SST activity) and 6 h (FFT activity). The reaction was stopped by injecting 2 mm^3 of the assay mixture directly into the HPAEC/PAD system. Alternatively, samples were centrifuged at 13000 ***g*** for 10 min and their supernatants were stored at −20 °C until quantitative analysis of sugars by HPAEC/PAD. SST activity was calculated by direct measurement of

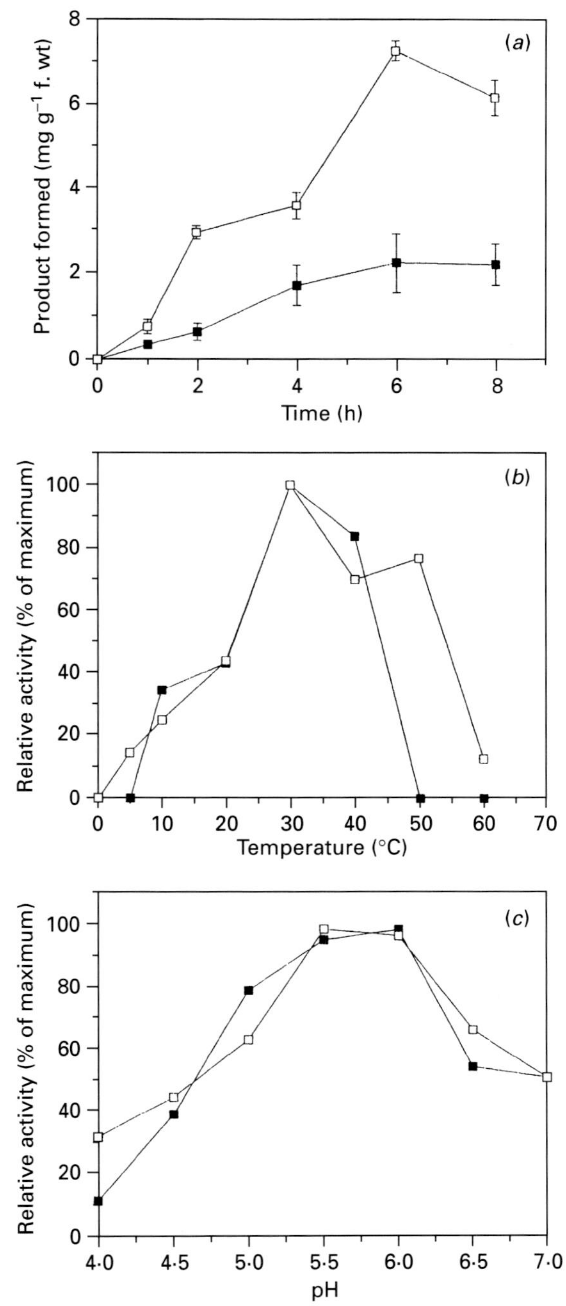

Figure 2. Optimization of fructan synthesis by cell-free extracts from tuberous roots of *Viguiera discolor* at the beginning of dormancy. (*a*) Time-course of formation of 1-kestose (■) and nystose (□) from 400 mol m^{-3} sucrose (SST activity) and 400 mol m^{-3} 1-kestose (FFT activity) incubated at 30 °C with the enzyme preparation equivalent to 1 g cm^{-3} fresh mass, (*b*) effect of temperature and (*c*) pH on SST (■) and FFT (□) activities in cell-free extracts incubated for 4 and 6 h, respectively with 400 mol m^{-3} sucrose and 400 mol m^{-3} 1-kestose. Data are means of duplicate determinations.

Table 1. *Effect of storage temperature on the stability of SST (sucrose: sucrose fructosyl transferase) and FFT (fructan:fructan fructosyl transferase) activities from tuberous roots of* Viguiera discolor *at the beginning of dormancy*

Storage temperature (°C)	Relative activity (% of maximum) SST			FFT		
	1 wk	2 wk	4 wk	1 wk	2 wk	4 wk
−80	—	76·19	60·84	—	100·00	100·00
−20	—	94·53	83·36	—	100·00	89·19
+5	51·00	44·19	44·20	100·00	83·16	82·40
+30	9·17	0·00	—	98·58	92·14	—
(+20−23)*	24·86	14·99	—	100·00	54·29	—

* Room temperature.
—, Not measured.
To prevent microbiological contamination toluene was included in the vials during storage at 30 °C and at room temperature.

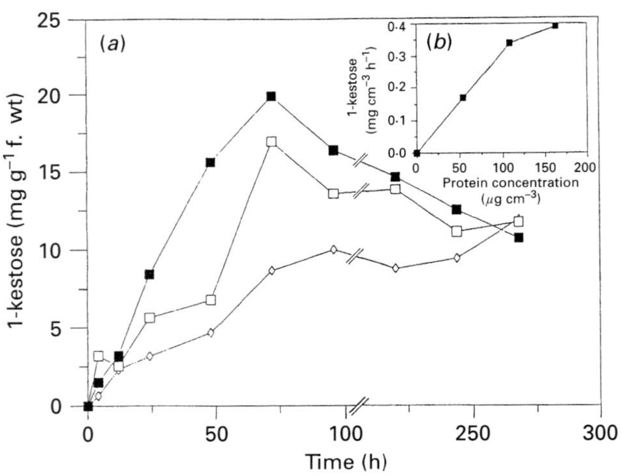

Figure 3. (*a*) Time-course of 1-kestose formation from 400 mol m^{-3} sucrose in McIlvaine buffer pH 5·5, at 30 °C, catalysed by different concentrations of the enzyme preparation from tuberous roots of *Viguiera discolor*, containing 54·25 (◇), 108·5 (□) and 162·7 (■) µg protein cm^{-3}. (*b*) Rate of fructan synthesis estimated by the production of 1-kestose from sucrose 400 mol m^{-3} catalysed by a dilution series of cell-free extracts containing different concentrations of protein. Data are means of duplicate determinations.

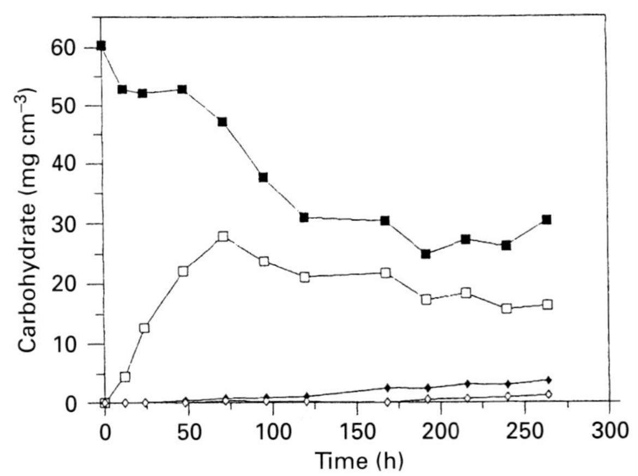

Figure 4. Time-course formation of 1-kestose (□), nystose (◆) and higher fructans (◇) from 400 mol m^{-3} sucrose (■). Incubation at 30 °C, pH 5·5 with an enzyme preparation from tuberous roots of *Viguiera discolor* equivalent to 162·7 µg protein cm^{-3}.

total trisaccharide production in the reaction mixture, whereas FFT activity was calculated by measurement of 1-kestose-dependent nystose production. The effect of pH was determined using McIlvaine buffer in the pH range 4·0–7·0 and the optimal assay temperature determined between 5 °C and 60 °C. To determine kinetic parameters for fructan synthesis, enzyme preparations were incubated for 4 and 6 h at 30 °C, pH 5·5, with seven sucrose and 1-kestose concentrations between 50 and 1000 mol m^{-3}. The apparent K_m and V_{max} were calculated from Lineweaver–Burk plots.

Aliquots of the enzyme preparation were maintained at −80 °C, −20 °C, 5 °C, 30 °C and at room temperature (20–23 °C), for investigation of enzyme stability. Samples of 100 mm^3 were taken over 7 d, mixed with 100 mm^3 of 800 mol m^{-3} sucrose or 1-kestose and incubated at 30 °C for 4 and 6 h respectively. The reaction was stopped by injecting 2 mm^3 of the assay mixture directly into the HPAEC system. The time-course of fructan production was investigated using sucrose or 1-kestose (800 mol m^{-3}) or nystose (400 mol m^{-3}). The enzyme preparation was mixed 1:1 (v/v) with the substrate prepared in McIlvaine buffer pH 5·5 and incubated at 30 °C for up to 264 h. The assays were performed in sterile sealed tubes, and a layer of toluene was present throughout the incubation period. Time zero and T 264 were examined microscopically, and samples stained with methylene blue. In both cases no evidence of the presence of micro-organisms was found. The substrates were also incubated with an enzyme preparation inactivated by boiling for 4 min.

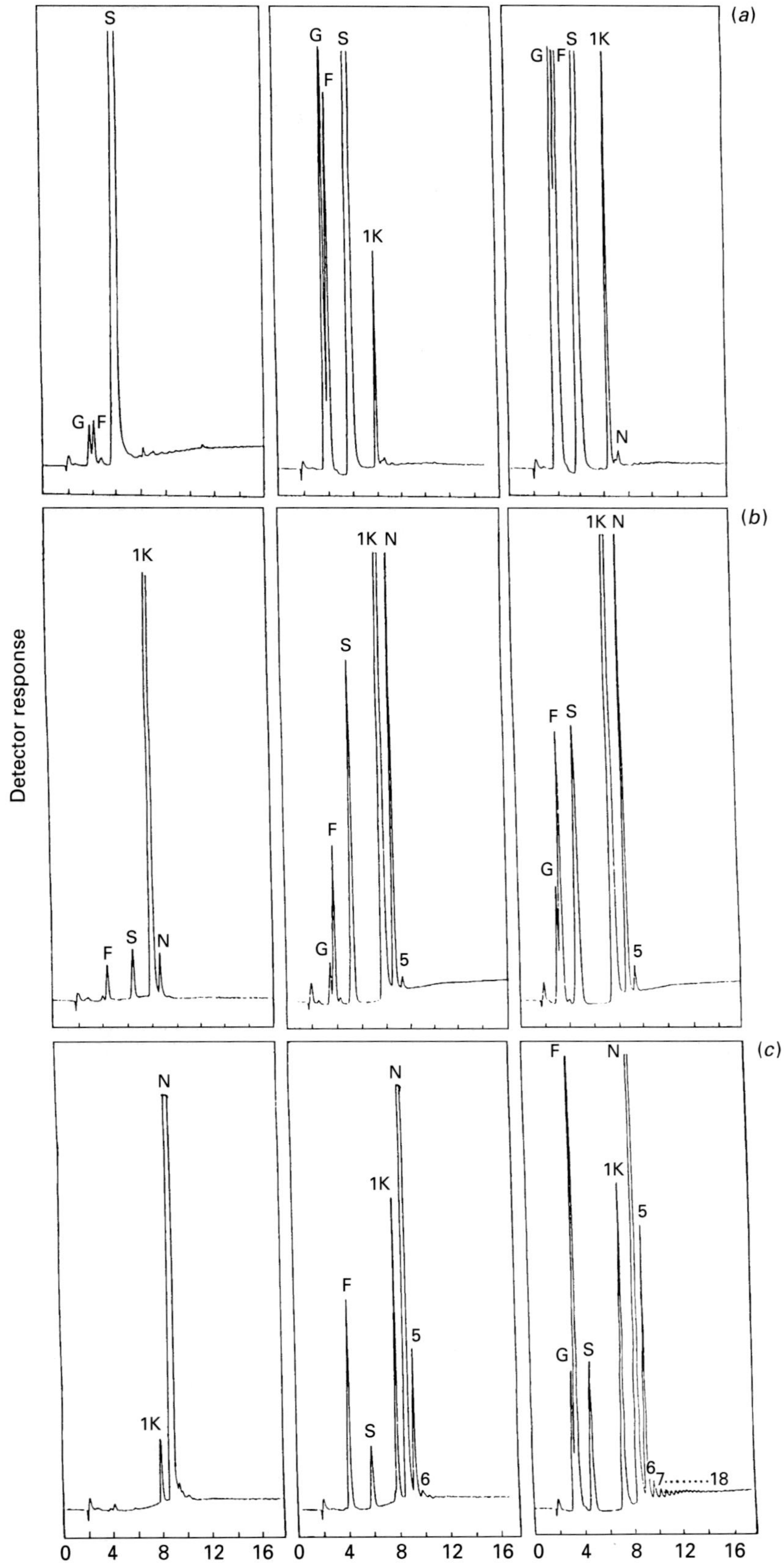

Figure 5. HPAEC/PAD analysis of fructans produced up to 96 h with enzyme preparations from tuberous roots of *Viguiera discolor* (108·5 μg cm^{-3}) incubated with 400 mol m^{-3} sucrose (*a*), 400 mol m^{-3} 1-kestose (*b*) and 200 mol m^{-3} nystose (*c*). Standards were those shown in Figure 1.

Samples of 2 mm^3 were collected at intervals over a 24-h period and the reaction was stopped by injecting directly into the HPAEC/PAD system.

To verify the effect of enzyme concentration on the products formed, samples of enzyme extract were diluted with McIlvaine buffer, pH 5·5 to provide a concentration range of 54·2–162·7 μg cm^{-3} protein (77·5–232·5 mg f. wt cm^{-3} equivalent) and assayed under standard conditions.

Protein concentrations in enzyme preparations were determined by the BioRad protein assay, which is based on the Bradford procedure (1976). Bovine serum albumin was used as a standard.

RESULTS AND DISCUSSION

As demonstrated for other systems, sucrose is the carbohydrate that initiates the synthesis of fructan. According to Cairns (1993) and Penson & Cairns (1994), to understand fructan biosynthesis it is necessary to show enzymatic synthesis of higher DP fructan from sucrose as the sole substrate *in vitro*.

In this study, enzyme preparations from tuberous roots of *V. discolor* harvested at the beginning of dormancy were able to synthesize fructan of high molecular mass using sucrose, 1-kestose or nystose as substrates (Fig. 1). HPAEC/PAD analysis revealed that fructans of up to DP 22 could be detected after 264 h incubation when nystose was used as substrate. Furthermore, the chromatographic pattern of the synthesized fructans resembled those found *in vivo* in terms of mobility and relative abundance (Fig. 1*c*).

Optimization and characterization of fructan-synthesizing activities

To differentiate SST from FFT activity, the time-courses of formation of 1-kestose and nystose in the reaction mixtures were analysed by HPAEC/PAD using, respectively, sucrose and 1-kestose as substrates. As shown in Figure 2*a*, product formation was linear with time up to 6 h of incubation for both the substrates. The optimal temperature for both activities was 30 °C (Fig. 2*b*) and the optimum pH range was between 5·5 and 6·0 (Fig. 2*c*). The latter data are consistent with the currently accepted vacuolar localization of fructan synthesis (Frehner, Keller & Wiemken, 1984).

The best temperature at which to store crude preparations of *V. discolor* in order to maintain SST activity was −20 °C, whereas for FFT either −80 °C or −20 °C maintained activity at high levels (90–100 %) for up to 4 wk (Table 1). Variable temperatures (20–23 °C) resulted in a significant loss of activity of both SST and FFT after 2 wk. Over the same period, storage, at 30 °C induced complete loss of SST activity while FFT retained more than 90 % of its activity. This could be a valuable treatment for the future purification of FFT from *V. discolor*.

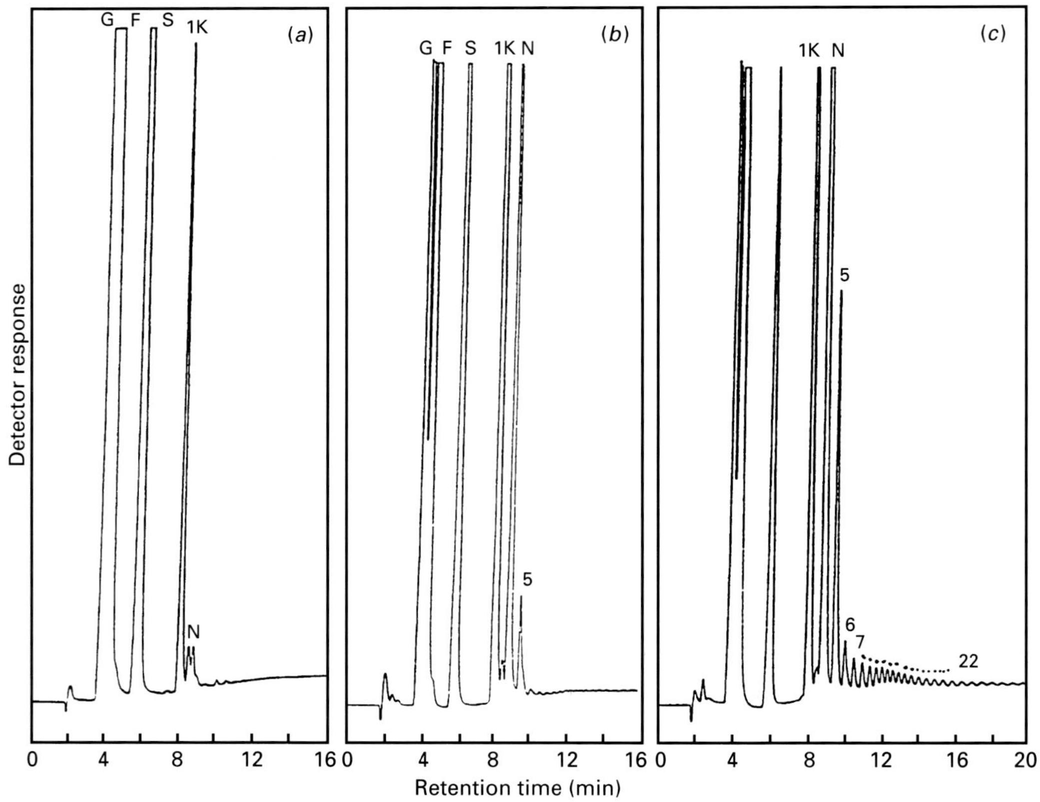

Figure 6. HPAEC/PAD analysis of fructans produced after 264 h incubation of enzyme preparation from tuberous roots of *Viguiera discolor* (108·5 μg protein cm^{-3}) with 400 mol m^{-3} sucrose (*a*), 400 mol m^{-3} 1-kestose (*b*) and 200 mol m^{-3} nystose (*c*). Peaks were identified as indicated in Figure 1.

Figure 3 shows that the rate of 1-kestose synthesis from sucrose increased with increasing enzyme concentration but that the profile of its formation was similar at enzyme concentrations > 100 μg cm^{-3}. When sucrose was the sole substrate, higher fructans (DP > 3) were detected only at higher enzyme concentrations and after long incubation periods (Fig. 4). Similarly, Cairns (1995) observed that enzymatic preparations from *Lolium temulentum* (Poaceae) and from fungal origin (pectolyase) synthesized higher fructo-oligosaccharides from sucrose only at higher enzyme concentrations (> 0·3 g f. wt equivalent cm^{-3} and 90 μg cm^{-3}, respectively). Reports from Shiomi (1992) and Lüscher & Nelson (1995) indicated that the synthesis of high-DP fructo-oligosaccharides both in *Asparagus* roots and in the leaf growth zone of tall fescue decreased at low enzyme concentrations. According to Cairns (1995), the use of low enzyme concentrations could complicate the correct interpretation of fructan biosynthesis *in vitro*. In this study we performed assays using the highest possible concentration of enzyme in order to promote the synthesis of higher-molecular-mass fructans.

In spite of the presence of pyridoxal, the continuous release of free fructose and glucose indicates that some invertase activity is also present in *V. discolor* extracts (Fig. 5*a*). Indeed, a number of authors reported the difficulty of separating SST activity from invertase activity, which can catalyse some fructosyl transfer at high sucrose concentrations, Lüscher & Nelson (1995). Furthermore, the large amounts of free fructose produced after long incubation periods (Fig. 6) indicate that enzyme preparation from tuberous roots of *V. discolor* has some hydrolytic activity as well.

V. discolor enzyme preparations were incubated with a range of sucrose and 1-kestose concentrations to determine the kinetic parameters of SST and FFT activities. The Lineweaver–Burk plots allowed determination of an apparent K_m of 173 mol m^{-3} for sucrose and a V_{max} of 0·52 mg of 1-kestose g^{-1} f. wt h^{-1} for SST activity and an apparent K_m of 592 mol m^{-3} for 1-kestose and a V_{max} of 1·48 mg of nystose g^{-1} f. wt h^{-1} for FFT activity.

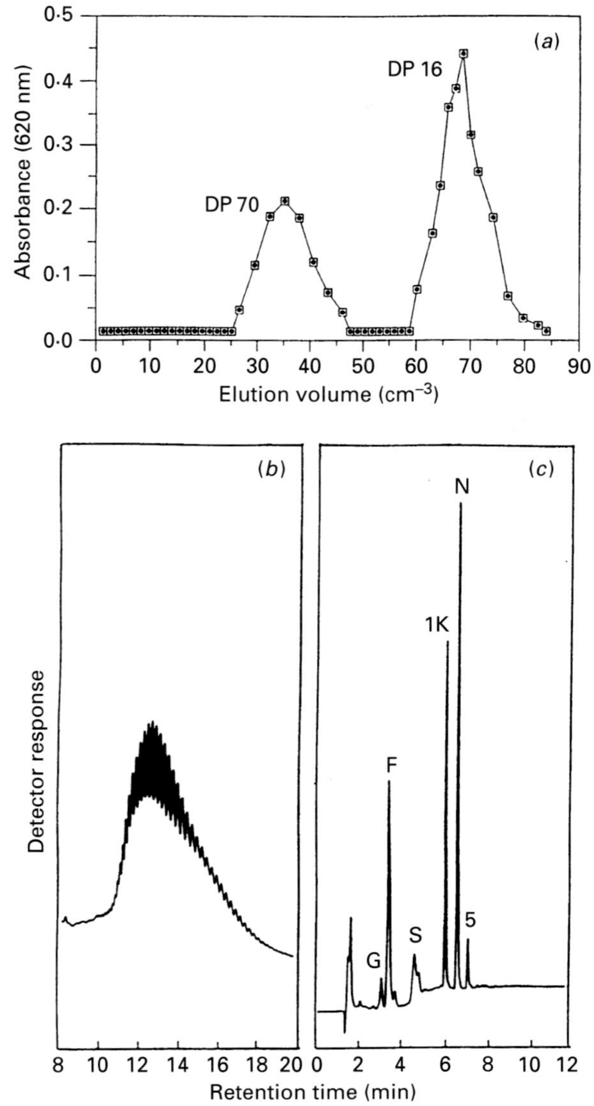

Figure 7. Gel-permeation chromatography on Bio-Gel P-10 of fructans produced after 2 wk incubation of cell-free extracts from tuberous roots of *Viguiera discolor* (108·5 μg protein cm^{-3}) with 200 mol m^{-3} nystose (*a*). Subfractions were analysed further by HPAEC/PAD on CarboPac PA-100 (mean DP 70; *b*) and CarboPac PA-1 (mean DP 16; *c*). G, glucose; F, fructose; S, sucrose; N, nystose; numbers represent the estimated DP (degree of polymerization).

Fructan biosynthesis catalysed by the enzyme extracts

In the presence of FFT, sucrose is thought to be a strong acceptor of fructosyl residues, thus contributing to a rapid formation of 1-kestose (Cairns, 1993). As shown in Figure 4, when sucrose was used as the sole substrate the concentration of 1-kestose rose sharply, reaching a maximum at 72 h (27 mg cm^{-3}–55 mol m^{-3}). As observed in other experiments (see below) the production of nystose started when the concentration of 1-kestose reached a threshold of *c.* 40 mol m^{-3}. It is interesting to note that, whereas the apparent K_m is approx. three times higher for sucrose than for 1-kestose, the V_{max} is approx. three times higher for 1-kestose. This might explain the need for high enzyme concentrations in order to obtain fructans of higher molecular mass *in vitro* (Fig. 3), since higher concentrations of 1-kestose (40 mol m^{-3}) were obtained in the assay mixture with sucrose as substrate under these conditions (Fig. 5*a*). The period of reaction seems to be relevant, since incubation of the enzyme preparation for 24 h with 1-kestose as substrate (Fig. 5*b*) yields only oligomers (DP 4 and 5), while longer incubation, up to 264 h, yields some higher polymers (Fig. 6*b*). Long incubation periods with 1-kestose or nystose, at high concentrations (400 mol m^{-3} and

200 mol m^{-3}, respectively) produced higher-molecular-mass material, which was assumed to be fructan since it reacted with specific reagents for fructose-containing carbohydrates and also showed a similar chromatographic profile to that of native inulin series fructan. Figures 5 and 6 also show that the higher the DP of the substrate, the higher the average DP which was obtained after incubation. The ratios of substrate to product, which might also indicate the affinity of SST and FFT for sucrose and 1-kestose respectively, were 2·72 (sucrose/1-kestose) and 1·04 (1-kestose/nystose). This suggests that FFT affinity for substrate is threefold lower than that of SST, which is consistent with the calculated kinetic parameters.

The substantial increase in free fructose during long incubation with 1-kestose and nystose (Figs 5*c*, 6*c*) indicated the presence of fructan hydrolytic activity in cell-free extracts from tuberous roots of *V. discolor*. As reported recently (Koops & Jonker, 1996), nystose is a substrate for the hydrolytic activity of purified SST from tubers of *H. tuberosus* rather than for synthesizing activity.

As shown in Figure 7, the enzyme preparation incubated with nystose (200 mol m^{-3}) for a long period catalysed the formation of fructo-oligosaccharides (Fig. 7*c*) and polysaccharides (Fig. 7*b*) with an average DP of 16 and 70, as determined by gel-permeation chromatography (Fig. 7*a*).

The results presented in this work indicate that developing tuberous roots of *V. discolor*, collected from plants at the beginning of dormancy, are suitable material for the study of fructan synthesis since they possess a biosynthetic system that is stable for very long incubation periods at relatively high temperatures (20–30 °C). These features permit the synthesis *in vitro* of fructans of high molecular mass (DP 70, see Fig. 7). It is possible that even longer incubation periods could produce fructan with even higher molecular mass, as fructans with DP values of 170 have already been found in this species (Isejima & Figueiredo-Ribeiro, 1993).

ACKNOWLEDGEMENTS

We would like to thank FAPESP for financial support to this work and CNPq for a fellowship to N.M.I. We also thank Dr Norio Shiomi for the kind gift of 1-kestose and nystose and MSc. Candida C. J. Vieira for critical reading of the manuscript.

REFERENCES

Bradford MM. 1976. A rapid and sensitive method for the quantitation of microgram quantities of protein utilizing the principle of protein-dye binding. *Analytical Biochemistry* **72**: 248–254.

Cairns AJ. 1992. Fructan biosynthesis in excised leaves of *Lolium temulentum* L. V. Enzymatic *de novo* synthesis of large fructans from the sucrose. *New Phytologist* **122**: 253–259.

Cairns AJ. 1993. Evidence for the *de novo* synthesis of fructan by enzymes from higher plants: a reappraisal of the SST/FFT model. *New Phytologist* **123**: 15–24.

Cairns AJ. 1995. Effects of enzyme concentration on oligofructan synthesis from sucrose. *Phytochemistry* **40**: 705–708.

Cairns AJ, Pollock CJ. 1988. Fructan biosynthesis in excised leaves of *Lolium temulentum* L. I. Chromatographic characterization of oligofructans and their labelling patterns following $^{14}CO_2$ feeding. *New Phytologist* **109**: 399–405.

Carvalho MAM, Dietrich SMC. 1993. Variation in fructan contents in the underground organs of *Vernonia herbacea* (Vell.) Rusby at different phenological phases. *New Phytologist* **123**: 735–740.

Edelman J, Jefford TG. 1968. The mechanism of fructosan metabolism in higher plants as exemplified in *Helianthus tuberosus* L. *New Phytologist* **67**: 517–531.

Figueiredo-Ribeiro RCL. 1993. Distribuição, aspectos estruturais e funcionais dos frutanos, com ênfase em plantas herbáceas do cerrado. *Revista Brasileira de Fisiologia Vegetal* **5**: 203–208.

Figueiredo-Ribeiro RCL, Isejima EM, Dias-Tagliacozzo GM, Carvalho MAM, Dietrich SMC. 1992. The physiological significance of fructan accumulation in Asteraceae from the Cerrado. *Ciência e Cultura* **43**: 443–446.

Frehner M, Keller F, Wiemken A. 1984. Localization of fructan metabolism in the vacuoles isolated from protoplasts of Jerusalem artichoke tubers (*Helianthus tuberosus* L.). *Journal of Plant Physiology* **116**: 197–208.

Isejima EM, Figueiredo-Ribeiro RCL. 1993. Fructan variations in tuberous roots of *Viguiera discolor* Baker (Asteraceae): the influence of phenology. *Plant Cell Physiology* **34**: 723–727.

Jermyn MA. 1956. A new method for the determination of kestohexoses in presence of aldohexoses. *Nature* **177**: 38–39.

Koops AJ, Jonker HH. 1994. Purification and characterization of the enzymes of fructan biosynthesis in tubers of *Helianthus tuberosus* Colombia. I. Fructan:fructan fructosyltransferase. *Journal of Experimental Botany* **45**: 1623–1631.

Koops AJ, Jonker HH. 1996. Purification and characterization of the enzymes of fructan biosynthesis in tubers of *Helianthus tuberosus* Colombia. II. Purification of sucrose:sucrose 1-fructosyltransferase and reconstitution of fructan synthesis *in vitro* with purified sucrose:sucrose 1-fructosyltransferase and fructan:fructan 1-fructosyltransferase. *Plant Physiology* **110**: 1167–1175.

Lüscher M, Frehner M, Nösberger J. 1993. Purification and characterization of fructan:fructan fructosyltransferase from Jerusalem artichoke (*Helianthus tuberosus*). *New Phytologist* **123**: 717–724.

Lüscher M, Nelson CJ. 1995. Fructosyltransferase activities in the leaf growth zone of tall fescue. *Plant Physiology* **107**: 1419–1425.

Penson SP, Cairns AJ. 1994. Fructan biosynthesis in excised leaves of wheat (*Triticum aestivum* L.): comparison of *de novo* synthesis *in vivo* and *in vitro*. *New Phytologist* **128**: 395–402.

Pollock CJ, Jones, T. 1979. Seasonal patterns of fructan metabolism in forage grasses. *New Phytologist* **83**: 9–15.

Pollock CJ, Suzuki M. 1986. Extraction and characterization of the enzymes of fructan biosynthesis in timothy (*Phleum pratense*). *Canadian Journal of Botany* **64**: 1884–1887.

Shiomi N. 1992. Content of carbohydrate and activities of fructosyl transferase and invertase in asparagus roots during the fructo-oligosaccharide and fructo-polysaccharide accumulating season. *New Phytologist* **112**: 421–423.

Shiomi N. 1993. Structure of fructo-polysaccharide (asparagosin) from roots of asparagus (*Asparagus officinalis* L.). *New Phytologist* **123**: 263–270.

Van den Ende W, Van Laere A. 1993. Purification and properties of an invertase with sucrose:sucrose fructosyltransferase (SST) activity from the roots of *Cichorium intybus* L. *New Phytologist* **123**: 31–37.

Vieira CCJ, Braga MR, Figueiredo-Ribeiro RCL. 1995. Fructans in callus of *Gomphrena macrocephala* St.-Hil. *Plant Cell Tissue and Organ Culture* **42**: 233–238.

New Phytol. (1997), **136**, 61–72

Fructan biosynthesis in excised leaves of *Lolium temulentum*

VII. Sucrose and fructan hydrolysis by a fructan-polymerizing enzyme preparation

By ANDREW J. CAIRNS[1]*, GRAHAM D. BONNETT[2], JOSEPH A. GALLAGHER[1], RICHARD J. SIMPSON[3] AND CHRISTOPHER J. POLLOCK[1]

[1] *Cell Biology Department, Institute of Grassland and Environmental Research, Plas Gogerddan, Aberystwyth SY23 3EB, UK*
[2] *Commonwealth Scientific and Industrial Research Organisation, Division of Tropical Crops and Pastures, Davies Laboratory, Private Mail Bag P.O. Aitkenvale, Queensland 4814, Australia*
[3] *Commonwealth Scientific and Industrial Research Organisation, Division of Plant Industry, Crop Adaptation Department, GPO Box 1600, Canberra, ACT 2601, Australia*

(*Received 24 July 1996*)

SUMMARY

A partially purified enzyme preparation from leaves of *Lolium temulentum* L. was previously shown to catalyse the net synthesis of oligofructans and polyfructans from sucrose. Here the same preparation is shown to catalyse the hydrolysis of both sucrose and oligofructans. The magnitude and properties of these hydrolytic activities have been determined. The significance of these catabolic activities for studies of fructan polymerization both *in vitro* and in tissues in a physiologically anabolic state are discussed.

The preparation hydrolysed 1-kestose, 6-kestose, neokestose, inulin oligosaccharides of low degree of polymerization (DP 4 and 5) and endogenous oligofructans from *L. temulentum*, with the concomitant release of monosaccharide. The preparation also released reducing sugar at low rates from high molecular weight inulin but had no detectable activity against bacterial levan. Simultaneous incubation of sucrose and Neosugar (a commercially available mixture of predominantly β-2, 1 linked tri-, tetra- and penta-saccharides) showed that sucrose was preferentially hydrolysed by the preparation, with Neosugar fructans being protected from hydrolysis at sucrose concentrations > 30 mol m^{-3}. The kinetic properties for hydrolysis of both sucrose and Neosugar were determined. For sucrose and Neosugar respectively, Michaelis constants at 30 °C and pH 6·0 were $7{\cdot}7 \pm 0{\cdot}5$ and $14{\cdot}1 \pm 1{\cdot}1$ mol m^{-3} (as terminal fructose) and maximum velocities were $6{\cdot}5 \pm 0{\cdot}1$ and $2{\cdot}7 \pm 0{\cdot}1$ mg g^{-1} fr. wt h^{-1} (equivalent to 10·0 and 4·2 nkat g^{-1} as reducing sugar release). Maximal temperatures for activity were 45 and 44 °C, and Arrhenius activation energies were 39·9 and 46·9 kJ mol^{-1}. Preincubations for 1 h at 49 and 48 °C caused 50% loss of activity in subsequent assays at 30 °C. The pHs for maximal activity for the two substrates were $5{\cdot}2 \pm 0{\cdot}1$ and $5{\cdot}5 \pm 0{\cdot}1$.

Using size exclusion chromatography (SEC), an activity catalysing the formation of fructan oligosaccharides and another catalysing sucrose hydrolysis, were not fully resolved, but exhibited distinct profiles of elution indicating M_r of 57 and 133 kD respectively. When assayed for the hydrolysis of Neosugar, the SEC eluate exhibited two peaks of activity indicative of M_r values of 57 and 133 kD.

Key words: Levan, invertase, Neosugar, oligofructans, sucrose.

* To whom correspondence should be addressed.
Abbreviations; DP, degree of polymerization; FSA, fructan synthetic activity, Invertase, β-fructofuranosidase (EC 3.2.1.26); MLP, Maximum Likelihood Program; SEC, size exclusion chromatography.

INTRODUCTION

Leaves of grasses contain enzymes which are capable of both the synthesis and degradation of the storage sugars, sucrose and fructan. It is not clear how these opposing reactions are integrated and regulated *in vivo*. To address this problem, the present study examines the properties of hydrolytic activities present in leaf tissue during sucrose and fructan accumulation.

We have reported the enzymatic polymerization of fructans from high concentrations of sucrose as the sole substrate, using concentrated, partially purified enzyme preparations from leaves of several grasses (Cairns, 1992*b*, 1993; Cairns & Ashton, 1993, 1994; Penson & Cairns, 1994; Cairns, 1995). These preparations also exhibit high rates of invertase activity (β-fructofuranosidase; EC 3.2.1.26), which competes with the fructan synthetic activity (FSA) for substrate and which might in part determine which fructans are synthesized, particularly in the trisaccharide fraction (Cairns & Ashton, 1991; Cairns, 1993). In addition, some invertases are known to hydrolyse fructan (Cairns & Ashton, 1991; Pollock & Cairns, 1991; Cairns, 1993; Simpson & Bonnett, 1993). For these reasons, the presence of invertase complicates the investigation of fructan synthesis in enzyme reactions (Cairns & Ashton, 1991; Cairns, 1993). In leaf tissue, acid invertase has been localized in the vacuole of mesophyll cells, which is also the site of sucrose accumulation (Wagner, Keller & Wiemken, 1983; Cairns, Winters & Pollock, 1989). This implies that invertase is inhibited *in vivo*. A role for the inhibition of acid invertase by free fructose in mesophyll vacuoles has recently been proposed (Walker, Winters & Pollock, 1996).

Our fructan polymerase preparations were obtained from leaves of high carbohydrate status in the phase of fructan accumulation (Cairns & Pollock, 1988*a*). Enzyme preparations from tissues in a similar physiological condition are reported to contain extractable fructan exohydrolase (FEH; EC 3.2.1.80), which hydrolyses fructan polymers with the release of fructose (Wagner & Wiemken, 1986; Wagner, Wiemken & Matile, 1986; Wagner & Wiemken, 1989; Simpson, Walker & Pollock, 1991; Smouter & Simpson, 1991; Bonnett & Simpson, 1993; Penson & Cairns, 1994; Henson & Livingston, 1996). A basal FEH activity of *c*. 6–12 nkat g^{-1} f. wt (3·9–7·8 mg h^{-1} g^{-1} as reducing sugar release) is a feature of most grass tissues examined, regardless of carbohydrate status, although extractable FEH is higher in the phase of fructan mobilization (Simpson & Bonnett, 1993). It is thus likely that FEH was also present in our polymerase preparations from excised grass leaves. Despite this, fructan products were stable and persisted in polymerase reactions for at least 10 h (Cairns, 1992*b*; Cairns & Ashton, 1993, 1994; Penson & Cairns, 1994). The *in vitro* activity of some of the FEH isoforms from *Lolium* spp. are known to be partly inhibited by the presence of sucrose at $\leqslant 20$ mol m^{-3} Simpson *et al.*, 1991; Bonnett & Simpson, 1993; Simpson & Bonnett, 1993) which might explain the *in vitro* persistence of the polymerase products. In common with invertase and sucrose, FEH is localized in the mesophyll vacuole (Wagner & Wiemken, 1986). Since this is also the site of fructan accumulation, FEH is presumably down-regulated *in vivo*, possibly by sucrose, to permit fructan accumulation.

The interpretation of data from *in vitro* polymerase systems in which a number of potential transformations of substrate and products occur simultaneously is problematic (Cairns & Ashton, 1991; Pollock & Cairns, 1991; Cairns, 1993, 1995). Hence it is necessary, for further progress in the investigation of enzymatic fructan synthesis, both *in vivo* and *in vitro*, to study concurrent hydrolytic processes, the role of fructose in the regulation of invertase and of sucrose in the inhibition of product degradation by FEH. As part of an ongoing study of the physiology and biochemistry of fructan polymerization, this paper characterizes the gross properties of the hydrolysis of sucrose and of a number of fructan substrates by the polymerase preparation from *Lolium temulentum* L.

MATERIALS AND METHODS

Plant tissue, induction of fructan biosynthesis and the preparation of enzyme extracts

Fully expanded 4th and 5th leaves of *Lolium temulentum* L. Ba. 3081 were illuminated continuously at 300 μmol m^{-2} s^{-1} and 20 °C for 24 h to induce fructan accumulation at a linear rate of 1·3 mg g f. wt h^{-1} (Cairns & Pollock, 1988*a*). Enzyme was extracted and partially purified by acid and salt precipitation as described by Cairns & Ashton (1994) and was re-suspended at a tissue equivalent concentration of 7–13 g f. wt cm^{-3} in McIlvaine buffer at pH 6·0 and 20 mM with respect to phosphate. Enzyme was stored as 0·5 cm^3 aliquots at −80 °C. The desalted enzyme preparations contained no detectable endogenous fructan, sucrose or monosaccharide (Cairns & Ashton, 1994).

Screening for the hydrolysis of a range of oligosaccharide substrates

The primary aim of the study was to characterize the hydrolytic properties in the polymerase preparation; hence most hydrolytic reactions were performed at pH 6·0, the optimum for fructan synthesis. The substrate concentration used for screening was 23 mg cm^{-3}, which approximated to the total amount of fructan generated by the polymerase from sucrose in a 5 h reaction at 30 °C (Cairns & Ashton, 1993, 1994). The extract (9·4 g f. wt cm^{-3}) was mixed 1:1 by volume with aqueous stock solutions of each substrate at 46 mg cm^{-3}. The reactions were incubated at 30 °C over a 5 h period. Samples of 50 mm^3 were removed at timed intervals and mixed with 2 cm^3 of the copper reagent of the Somogyi-Nelson reducing sugar assay (Hestrin, Feingold & Schramm, 1955; measured pH 9·9) to stop the reaction. Fructose and glucose gave an equivalent, linear molar colour-yield up to 1·0 A_{500nm} (from 295 μg) with this assay. Additional samples (100 mm^3) were removed at 0 h and 5 h and the reaction was stopped by heating at 100 °C for 4 min. After processing by ion-exchange chromatography (Cairns, 1992*a*), the neutral sugars from 0 h and 5 h samples were compared by TLC (Cairns & Pollock, 1988*a*). Apparent DP of fructans was determined on TLC by comparison with the mobilities of oligoinulins from *Helianthus tuberosus* L. The trisaccharide isomers of *L. temulentum* fructan were identified previously (Cairns & Pollock, 1988*a*, *b*).

The substrates were sucrose, Neosugar (a commercially available oligofructan mixture; Meiji Seika Keisha Ltd., Kawasaki, Japan; Hidaka, Hirayama & Sumi, 1988), high molecular weight inulin (from *Dahlia*; Sigma), and levan (from *Erwinia herbicola*; Sigma). Inulin oligosaccharides from roots of *H. tuberosus* (oligoinulin; DP 3–18, by TLC; 25·7% (w/w) sucrose by HPLC) were kindly provided by Dr Ian Sims (University of Melbourne, Australia). The total endogenous neutral soluble carbohydrate from excised, induced 4th and 5th leaves of *L. temulentum* (DP 3–*c.* 30; Pollock, 1986) was prepared in bulk as described by Cairns (1989). Sucrose and monosaccharides were removed from *L. temulentum* fructan by HPLC before use in enzyme reactions (Cairns & Pollock, 1988*b*). The tissue used as a source of endogenous substrate was physiologically identical to that used as the source of the enzyme.

Simultaneous incubation of sucrose and Neosugar fructan

Sucrose and Neosugar, each at a final concentration of 23 mg cm^{-3}, were simultaneously incubated with the extract at pH 6·0, 30 °C (standard pH and temperature for subsequent reactions) and a final enzyme concentration of 4·7 g f. wt equiv. cm^{-3}. In the first experiment, rates of reducing sugar release were determined during a 3 h incubation and were compared with the rates of release for each sugar as sole substrate. In the second experiment, samples (100 mm^3) were removed at timed intervals during a 5 h incubation, stopped by heating, and processed to obtain the neutral fraction. The persistence of sucrose and Neosugar in the samples was determined by HPLC (Cairns & Pollock, 1988*a*).

Determination of the characteristics of sucrose and Neosugar hydrolysis

For the determination of the properties of sucrose and Neosugar hydrolysis, the stock enzyme solution was diluted tenfold with the McIlvaine buffer (pH 6·0, 20 mol m^{-3} phosphate) immediately before use. The diluted enzyme was mixed 1:1 (v/v) with buffered substrate, resulting in a final enzyme concentration of 0·47 g f. wt equivalent cm^{-3}. In all experiments, rates of reducing sugar release were calculated by linear regression from a minimum of six timed samples during a 120 min time-course for each treatment. Correlation coefficients in all regression analyses were > 0·97. Because Neosugar contains oligoinulins (mainly β-2, 1 linked) of DP 3, 4 and 5 in the mass ratio 3·7:4·5:1, respectively (manufacturer's specification), it was possible to calculate a mean molar concentration of terminal fructose moieties for the mixture. Neosugar was used, assuming a mean molecular mass of 603, and contained 4% (w/w) of reducing compounds determined with the Somogyi-Nelson reagent against standard fructose.

The effect of substrate concentration on reducing sugar release was determined in the range 1–100 mol m^{-3} (terminal fructose) for both substrates. Kinetic parameters were estimated using a logistic fit in the maximum likelihood program (MLP, Ross, 1991, version 3.01). The effect of pH was determined using a McIlvaine buffer series in the range 3–8 (Hale, 1965); rate data were analysed by a Gaussian fit in the MLP to calculate the pH at maximal activity.

The effect of temperature on enzyme stability was determined by preincubating the diluted enzyme at temperatures between 5–70 °C for 1 h on a temperature gradient bar (Cairns & Ashton, 1991). Residual activity was assayed as reducing sugar release from 100 mol m^{-3} substrate under standard conditions. The data were analysed using a logistic fit in the MLP to determine the temperature causing a 50% loss of activity. The effect of temperature on rates of hydrolysis was determined by mixing substrate and enzyme on ice, transfer to positions on the temperature gradient bar in the range 5–70 °C and determination of the time-course of reducing sugar release. Temperature had a negligible effect on the pH of this buffer system (Cairns *et al.*, 1995). Above 45 °C, Neosugar, but not sucrose, was susceptible to hydrolysis in the absence of enzyme. This non-enzymatic hydrolysis accounted for roughly 10% of the rate of reducing sugar release in enzyme reactions at 60 °C. To control for non-enzymatic hydrolysis, the Neosugar experiment was repeated in the absence of the enzyme, and rates of reducing sugar release at each temperature were subtracted from the rates determined in the enzyme reactions. The variation in submaximal rate with temperature (5–40 °C) was analysed by both exponential regression and by Arrhenius analysis to determine activation energies for the hydrolysis of the two substrates (Nobel, 1988).

Behaviour of the polymerase and sucrose–hydrolase activities during size exclusion HPLC (SEC) of protein

Aliquots of an undiluted extract (0·2 cm^3; equivalent to 1·5 g f. wt) were chromatographed on a 300 × 7·8 mm Bio-Sil TSK-250 SEC column (Biorad) at room temperature with McIlvaine buffer pH 6·0 (78 mol m^{-3} with respect to phosphate) as mobile phase and at a flow rate of 1·0 cm^3 min^{-1}. The 280 nm absorbance of the eluate was monitored continuously and 0·25 cm^3 fractions were collected on ice. Fractions were mixed 1:1 (v/v) with 2000 mol m^{-3} aqueous sucrose, and incubated at 40 °C (Cairns & Ashton, 1994). After 2 h, 0·1 cm^3 subsamples of each reaction were screened for reducing sugar release by mixing in microtitre plate wells, with 0·1 cm^3 of 0·2 % (w/v) triphenyl tetrazolium chloride (TTC) in 1000 mol m^{-3} NaOH. Red formazan formed in fractions which released reducing sugar from sucrose. These fractions were incubated for a further 12 h and the reaction was stopped by heating at 100 °C for 4 min. Fructan and monosaccharide products of each active fraction were determined by HPLC and the pooled eluate containing fructan was concentrated by freeze-drying and further fractionated by TLC (Cairns & Pollock, 1988*a*, *b*). The M_r of fructose releasing and fructan-synthesizing fractions were determined by comparison of the elution patterns with those of gel filtration standards of known molecular mass (Biorad).

Behaviour of Neosugar and sucrose hydrolysing activities during SEC

Protein equivalent to 2·7 g f. wt in 0·2 cm^3 was chromatographed as described above, with the following exceptions: fractions of 0·2 cm^3 were collected, one half of which was mixed 1:1 (v/v) with 200 mol m^{-3} sucrose, the other 1:1 (v/v) with 160 mol m^{-3} Neosugar. Both aliquots of each fraction were incubated at 37 °C and rates of reducing sugar release were determined. The elution behaviour of fractions which hydrolysed Neosugar and sucrose were compared with those of M_r standards.

RESULTS

Hydrolysis of sucrose and fructan

Figure 1 summarizes the release of reducing sugar from a range of oligosaccharide substrates. Sucrose was hydrolysed at a linear rate of 5·2 mg g^{-1} h^{-1}, being completely degraded within 80 min. Reducing sugar release from Neosugar was linear during the initial 60 min at 2·0 mg g^{-1} h^{-1} and tended to an asymptote after longer periods of incubation. Oligoinulin hydrolysis was biphasic, exhibiting an initial (0–10 min) fast rate of 5·2 mg g^{-1} h^{-1} and a subsequent slower linear rate of release of reducing sugar of 0·4 mg g^{-1} h^{-1}. *L. temulentum* fructan was also degraded, with an initial short period of rapid hydrolysis (0–10 min) followed by a slower linear rate, similar to the second phase of hydrolysis of the oligoinulins (0·4 mg g^{-1} h^{-1}). Inulin was relatively resistant, exhibiting a rate of hydrolysis $< 0·1$ mg g^{-1} h^{-1}. Activity against levan was at the limit of detection.

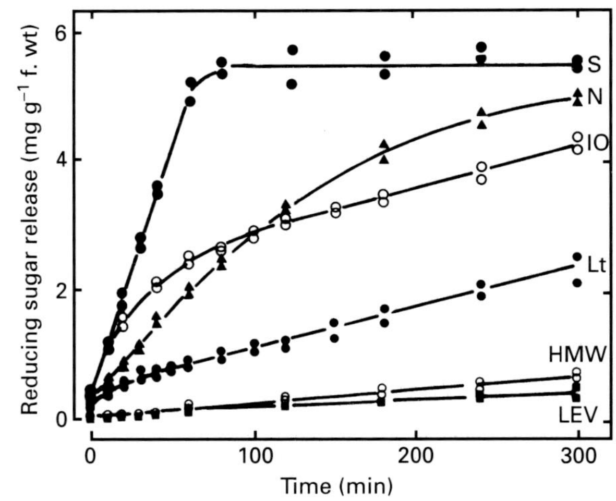

Figure 1. Time-courses of release of reducing sugar from sucrose and fructan substrates by the enzyme preparation. Reactions were performed at pH 6·0 and at 30 °C, contained 23 mg cm^{-3} of substrate and enzyme equivalent to 4·68 g f. wt cm^{-3}. Data are from two independent determinations for each substrate. Substrates were sucrose (S), Neosugar (N), inulin oligosaccharides (IO), total fructan from leaves of *Lolium temulentum* (Lt), high molecular weight inulin (HMW) and bacterial levan (LEV).

Qualitative analysis by TLC (Fig. 2), showed that, of the β-2, 1 linked sugars in both the oligoinulin and Neosugar reactions, 1-kestose and tetrasaccharide were the fructans most susceptible to hydrolysis. Inulins of DP 5 and 6 stained less intensely after 5 h of incubation, indicating limited hydrolysis (Fig. 2; lanes OI and N). Of the fructans from *L. temulentum*, sugars in the range of apparent DP 3–14 (Fig. 2; lanes Lt) were degraded by the extract. Inulin was hydrolysed to release mainly monosaccharide, and small amounts of oligosaccharides of DP 9–11 with similar mobilities to oligoinulins (Fig. 2, compare lanes HMW with IO). TLC showed no hydrolysis products from bacterial levan.

Simultaneous incubation of sucrose and Neosugar

When equal masses of Neosugar and sucrose were co-incubated with the extract, a biphasic pattern of

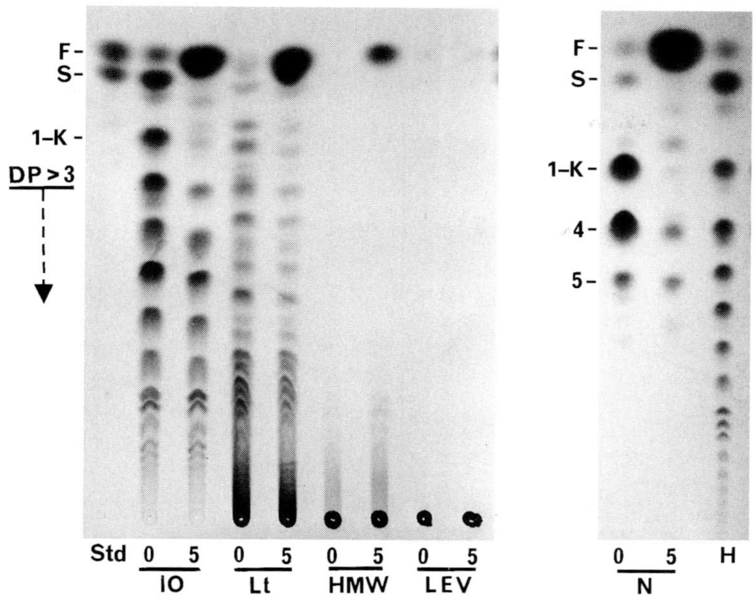

Figure 2. Qualitative analysis of substrate degradation. Neutral soluble sugars from time zero and 5 h samples are compared by TLC. Inulin oligosaccharides (IO), total fructan from leaves of *Lolium temulentum* (Lt), high molecular weight inulin (HMW), bacterial levan (LEV) and Neosugar (N) are compared with standard fructose (F) sucrose (S) and inulin oligosaccharides from roots of *Helianthus tuberosus* (H and IO: $t = 0$). 1-K and DP > 3 denote 1-kestose and fructans of degree of polymerization greater than three, respectively.

hexose release resulted (Fig. 3*a*). The initial phase of hydrolysis was linear, with a rate and duration identical to those of sucrose as sole substrate (5·2 mg g^{-1} h^{-1}, 60 min). This initial rate of reducing sugar release was lower than the theoretical additive rate obtained for Neosugar and sucrose incubated as sole substrates (6·75 mg g^{-1} h^{-1}). The second phase began after 60 min, coinciding with consumption of > 95 % of the sucrose and was characterized by a curve tending to an asymptote. The second phase was similar to the curve obtained with Neosugar as sole substrate. When the reactions were followed by substrate depletion (Fig. 3*b*), sucrose was found to be hydrolysed rapidly in the initial 45 min of reaction, whilst the Neosugar fructans were not hydrolysed in this period. Fructans were degraded only after sucrose concentration in the reaction fell to < 10 mg cm^{-3} (29 mol m^{-3}).

Kinetic properties for sucrose and Neosugar hydrolysis

Kinetic data for sucrose and for Neosugar hydrolysis are presented in Figure 4. Maximum velocities at 30 °C and pH 6·0 were calculated at 6·5 ± 0·1 and 2·7 ± 0·1 mg g^{-1} f. wt. h^{-1} (10·0 and 4·2 nkat g^{-1}) whilst values for apparent K_m were 7·7 ± 0·5 and 14·1 ± 1·1 mol m^{-3} respectively (values ± SE). Figure 5 shows the effect of pH on sucrose and Neosugar hydrolysis. Both substrates produced roughly symmetrical bell-shaped curves with optima at 5·2 ± 0·1 5·5 ± 0·1, respectively.

The preincubation temperatures causing 50 % loss of activity were 47·9 °C ± 0·5 for sucrose and 46·1 ± 0·7 °C for Neosugar (Fig. 6*a*). The correlation coefficient for both curves was > 0·97. The activity against both substrates increased exponentially with temperature in the range 5–40 °C. By regression, the line equations were, for sucrose; $v = 0{\cdot}107\, e^{0{\cdot}054t}$; $n = 9$; $r^2 = 0{\cdot}993$, and for Neosugar, $v = 0{\cdot}240\, e^{0{\cdot}067t}$; $n = 11$; $r^2 = 0{\cdot}988$ (where v = velocity in mg g^{-1} h^{-1} and t = temperature °C). Arrhenius analysis gave continuous negative linear relationships allowing calculation of activation energies of 39·9 and 46·9 kJ mol^{-1} respectively for the two substrates. Sharply defined optimum temperatures for hydrolysis of 45 °C and 44 °C were determined respectively for sucrose and Neosugar (Fig. 6*b*). The optimum determined for Neosugar hydrolysis was not influenced by the subtraction of non-enzymatic reducing sugar release.

SEC of the protein extract

The synthesis of oligofructan of DP > 3 from sucrose requires the use of high enzyme concentrations. To maximize product formation in column fractions, a relatively large sample volume of 0·2 cm^3 of concentrated enzyme preparation was used. To compensate for the effects of dilution (Cairns, 1995) during chromatography, reactions were performed at 40 °C for 12 h to enhance the formation of polymers (Cairns & Ashton, 1994). Fructose release from sucrose (Fig. 7c) and FSA (Fig. 7*c*, *d*) showed distinct, though overlapping patterns of elution, indicating relative molecular masses of 133 and 57 kDa respectively. Recoveries were 82 % and 77 % for sucrose hydrolysis and fructan synthesis. Incubation of the column fractions with a relatively low concentration of sucrose (100 mol m^{-3}) produced a

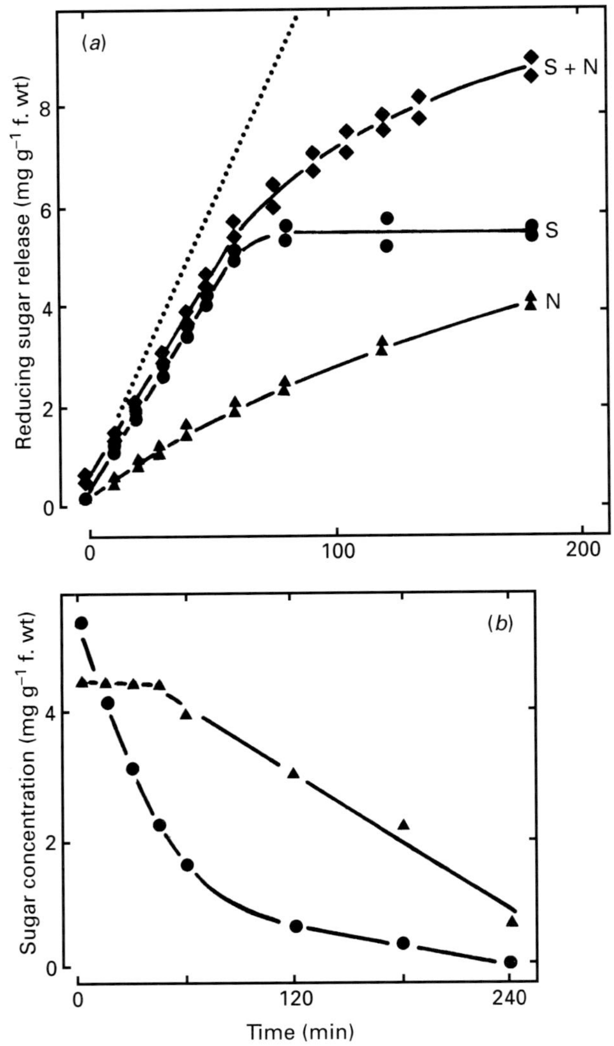

Figure 3. (*a*) Reducing sugar release from sucrose alone (S), Neosugar alone (N) and from a mixture of sucrose and Neosugar (S+N), by the enzyme preparation under the standard conditions. The discontinuous line represents the theoretical additive rate of release of the initial rate for S only, added to that for N only. (*b*) Preferential degradation of sucrose in a mixture of sucrose and Neosugar. Sucrose (●---●) and total Neosugar oligosaccharides (▲---▲) were determined by HPLC. The reaction contained an initial concentration of *c.* 23 mg cm^{-3} of each substrate and was performed under standard conditions.

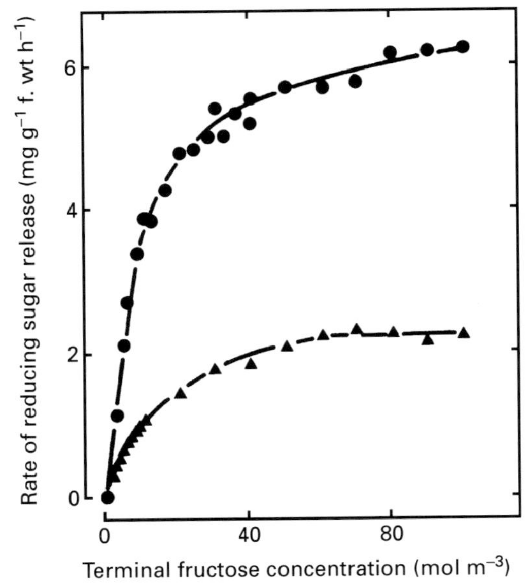

Figure 4. Effect of substrate concentration on rates of reducing sugar release from sucrose (●---●) and Neosugar (▲---▲). Substrate concentration was expressed as the molar concentration of terminal fructose moieties. Each point is a rate calculated from a regression line where $n = 6$ and $r^2 > 0{\cdot}97$.

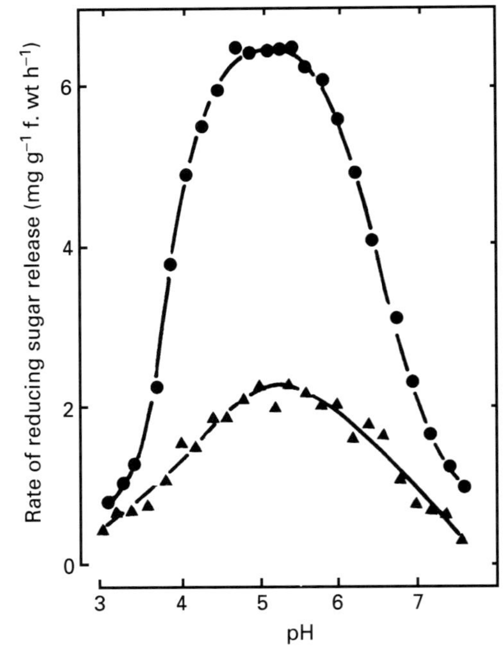

Figure 5. Effect of pH on reducing sugar release from 100 mol m^{-3} sucrose (●---●) and 80 mol m^{-3} Neosugar (▲---▲). Each point is a rate calculated from a regression line where $n = 6$ and $r^2 > 0{\cdot}97$.

single peak of reducing sugar release, with an apparent $M_r = 133$ kDa (Fig. 8). A peak of activity against Neosugar was also found at 133 kDa together with a second peak of activity at 57 kDa. Recoveries of activity in the eluate was 83 % and 85 % for the hydrolysis of sucrose and Neosugar (based upon the respective activities applied to the column).

DISCUSSION

Hydrolysis of substrates

In view of the effects of substrate concentration on the rates of sucrose and neosugar hydrolysis (Fig. 4), their asymptotic time-courses of hydrolysis (Fig. 1: S, N) are explicable in terms of the respective affinities of the preparation for these substrates, their relative rates of hydrolysis and progressive limitation by substrate utilization during the course of the reaction. The oligoinulin preparation contained sucrose (Fig. 2, lanes IO) equivalent to 5·9 mg cm^{-3} in the reaction. This amount of sucrose would be completely hydrolysed within 14·5 min at the rate of sucrose hydrolysis determined above. The biphasic

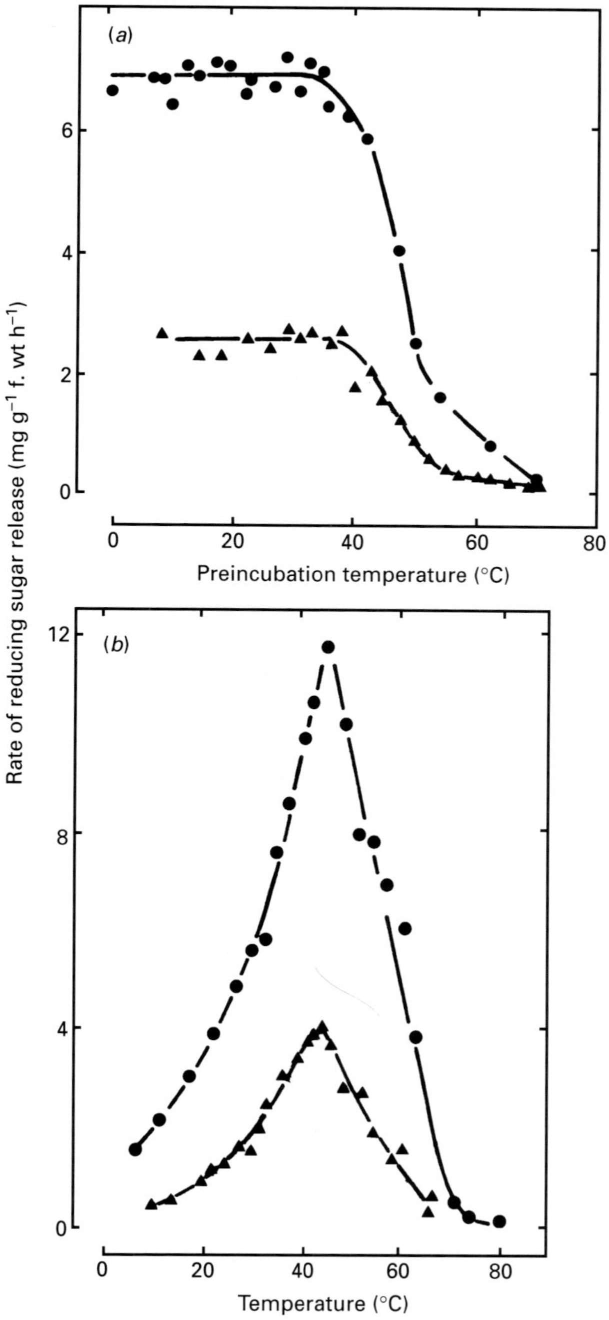

Figure 6. Effect of temperature on reducing sugar release from sucrose (●---●) and Neosugar (▲---▲). (*a*) Temperature-stability characteristics. The enzyme was preincubated for 1 h at a range of temperatures prior to the assay of residual hydrolytic activity against the two substrates. (*b*) The effect of temperature on the rate of reducing sugar release from sucrose and Neosugar. Each point is a rate calculated from a regression line where $n = 6$ and $r^2 > 0{\cdot}97$.

time-course of hydrolysis of oligoinulin suggested an initial complete degradation of sucrose at high rate and a subsequent hydrolysis of fructan at a lower, constant rate. This proposal was confirmed in the simultaneous incubation experiments (discussed below). The endogenous fructan of *L. temulentum* was also degraded with a short initial high rate of reducing sugar release, followed by a lower linear rate of hydrolysis (Fig. 1, Lt). This pattern was consistent with the low but detectable sucrose concentration in this substrate (Fig. 2, lane Lt, 0 h). TLC showed that fructan species in the range of apparent DP $\leqslant$ 16 were broken down with the release of monosaccharide (Fig. 2, lanes Lt). There was no evidence of the hydrolysis of the *L. temulentum* fructan of apparent DP > 16. This pattern parallels the early phase of fructan degradation by *L. temulentum* during reserve mobilization *in vivo* (Simpson *et al.*, 1991). Since the trisaccharide and oligofructan fractions of *L. temulentum* contain β-2,1, β-2,6 and 6-G glycosidic linkages (Cairns & Pollock, 1988*a*; Sims, Pollock & Horgan, 1992), activities hydrolysing all of these linkage types were present in the extract.

Activity against Neosugar and oligoinulin demonstrated the ability to cleave β-2,1 glycosidic linkages of small oligofructans (DP 3–5; Fig. 2, lanes N and IO). However, when inulin (also β-2,1 linked) was used as substrate, low rates of fructose release were detected. The results do not allow distinction between two possible explanations: (i) the lower molar concentration of terminal fructose residues per unit weight available for reaction, or, (ii) a lower affinity of the hydrolase activity for longer-chain substrates. The formation of inulooligosaccharides and monosaccharide as the products of inulin hydrolysis (Fig. 2, HMW, lane 5 h) indicate exo-hydrolase rather than endo-hydrolase activity. The minimal activity against bacterial levan (β-2,6 linked; Fig. 1; LEV) might also be a consequence of the low concentration of termini or of steric effects of chain length. Negligible activity against bacterial levan is consistent with other reports of FEH enzymes from grasses (Wagner & Wiemken, 1986; Henson, 1989; Bonnett & Simpson, 1993).

Protection of fructan from degradation in the presence of sucrose

Screening for activity against a number of substrates showed the hydrolysis of the three glycosidic bonds found in grass fructan and a marked susceptibility of small fructans to hydrolysis. This is of particular importance because the persistence and accumulation of smaller fructan intermediates is crucial to the subsequent building of higher polymers, both *in vitro* and *in vivo*. The hydrolytic nature of the preparation towards oligofructan was in marked contrast to our previous reports where oligofructans were synthesized, and persisted for at least 10 h in the presence of these hydrolytic activities (Cairns, 1992*b*; Cairns & Ashton 1993, 1994; Penson & Cairns, 1994).

The biphasic time-course of oligoinulin hydrolysis (Fig. 1) suggested that sucrose hydrolysis preceded that of oligofructan. Experiments were devised to test this possibility. Neosugar was chosen as substrate

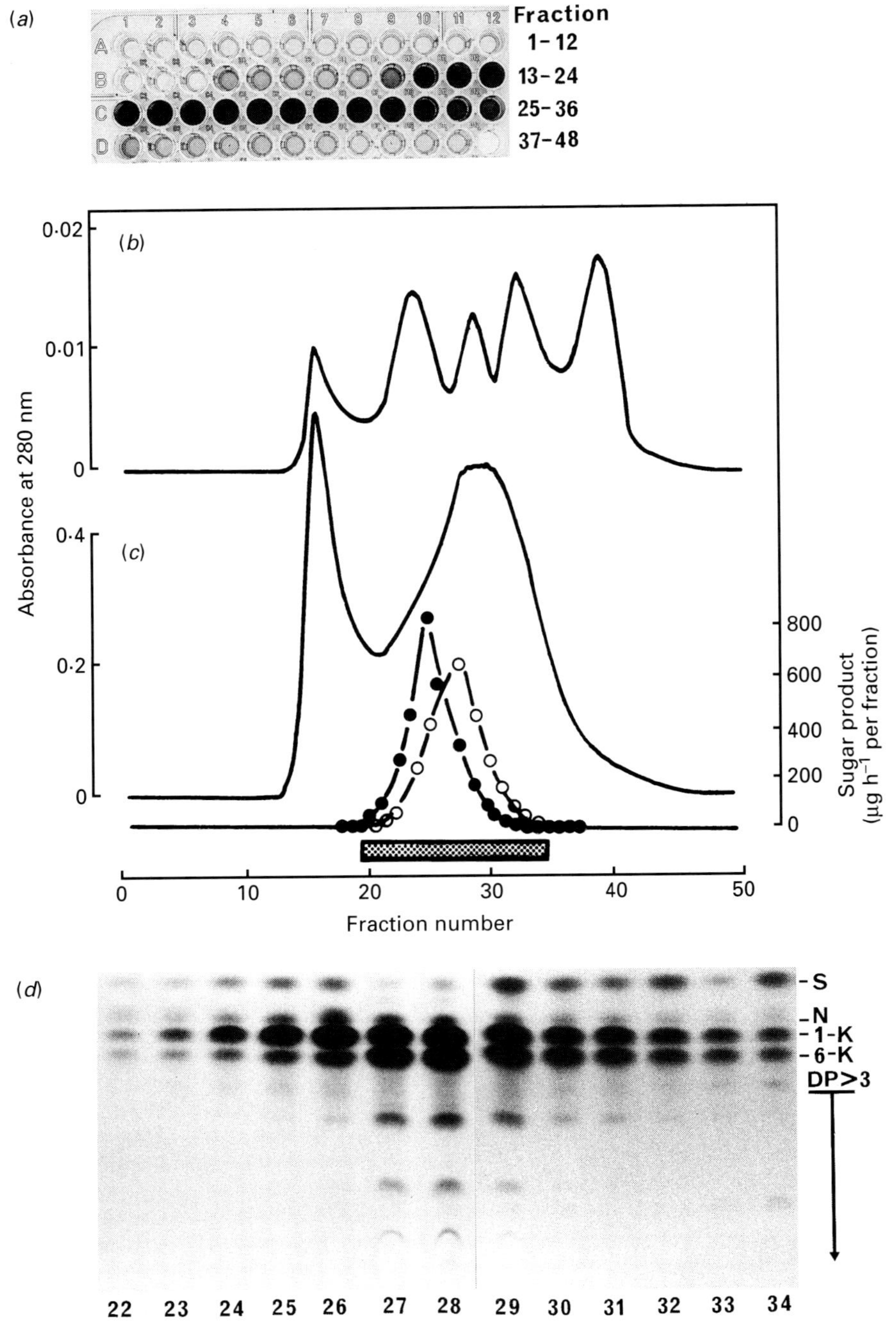

Figure 7. Size exclusion chromatography of the extract by HPLC on Bio-Sil TSK-250. Protein equivalent to 1·5 g f. wt was separated and collected as 0·25 cm^{-3} fraction. Fractions were incubated for long periods at high sucrose concentration to enhance oligofructan synthesis. Fructan synthesis and sucrose hydrolysis were measured by HPLC. (*a*) Screening for sucrose-active fractions using TTC. (*b*) Size calibration of the column using (left to right) thyroglobulin (670 kDa), immunoglobin G (158 kDa) ovalbumin (44 kDa); Myoglobin (17 kDa) and vitamin B_{12} (1·35 kDa). (*c*) Chromatography of the partially purified protein showing the elution profiles of u.v. absorbance (continuous line), invertase activity (●---●) and FSA (○--○). The shaded bar indicates sucrose active fractions from the microtitre plate screen. (*d*) Fructan synthesis in column fractions shown by TLC. N and G-K denote neokestose and 6-kestose respectively. Other labels are as Figure 2.

because of its simple, low-DP fructan composition, its negligible sucrose content and because of its availability in quantity. When Neosugar was simultaneously incubated with sucrose (Fig. 3*a*), a biphasic pattern of reducing sugar release was obtained. The rate in the first phase was identical to that of sucrose alone. The second phase of reducing sugar release, which resembled that of Neosugar alone, did not commence until the bulk of the sucrose had been degraded. The results were consistent with a preferential hydrolysis of sucrose, before the hydrolysis of fructan. Direct determination of the persistence of sucrose and Neosugar by HPLC, confirmed this view, showing the 'protection' of fructan by the presence of sucrose above a threshold concentration of 29 mol m^{-3} (Fig. 3*b*).

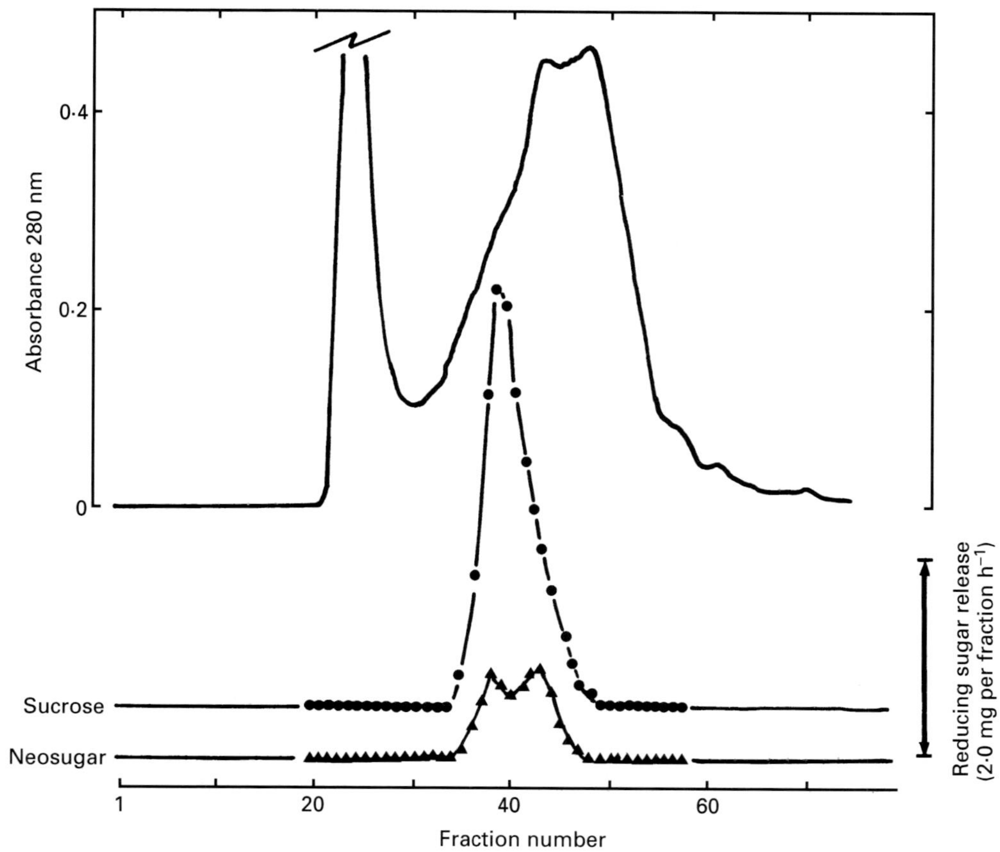

Figure 8. Distribution of sucrose (●---●) and Neosugar (▲---▲) hydrolysis during SEC chromatography. Protein equivalent to 2·75 g f. wt was separated and collected as 0·2 cm^3 fractions which were incubated for short periods at low substrate concentration to distinguish hydrolysis from polymerization. The continuous line represents the u.v. profile of the eluate.

Table 1. *Summary of the properties of oligofructan (Neosugar) and sucrose hydrolysis compared with those of fructan synthesis by the same enzyme preparation from induced leaves of* Lolium temulentum

Parameter	Neosugar hydrolysis	Sucrose hydrolysis	Fructan synthesis*
K_m at 30 °C ($mol\ m^{-3}$)	14.1±1.1	7.7±0.5	Not saturable at 1500 mol m^{-3}
V_{max} at 30 °C ($mg\ g^{-1}\ h^{-1}$)	2.7±0.1	6.5±0.1	1.5±0.1 at 1500 mol m^{-3}
pH for maximal activity	5.5±0.1	5.2±0.1	6.0
Temperature for maximal activity (°C)	44	45	43
Temperature for 50% loss of activity (°C)	48	49	32
Apparent M_r by gel filtration (kDa)			
I	133	133	57
II	57	—	—
Activation energy ($kJ\ mol^{-1}$)	46.9	39.9	36.9

* Data from Cairns & Ashton (1994).

Two possible mechanisms may be envisaged to explain these results: (i) that invertase initially and preferentially acts on sucrose, and only later on fructan; (ii) that a sucrose-inhibited FEH was present, the FEH becoming active only after extensive sucrose degradation by invertase. The SEC results (Fig. 8) indicate the possibility of both an invertase-associated and an invertase-independent FEH activity; hence both mechanisms might be involved. Regardless of the mechanism, the results can explain why fructan persisted in our previous fructan polymerase reactions. The high concentrations of sucrose required for fructan synthesis (Cairns, 1992*a*; Cairns & Ashton, 1993, 1994) would inhibit fructan degradation and would permit the accumulation of intermediate oligofructans and larger fructan products.

Properties of sucrose and fructan hydrolysis

The characteristics of the hydrolases are summarized in Table 1 and are generally consistent with reported values for invertase and FEH from other grasses, though the temperature optima were higher than those generally reported (Simpson & Bonnett, 1993; Bonnett & Simpson, 1993; Henson & Livingston, 1996). Simpson & Bonnett (1993) expressed the view that many published values for these temperature parameters are underestimates. The large number of samples (facilitated by use of the gradient bar) and the consistent linear rates obtained for each treatment suggest that our values are accurate.

The maximal rate of Neosugar hydrolysis was, at 2·7 mg g^{-1} h^{-1}, lower than the range of 4–8 mg g^{-1} h^{-1} for fructan hydrolysis by FEH in previous reports (Simpson & Bonnett, 1993). This might reflect the fructan-synthetic status of the tissues used. Generally FEH was studied in fructan-degrading tissues which might be expected to have higher rates of FEH. The enzymes studied here were deliberately derived from tissue in the phase of net synthesis. The lower FEH activity implies a small degree of coarse down-regulation of FEH during fructan synthesis, though the rate remains significant in the context of the rate of accumulation of fructan (discussed below).

Comparison of the hydrolase activities with enzymatic fructan synthesis

For a complete understanding of the factors affecting fructan synthesis *in vitro*, and to assist in purification of the polymerase(s), a knowledge of the characteristics of the hydrolytic activities in the extract is of value. The properties of the FSA, sucrose hydrolysis and Neosugar hydrolysis are compared in Table 1. The invertase and fructan hydrolase activities are generally similar in gross physical properties but differ significantly from FSA in a number of important respects. By virtue of the large difference in substrate affinities, hydrolytic activities can be assayed with little interference from the polymerase. The polymerase is barely active at 100 mol m^{-3} sucrose (< 0·1 mg g^{-1} h^{-1}, Cairns & Ashton, 1994) and produces little detectable product at *c*. 0·5 g f. wt equivalent cm^{-3} in short reactions (< 2 h) even at high substrate concentration (Cairns *et al.* 1989; Cairns, 1995). By contrast, sucrose and fructan hydrolysis are measurable at low substrate and low enzyme concentration. Hence the properties determined for the hydrolases were not affected significantly by polymerase-mediated hexose release. The converse is not true: with the exception of the protection of synthesized fructan by the sucrose substrate, none of the properties of the hydrolytic activities can be exploited to favour the polymerase reaction in partially purified preparations.

Separation of activities by SEC

The elution behaviour of FSA and sucrose hydrolysis were distinct on SEC, with M_r of 57 and 133 kDa respectively (Fig. 7*c*, *d*). This is the first evidence that these two reactions are catalysed by distinct proteins in *L. temulentum*, and it provides a basis for the further purification of the polymerase(s). These molecular masses are similar to those reported for invertase (64, 116 and 155 kDa: Obenland *et al.*, 1993) and an invertase-6-SFT activity from barley leaves (67 kDa; Duchateau *et al.*, 1995). The pattern of oligofructans synthesized by fractions 27–29 (Fig. 7*d*) was identical to the early stages of the optimized FSA reaction (Cairns, 1992*b*; Cairns & Ashton, 1994) and to the restricted range of oligosaccharides formed when FSA is diluted (Cairns, 1995). The small size range of the fructan product (apparent DP 3–6) was therefore assumed, despite the long incubation period, to be due to the dilution of the FSA during chromatography and assay. This was in the order of approx. 25-fold (to an average of 0·06 g f. wt equivalent cm^{-3}) which is too low for efficient polymerization (Cairns, 1992, 1995; Cairns & Ashton, 1993, 1994; Penson & Cairns, 1994), and underlines the difficulties associated with the purification of enzyme(s) capable of the functional polymerization of fructan. The recovery of oligofructan synthesis (DP > 3) in column fractions shows that enzyme(s) of M_r 57 kDa are responsible for both trisaccharide and higher oligofructan synthesis. If more than one distinct enzyme is required for the synthesis of fructan of DP > 3 in *L. temulentum* (Edelman & Jefford, 1968; Simmen *et al.*, 1993; Sprenger *et al.*, 1995), the contributing enzymes must have co-eluted.

When SEC fractions were assayed at low sucrose concentration, in order to minimize interference by the polymerase (discussed above), a single peak of activity with an apparent M_r of 133 kDa was detected. Roughly half the total activity against Neosugar also had M_r of 133 kDa, and the same enzyme might be responsible for degrading both substrates. The remainder of the activity against Neosugar exhibited an M_r of 57 kDa, coincident with that of the FSA. This release of reducing sugar shows that the FSA fraction is both distinct from invertase and capable of the net hydrolysis of Neosugar fructans under these conditions of assay. The M_r of 57 kDa is identical to that of the FEH of

Dactylis glomerata (Yamomoto & Mino, 1985) and the results closely parallel those reported for *L. rigidum* where an invertase of M_r 132 kDa was separated from an FEH of M_r 68 kDa using SEC (Bonnett & Simpson, 1993). The values for *L. temulentum* fructan hydrolase and invertase fall within the range of 57–68 kDa and 40–155 kDa reported for FEH and acid invertases, respectively, from grasses (Simpson & Bonnett, 1993; Obenland *et al.*, 1993; Henson & Livingston, 1996).

Physiological significance

The results raise a number of issues concerning the function and regulation of hydrolytic activities in tissues engaged in sucrose and fructan accumulation. Leaves which accumulate sucrose at 1·6 mg g^{-1} f. wt h^{-1} at 20 °C (Cairns & Pollock, 1988*a*), are shown here to contain extractable acid invertase which is capable of the hydrolysis of sucrose with a rate, at 20 °C, of 3·6 mg g^{-1} f. wt h^{-1} (Fig. 6*b*). The affinity of the invertase for sucrose (K_m = 7·7 mol m^{-3}) is high in the context of the estimated sucrose concentration of mesophyll vacuoles during fructan synthesis (⩽ 100 mol m^{-3}; Cairns *et al.*, 1989). Similarly, at 20 °C the tissue accumulates fructan at 1·3 mg g^{-1} f. wt h^{-1} (Cairns & Pollock, 1988*a*), yet is shown here to contain extractable FEH with an activity at 20 °C of 0·9 mg g^{-1} f. wt h^{-1} against 1-kestose and inulin tetrasaccharide (Fig. 6*b*). The accumulation of such small fructans during synthesis is crucial to the formation of higher polymers.

The established model for fructan synthesis proposes that sucrose, fructan, invertase, FEH and the enzyme(s) of fructan synthesis are co-compartmentalized in the vacuole of grass leaf mesophyll cells (Wagner *et al.*, 1983; Cairns *et al.*, 1989; Cairns & Pollock, 1991). However, sucrose and fructan accumulate in the tissue, despite the presence of significant extractable activities of invertase and FEH (Cairns & Pollock, 1988*a*, *b*). This implies that both enzymes are inhibited *in vivo*. Since vacuolar sucrose concentration was *c.* 100 mol m^{-3} during fructan synthesis in leaves of *L. temulentum* (Cairns *et al.*, 1989), sucrose inhibition of FEH (effective above 29 mol m^{-3}; Fig. 3*b*) provides a mechanism by which fructan hydrolysis might be prevented *in vivo*, permitting trisaccharide accumulation and subsequent higher fructan synthesis. Purified acid invertase in *L. temulentum* is inhibited by fructose *in vitro*. The *K*i of 2·5 mol m^{-3} is sufficiently low to suggest that invertase inhibition *in vivo* by fructose is a possibility (Walker *et al.*, 1996). However, rates of sucrose hydrolysis observed *in vitro* in this study were linear, even though product concentrations were in the order of 30 mM (Fig. 1). Distribution of specific isoforms of acid invertase between tissue types has been detected in barley leaves (Obenland *et al.*, 1993). Contamination of this preparation by invertase isoforms not normally present in mesophyll vacuoles is a possibility and might explain the inconsistent inhibition kinetics. Delineation of the detailed mechanisms regulating fructan biosynthesis in the presence of hydrolytic activities will require the determination of the spatial distribution of the relevant enzymes.

It is clear that sucrose is the central metabolite in fructan synthesis. Its accumulation triggers the onset of fructan synthesis, it provides the substrate for polymerization (Cairns & Pollock, 1991) and there is evidence that it stabilizes the enzyme(s) of fructan synthesis against proteolytic degradation, a proposed feature of the metabolic control of fructan metabolism (Obenland *et al.*, 1991, Cairns & Ashton, 1994). The presence of sucrose *in vitro* stabilizes small fructans against hydrolysis and is crucial to their accumulation. This in turn allows the elaboration of larger polymers. It might also be an important feature of the process *in vivo*.

ACKNOWLEDGEMENTS

We thank Dr Ian Sims, University of Melbourne, for the gift of inulooligosaccharides and Jenny Ashton for excellent technical assistance. We are grateful to the British Council (Australia) and CSIRO for provision of travel and subsistence funding to A. J. C. This research was supported by the Biotechnology and Biological Sciences Research Council (UK).

REFERENCES

Bonnett GD, Simpson RJ. 1993. Fructan hydrolase activities from *Lolium rigidum* Gaudin. *New Phytologist* **123**: 443–451.

Cairns AJ. 1989. Fructan biosynthesis in excised leaves of *Lolium temulentum* L. IV. Cell-free ^{14}C labelling of specific oligofructans at low sucrose concentration. *New Phytologist* **112**: 465–473.

Cairns AJ. 1992*a*. A reconsideration of fructan biosynthesis in storage roots of *Asparagus officinalis*. *New Phytologist* **120**: 463–473.

Cairns AJ. 1992*b*. Fructan biosynthesis in excised leaves of *Lolium temulentum* L. V. Enzymatic *de novo* synthesis of large fructans from sucrose. *New Phytologist* **122**: 253–259.

Cairns AJ. 1993. Evidence for the *de novo* synthesis of fructan by enzymes from higher plants; a re-appraisal of the SST/FFT model. *New Phytologist* **123**: 15–24.

Cairns AJ. 1995. Effects of enzyme concentration on oligofructan synthesis from sucrose. *Phytochemistry* **40**: 705–708.

Cairns AJ, Ashton JE. 1991. The interpretation of *in vitro* measurements of fructosyl transferase activity: an analysis of patterns of fructosyl transfer by fungal invertase. *New Phytologist* **118**: 23–34.

Cairns AJ, Ashton JE. 1993. Species-dependent patterns of fructan synthesis by enzymes from excised leaves of oat, wheat, barley and timothy. *New Phytologist* **124**: 381–388.

Cairns AJ, Ashton JE. 1994. Fructan biosynthesis is excised leaves of *Lolium temulentum* L. VI. Optimization and stability of enzymatic fructan synthesis. *New Phytologist* **126**: 3–10.

Cairns AJ, Howarth, CJ, Pollock CJ. 1995. Characterisation of acid invertase from the snow mould *Monographella nivalis*: a mesophilic enzyme from a psychrophilic fungus. *New Phytologist* **130**: 391–400.

Cairns AJ, Pollock CJ. 1988*a*. Fructan biosynthesis in excised leaves of *Lolium temulentum* L. I. Chromatographic characterisation of oligofructans and their labelling patterns following $^{14}CO_2$ feeding. *New Phytologist* **109**: 399–405.

Cairns AJ, Pollock, CJ. 1988*b*. Fructan biosynthesis in excised leaves of *Lolium temulentum* L. II. Changes in fructosyl transferase activity following excision and application of inhibitors of gene expression. *New Phytologist* **109**: 407–413.

Cairns AJ, Winters A, Pollock CJ. 1989. Fructan biosynthesis in excised leaves of *Lolium temulentum* L. III. A comparison of the *in vitro* properties of fructosyl transferase activities with the characteristics of *in vivo* fructan accumulation. *New Phytologist* **112**: 343–352.

Duchateau N, Bortlik K, Simmen U, Wiemken A, Bancal P. 1995. Sucrose: fructan 6-fructosyltransferase, a key enzyme for converting carbon from sucrose to fructan in barley leaves. *Plant Physiology* **107**: 1249–1255.

Edelman J, Jefford TG. 1968. The mechanism of fructosan metabolism in higher plants as exemplified in *Helianthus tuberosus* L. *New Phytologist* **67**: 517–531.

Hale LJ. 1965. Biological Laboratory Data. *Methuen's Monographs on Biological Subjects*. London: Chapman & Hall.

Henson CA. 1989. Purification and properties of barley stem fructan hydrolase. *Journal of Plant Physiology* **134**: 186–191.

Henson CA, Livingston DP. 1996. Purification and characterisation of an oat fructan exohydrolase that preferentially hydrolyses β-2,6 fructans. *Plant Physiology* **110**: 639–644.

Hestrin S, Feingold DD, Schramm M. 1955. Hexoside hydrolases. In: Colowick SP, Kaplan NO, eds. *Methods in Enzymology, vol.* New York Academic Press, 231–257.

Hidaka H, Hirayama M, Sumi N. 1988. A fructo-oligosaccharide-producing enzyme from *Aspergillus niger* ATCC 20611. *Agricultural and Biological Chemistry* **52**: 1181–1187.

Nobel PS. 1988. Principles underlying the prediction of temperature in plants, with special reference to desert succulents. In: Long SP, Woodward FI, eds. *Symposia of the Society for Experimental Biology XXXXII*, 1–23.

Obenland DM, Simmen U, Boller T, Wiemken A. 1991. Regulation of sucrose-sucrose-fructosyl transferase in barley leaves. *Plant Physiology* **97**: 811–813.

Obenland DM, Simmen U, Boller T, Wiemken A. 1993. Purification and characterisation of three soluble invertases from barley (*Hordeum vulgare* L.) leaves. *Plant Physiology* **101**: 1331–1339.

Penson SP, Cairns AJ. 1994. Fructan biosynthesis in excised leaves of wheat (*Triticum aestivum* L.): a comparison of *de novo* synthesis *in vivo* and *in vitro*. *New Phytologist* **128**: 395–402.

Pollock CJ. 1986. Fructans and the metabolism of sucrose in vascular plants. *New Phytologist* **104**: 1–24.

Pollock CJ, Cairns AJ. 1991. Fructan metabolism in grasses and cereals. *Annual Review of Plant Physiology and Molecular Biology* **42**: 77–101.

Ross GJS. 1991. *Maximum likelihood program. Release* 3.09. Numerical Algorithms Group, Oxford.

Simmen U, Obenland D, Boller T, Wiemken A. 1993. Fructan synthesis in excised barley leaves. Identification of two sucrose-sucrose fructosyl transferases induced by light and their separation from constitutive invertases. *Plant Physiology* **101**: 459–468.

Sims IM, Pollock CJ, Horgan R. 1992. Structural analysis of oligomeric fructans from excised leaves of *Lolium temulentum*. *Phytochemistry* **31**: 2987–2992.

Simpson RJ, Bonnett GD. 1993. Fructan exohydrolase from grasses. *New Phytologist* **123**: 453–469.

Simpson RJ, Walker RP, Pollock CJ. 1991. Fructan exhohydrolase activity of leaves of *Lolium temulentum* L. *New Phytologist* **119**, 499–507.

Smouter H, Simpson RJ. 1991. Fructan metabolism in leaves of *Lolium rigidum* Gaudin. II. Fructosyl transferase, invertase and fructan hydrolase activity. *New Phytologist* **119**: 517–526.

Sprenger N, Bortlik K, Brandt A, Boller T, Wiemken, A. 1995. Purification, cloning, and functional expression of sucrose: fructan 6-fructosyl transferase, a key enzyme of fructan synthesis in barley. *Proceedings of the National Academy of Sciences, USA* **92**: 11652–11656.

Walker RP, Winters AL, Pollock CJ. 1996. Purification and characterization of invertases from leaves of *Lolium temulentum* L. *New Phytologist* **135**, 259–266.

Wagner W, Keller F, Wiemken A. 1983. Fructan metabolism in cereals: induction in leaves and compartmentation in protoplasts and vacuoles. *Zeitschrift für Planzenphysiologie* **112**: 359–372.

Wagner W, Wiemken A. 1986. Properties and subcellular localisation of fructan hydrolase in the leaves of barley. *Journal of Plant Physiology* **123**: 492–439.

Wagner W, Wiemken A. 1989. Fructan metabolism in expanded primary leaves of barley (*Hordeum vulgare* L. cv. Gerbel): change upon ageing and spatial organisation along the leaf. *Journal of Plant Physiology* **134**: 237–242.

Wagner W, Wiemken A, Matile Ph. 1986. Regulation of fructan metabolism in leaves of barley (*Hordeum vulgare* L. cv. Gerbel). *Plant Physiology* **81**: 444–447.

Yamomoto S, Mino Y. 1985. Partial purification and properties of phleinase induced in stem base of orchard grass after defoliation. *Plant Physiology* **78**: 591–595.

New Phytol. (1997), **136**, 73–79

Fructan and fructan-metabolizing enzymes in the growth zone of barley leaves

BY ADRIAN ROTH, MARCEL LÜSCHER, NORBERT SPRENGER, THOMAS BOLLER AND ANDRES WIEMKEN*

Department of Botany, University of Basel, Hebelstrasse 1, CH-4056 Basel, Switzerland

(*Received 24 July 1996*)

SUMMARY

To study the role of fructan in leaf development of barley seedlings, fructan contents and the activities of enzymes involved in fructan metabolism were measured in segments of the third leaf at the stage of maximal growth. Fructans amounted to more than 10% of the d. wt in the growth zone at the base of the developing leaf and decreased sharply in the zone above. The specific activities of sucrose:sucrose 1-fructosyltransferase (1-SST) and sucrose:fructan 6-fructosyltransferase (6-SFT), two key enzymes in fructan synthesis, also peaked in the growth zone but decreased more slowly in the zone above. By contrast, the specific activities of invertase and fructan exohydrolase (FEH) were low in the growth zone at the leaf base and had their peak just above it. The specific activity of fructan:fructan 1-fructosyltransferase (1-FFT) was approx. constant over the whole leaf. These results indicate that transitory accumulation of fructans in the growth zone plays a key role in the development of barley leaves.

Key words: Fructan, fructan-metabolizing enzymes, growth zone, *Hordeum vulgare* (barley), partitioning.

INTRODUCTION

Fructans, polymers of fructofuranosyl units starting with sucrose, are the main storage carbohydrates in the vegetative parts of cool-season grasses such as wheat, barley, rye, oat and many forage grasses (Hendry, 1993). They play an important role in buffering temporary imbalances between photosynthetic carbon supply and carbon demand for growth and development. Thus, fructans are accumulated in leaves and underground organs of many grasses during autumn, when low temperatures restrict growth but still allow photosynthesis. In spring, these fructan reserves are rapidly mobilized, allowing early growth and differentiation of assimilatory organs (Hendry, 1993). Later in the year, a surplus of photosynthate is often produced in fully grown cereal plants immediately after anthesis, when demand for assimilates is low in the early stages of seed development. During this period, assimilates are temporarily stored in the form of fructan in the stems, and they are re-mobilized during grain filling when the need for assimilates frequently exceeds current supply by photosynthesis (Schnyder, 1993).

Fructan metabolism also appears to play a major role in carbohydrate partitioning in growth zones, particularly in rapidly expanding tissues. Thus, fructan is temporarily deposited in wheat kernels for a short period after anthesis, until the milky stage, when the most rapid volumetric growth occurs (Schnyder *et al.*, 1988). Similarly, fructan stores are built up in the growth zone of grass leaves (Spollen & Nelson, 1988; Schnyder & Nelson, 1989). Fructan metabolism is known to take place in the vacuole (Wagner, Keller & Wiemken, 1983), the cell compartment primarily involved in cell expansion, and might also play a role in osmotic adjustment.

Rapid accumulation of large quantities of fructan can be induced in fully grown barley leaves by excision followed by continuous illumination, or by feeding with sugars (Wagner *et al.*, 1983). This model system has been used extensively to study fructan physiology and enzymology (Wiemken, Sprenger & Boller, 1995). Based on these studies, our current view of fructan synthesis in barley leaves is as follows: sucrose:sucrose 1-fructosyltransferase (1-SST), the enzyme first induced (Simmen *et al.*, 1993), is responsible for triggering fructan synthesis by producing isokestose, the 'starter molecule' for fructan synthesis. Thereafter, sucrose:fructan 6-fructosyltransferase (6-SFT) is induced and becomes the main enzyme responsible for diverting carbon from sucrose to fructan (Duchateau *et al.*, 1995). A cDNA encoding 6-SFT has recently been cloned (Sprenger *et al.*, 1995), whose preferred fructosyl-acceptor substrate is isokestose, and therefore the first product formed by the concerted action of 1-

* To whom correspondence should be addressed.

SST and 6-SFT is the branched tetrasaccharide bifurcose. Bifurcose is subsequently extended by the repeated transfer of fructosyl residues from sucrose, catalysed by 6-SFT, yielding a β-2→6 linked fructosyl chain attached to isokestose. Further modification of the fructan molecules formed by these two key enzymes is generated by fructan:fructan 1-fructosyltransferase (1-FFT, Jeong & Housley, 1989) and fructan exohydrolase (FEH, Henson, 1989). In principle, this set of enzymes is sufficient for the synthesis of all the mixed-linkage type fructans which occur in barley and wheat (Carpita, Housley & Hendrix, 1991), which are collectively called graminans and contain mainly β-2→6 and a few β-2→1 linkages (Lewis, 1993).

In the present study, we show that 1-SST and 6-SFT are also highly active in the growth zones of elongating barley leaves, indicating that the enzymes might play an important role in assimilate partitioning during barley leaf development. This is of particular interest since it has been found recently that the fructosyltransferases in the growth zone of tall fescue leaves are quite different from barley 1-SST and 6-SFT (Lüscher & Nelson, 1995).

MATERIALS AND METHODS

Plant material

Barley seedlings (*Hordeum vulgare* L. cv. Express) were grown in plastic trays in a glasshouse with a light period of 14 h at ambient temperature (*c.* 22 °C day and 15 °C night). The substrate was a mixture of farmland soil, vermiculite, sand and fertilized peat (3:2:1:1 (v/v)). The plants were harvested at the three-leaf stage, when the tip of the third leaf was just emerging from the sheath of the two older leaves, about 14 d after sowing (Fig. 1).

Determination of growth

Leaf growth was determined by measuring leaf length daily. The elongation zone of the third leaf, the fastest-growing leaf, was localized as described by Schnyder & Nelson (1987). Briefly, 20 equidistant marker holes were pierced through the tiller with a fine steel wire, starting from the ligule at the leaf base. After 24 h, the tertiary leaf was excised and the distances between the marker holes were measured under a dissection microscope. Elongation of each segment was calculated as a percentage of the original distance between the holes.

Carbohydrate analysis

Plants were harvested and the first, second and third leaf were separated. The sheaths of the first and the second leaves, as well as the whole growing third leaf which had not yet formed a sheath (length *c.* 7 cm) were cut into 10 equal segments (Fig. 1). The blades of the first and second leaves were cut into segments of 1-cm length (Fig. 1). The segments were immediately frozen in liquid nitrogen and lyophilized, and their d. wt determined.

For extraction of soluble carbohydrates, 300 μl of 95% (v/v) ethanol were added to each of the

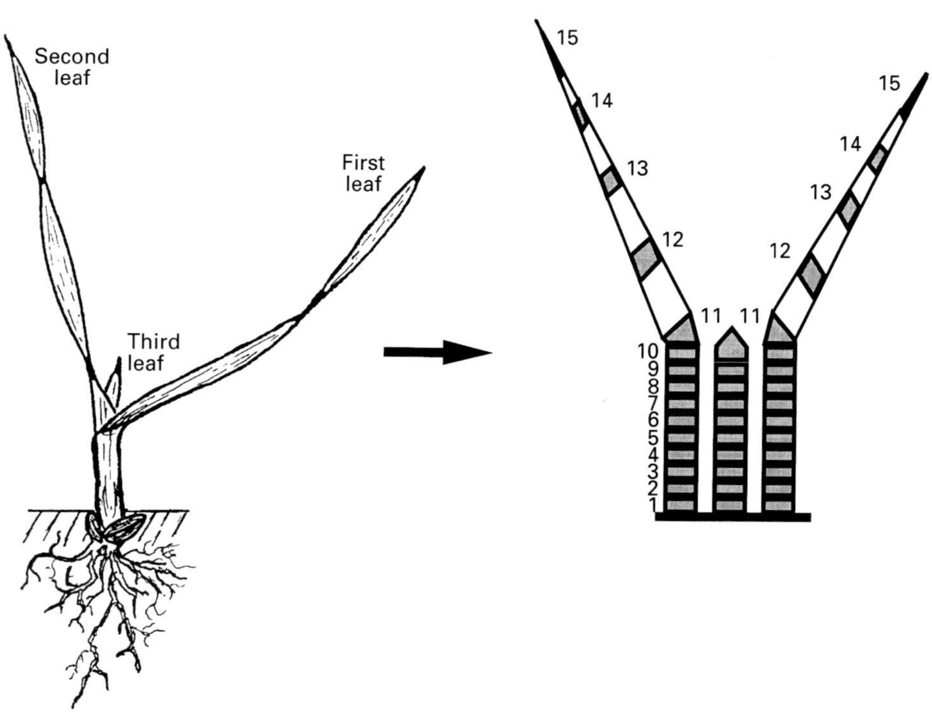

Figure 1. Schematic view of a barley seedling at the two-leaf stage, about 14 d after sowing. The sheaths of the second and the first leaves as well as the whole third leaf were cut into 10 equal segments (1–10). From the blades of the first and second leaves, representative 1-cm segments were cut as indicated (11–15).

lyophilized samples. After incubation at room temperature for 5 min, the plant material was ground in a mortar and pestle with a spatula of silica sand. The slurry was heated for 5 min at 80 °C and centrifuged at 11 500 ***g*** for 5 min. The supernatant was transferred into a fresh tube, and the pellet was re-extracted twice with 300 μl water under the same conditions but at 85 °C. The combined supernatants were centrifuged once more at 11 500 ***g*** for 7 min, freeze-dried and dissolved in 20 μl of water. Two μl were spotted onto a TLC silica gel plate (Merck, Darmstadt, FRG) and developed twice with a mixture of acetone/water (90/10 (v/v)).

Spots corresponding to glucose, fructose, sucrose and fructan were scraped off the plate, dissolved in 20 μl of water and quantified as described previously (Wagner *et al.*, 1983). Glucose was measured with the glucose oxidase kit (Boehringer, Mannheim, FRG). Starch was determined in pellets after extraction of soluble carbohydrates, following the procedure of Wagner *et al.* (1983).

Enzyme extraction

Sections cut from the third leaves of *c.* 15 plants were frozen in liquid nitrogen and ground to a powder. Proteins were extracted with two volumes of 50 mM citric acid (Na^+), pH 5·75, containing 2 mM EDTA and 2 mM dithiothreitol. The slurry was centrifuged at 11 500 ***g*** for 2 min. Ammonium sulphate was added to the supernatant to 70% saturation. Proteins were precipitated for 1 h at 4 °C and collected by centrifugation (2 min, 11 500 ***g***). The pellet was re-suspended in 50 mM MES (Na^+), pH 5·8, and desalted using Biogel P4 (Biorad) in the same buffer. The resulting protein solutions were used for enzyme assays.

Enzyme assays

Aliquots of 10 μl of desalted protein extracts were incubated at 25 °C for 30 min with 10 μl of 50 mM MES-Na^+, pH 5·8, containing the appropriate substrates. The following substrates (end concentration) were used: 100 mM sucrose for 1-SST and invertase; 50 mM sucrose and 50 mM isokestose for 6-SFT; 100 mM isokestose for 1-FFT; and a 5% (w/v) solution of a fructan mixture purified from barley growth zones (see below) for FEH. The assays were stopped by heating at 95 °C for 3 min. After centrifugation at 11 500 ***g*** for 3 min, aliquots of the supernatant were analysed for carbohydrates by anion-exchange HPLC with pulsed amperometric detection (Simmen *et al.*, 1993). Activity of 1-SST was defined as the production of isokestose, invertase activity as the production of fructose, 1-FFT activity as the production of nystose, 6-SFT activity as the production of bifurcose and FEH activity as the production of fructose, using the substrates indicated above. Enzyme activities are expressed as specific activities. Protein concentration was determined according to Bradford (1976) using bovine serum albumin as the standard.

Preparation of the substrate for fructan exohydrolase measurements

Sliced growth zones of young leaves from barley (800 mg f. wt) containing large amounts of fructans were ground to a fine powder in liquid nitrogen and extracted twice with 1 ml of boiling 80% ethanol, twice with 1 ml of 40% ethanol and twice with 1 ml of water. The combined extracts were evaporated under reduced pressure, re-dissolved in 1·5 ml of water and separated on a TSK-40W size exclusion column (100 cm × 3 cm, Merck, Darmstadt, FRG) with water as the mobile phase. Fractions (8 ml) were collected; those containing fructans with a degree of polymerization (DP) of 4–15 were pooled and lyophilized.

Isoelectric focusing (IEF) gel electrophoresis

The proteins of each segment of the third leaf were extracted as described above, except that the precipitated protein was desalted in 20 mM histidine-(Cl)-buffer, pH 5·75, and separated by electrophoresis on IEF gels.

Isoelectric focusing and staining of the gels for activities consuming sucrose and producing hexoses were performed as described by Lüscher & Nelson (1995). One lane of the IEF gel, which had been loaded with a protein extract from a whole third leaf, was cut away from the gel before activity staining and was sliced into 2-mm transverse segments. These segments were incubated in 100 mM MES-(Na^+)-buffer, pH 5·75 with 200 mM sucrose for 1 h at 25 °C. Aliquots of these reactions were analysed for sugars by anion exchange HPLC (Simmen *et al.*, 1993).

RESULTS

Growth rates of the leaves of barley seedlings and localization of the elongation zone of the growing third leaf

Seedlings were always harvested at the two-leaf stage, which was reached *c.* 14 d after sowing (Fig. 1). The daily growth rates of the leaves of 12–17-d-old seedlings is shown in Figure 2. The first leaf was fully expanded, whereas the second leaf had already differentiated into sheath and blade but was still growing slowly. The third leaf of 14-d-old seedlings was just emerging from the envelope formed by the sheaths of the two older leaves. It consisted only of a blade, and had a high growth rate that decreased rapidly during the following days (Fig. 2). The

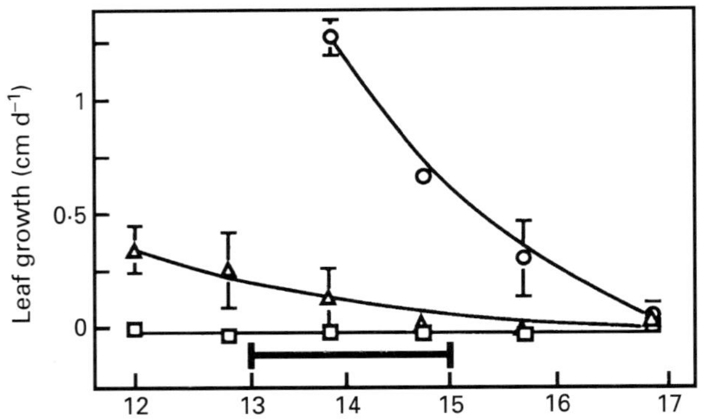

Figure 2. Leaf growth of barley seedlings at the two-leaf stage. The lengths of the first (□), second (△), and third leaves (○) were measured daily for 5 d. The crossbar indicates the harvesting time for all analyses reported, about 14 d after sowing.

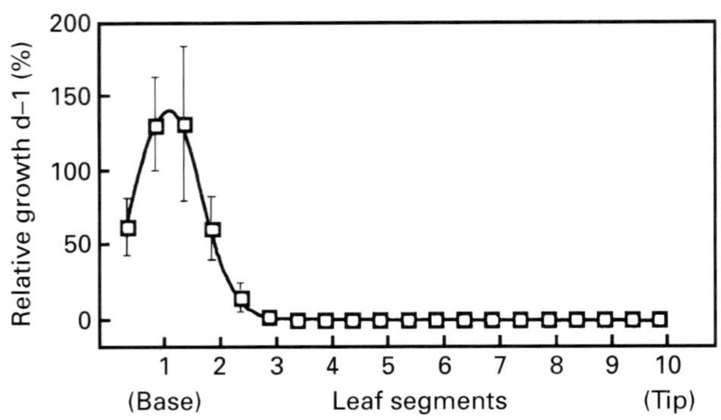

Figure 3. Growth of the third leaf of barley in a 24-h period. Twenty equidistant holes were pierced from the small basal thickening of the tiller to the tip. After 24 h, the plants were harvested and the third leaf excised. The distance between adjacent holes was determined and expressed as the percentage of the original distance.

elongation zone of the third leaf was confined to the leaf base region (Fig. 3), comprising the first 2 cm distal to the ligule (segments 1–3 in Fig. 1).

Distribution of non-structural carbohydrates in the leaves of barley seedlings

The plants were harvested, cut into segments and analysed for water-soluble carbohydrates and starch (Fig. 4). Fructan was found in all leaves, but by far the highest content (up to 120 mg g^{-1} d. wt) was in the basal part of the rapidly expanding third leaf (segments 1–3) where a comparatively small amount of starch (up to 20 mg g^{-1} d. wt) was also found. Sucrose remained constant over the whole third leaf (*c.* 4 mg g^{-1} d. wt) whilst hexoses were clearly reduced at the leaf base. The second leaf also contained fructan at the base but the level was only half that in the third leaf. In this second leaf, the levels of sucrose and hexoses were markedly increased at the leaf base, where fructan levels were highest. Starch was present at low levels throughout the sheath (segments 1–10) but accumulated in the leaf blade.

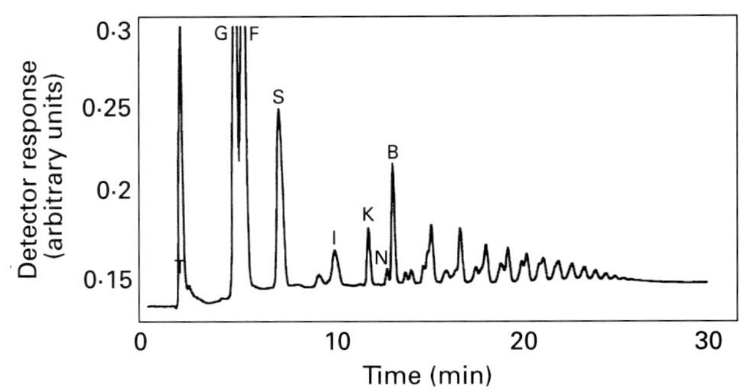

Figure 5. Anion-exchange HPLC analysis of water-soluble carbohydrates from the growth zone of the third leaf. Carbohydrates were identified by comparison with standards of known retention time. Trehalose (T) served as an internal standard. Identified sugars were glucose (G), fructose (F), sucrose (S), isokestose (I), kestose (K), nystose (N) and bifurcose (B).

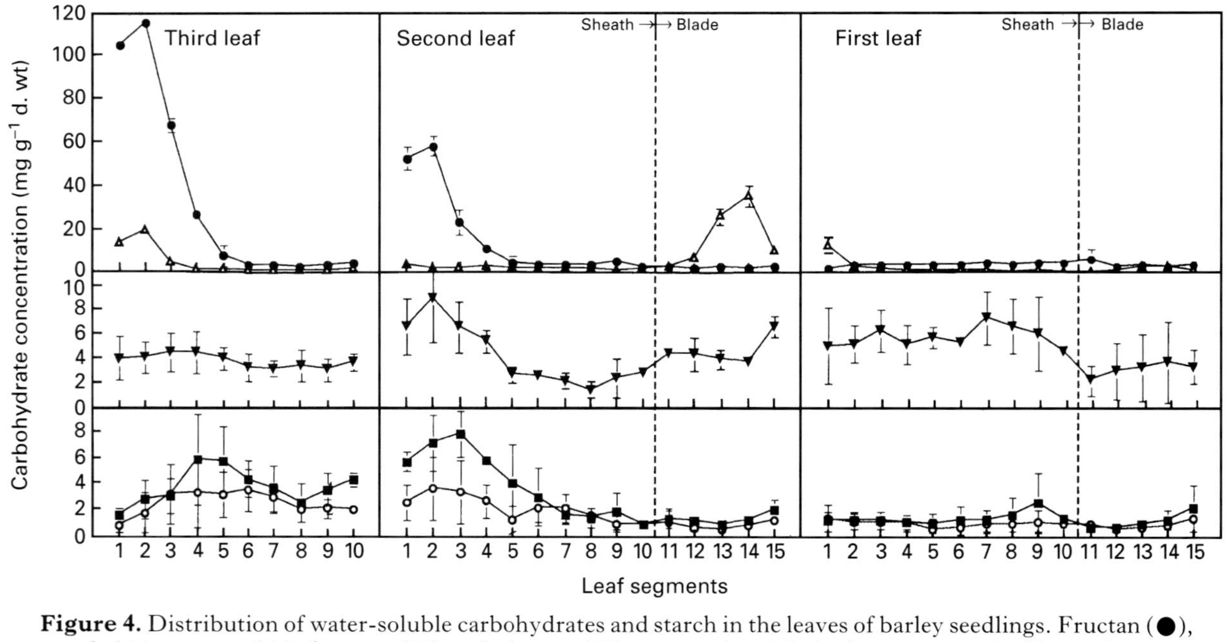

Figure 4. Distribution of water-soluble carbohydrates and starch in the leaves of barley seedlings. Fructan (●), starch (△), sucrose (▲), fructose (■) and glucose (○) were analysed in leaf segments as in Figure 1.

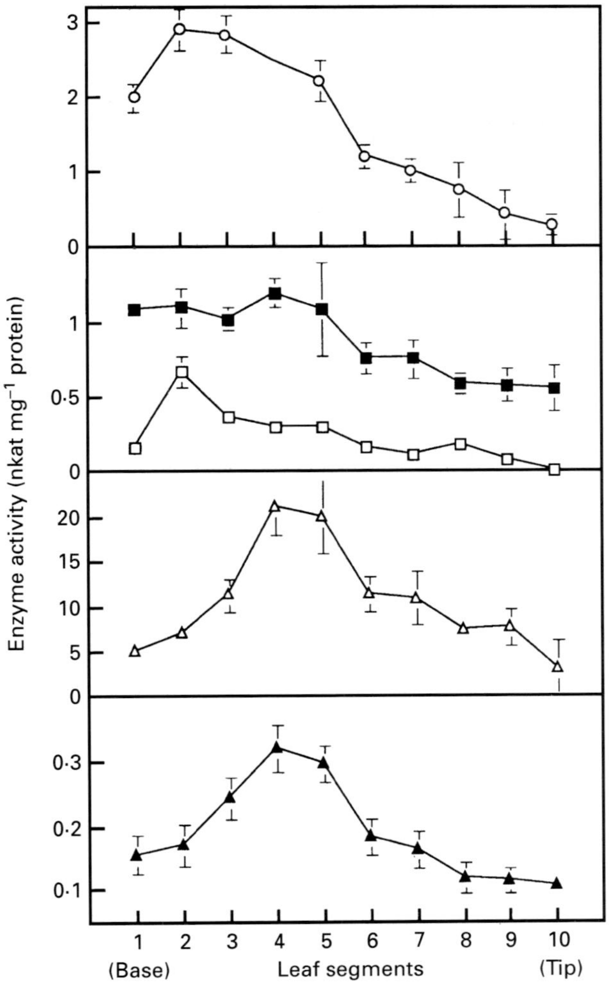

Figure 6. Distribution of enzymatic activities in the third leaf of barley seedlings. The third leaf was cut into 10 equal segments (see Fig. 1). Activities of 1-SST (○), 6-SFT (□), invertase (△), 1-FFT (■), and FEH (▲) were determined as described above.

The first leaf contained only minute amounts of fructan at the base whilst the content of sucrose was relatively high, at least in the sheath.

Extracts prepared from the growth zone of the third leaf (corresponding to segments 1–3 in Fig. 1) were analysed by anion exchange HPLC (Fig. 5). The pattern of fructan species was similar to that found in excised, illuminated barley leaves (Simmen *et al.* 1993). The extracts contained the trisaccharides kestose and isokestose, as well as a relatively large amount of the branched tetrasaccharide bifurcose and a small quantity of the linear tetrasaccharide nystose. A series of larger fructooligosaccharides was also present, yielding steadily smaller peaks with increasing DP.

Distribution of enzyme activities involved in fructan metabolism in the third leaf of barley seedlings

The third leaf of 14-d-old barley seedlings was cut into 10 segments and analysed for enzyme activities involved in fructan metabolism (Fig. 6). All enzyme activities were expressed on a protein basis; protein concentrations, calculated on a d. wt basis, remained virtually constant throughout the leaf (data not shown). The activity of 1-SST was greatest (3 nkat mg^{-1} protein) in the second segment where the accumulation of fructan (Fig. 4) and elongation rates (Fig. 3) were maximal. The activity decreased only gradually, extending into the fourth and fifth segments, where fructan contents were already very low. The highest activity of 6-SFT was also found in segment 2, but the activity of this enzyme was much lower than in excised and illuminated barley leaves (Sprenger *et al.*, 1995). By contrast, the maximal activity of invertase was located in segments 4 and 5; this might have contributed to isokestose production in these segments since barley invertases are known to exhibit some fructosyltransferase activity (1-SST activity) besides their main β-fructosidase activity (Obenland *et al.*, 1993). Fructan-hydrolysing activity (FEH) had a maximum in the same region of the leaf as invertase (segments 4 and 5). The activity of 1-FFT was constant (1·2 nkat mg^{-1} protein) in the first five segments, then declined slowly towards the tip (0·6 nkat mg^{-1} protein).

An IEF gel loaded with proteins extracted from whole third leaves was run, subsequently sliced, and incubated with 200 mM sucrose. The products formed by each gel slice were analysed by HPLC. They indicate the position of the different enzymes in the IEF gel (Fig. 7*a*). In this single-step separation procedure, 1-SST (which produced isokestose and virtually no fructose), the main invertases (which produced fructose and little isokestose), and the 6-SFT (which produced kestose and fructose as known from previous work with the purified enzyme (Duchateau *et al.*, 1995)), could be clearly differentiated (Fig. 7*a*).

The distribution of 1-SST, 6-SFT and invertases was determined in segments of the third leaf after separation by IEF gel electrophoresis, using the production of reducing sugars from sucrose (both glucose and fructose by invertases and mostly glucose by 1-SST and 6-SFT) in the gel as an assay (Fig. 7*b*). The main leaf invertases (pI *c.* pH 4·5) were present in the region above the growth zone and were most active in segments 5–8 (corresponding to the middle and upper part of the leaf); remarkably, in segment 1 there were practically no invertases. By contrast, the highest activities of 1-SST (pI *c.* pH 4·8) and of 6-SFT isozymes (pI *c.* pH 5·0), were found at the leaf base in segments 1–3 (Fig. 7*b*). Two additional minor activities were observed on the IEF gel (pI *c.* pH 4·9 and pH 5·05, respectively), primarily in segments from the leaf base. These could not be attributed to a specific enzyme activity. In qualitative terms, these results are consistent with those of the data of Fig. 6, which was conducted with a separate batch of leaves, although the peak of invertase activity was shifted slightly towards the more distal zone.

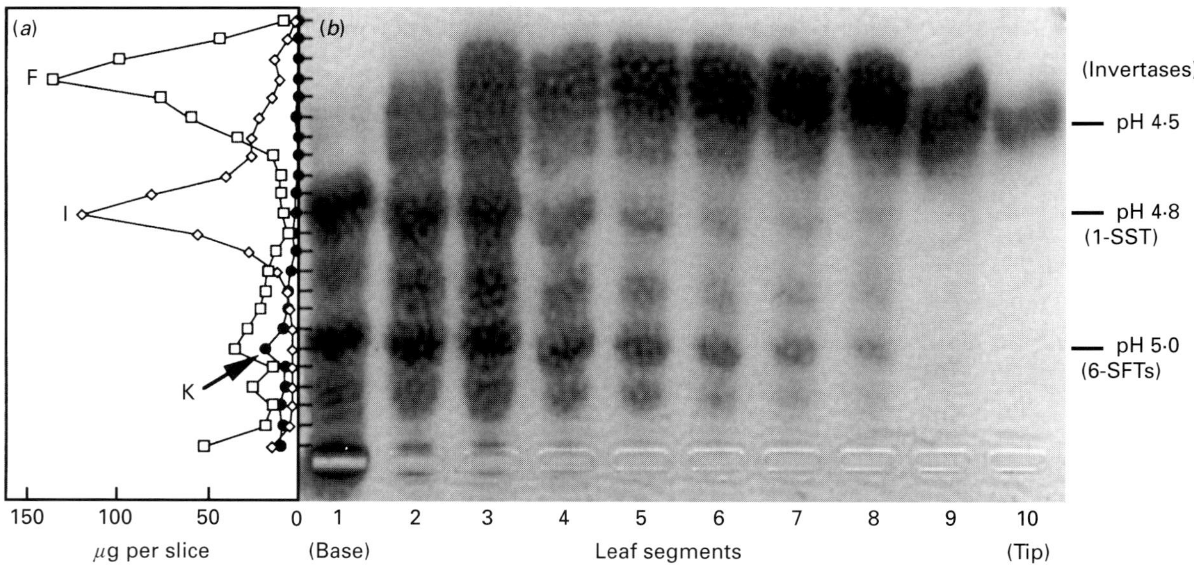

Figure 7. Distribution of enzymes in the third leaf of barley seedlings, analysed by IEF. (*a*) A protein extract from a whole tertiary leaf was subjected to IEF, and the gel was sliced into 2-mm transverse segments. The segments were incubated in 100 mM MES-(Na^+)-buffer, pH 5·75 and 200 mM sucrose for 1 h at 25 °C. Aliquots of the reaction mixtures were analysed by anion-exchange HPLC for fructose (F), isokestose (I) and kestose (K) production. (*b*) The third leaf was cut into 10 equal segments (see Fig. 1). Proteins of each segment were separated on an IEF gel; the whole gel was incubated with sucrose and subsequently stained for reducing sugars.

DISCUSSION

As is the case in several grasses (Spollen & Nelson, 1988; Schnyder & Nelson, 1989, and references therein), large fructan stores are deposited temporarily in the growth zone at the base of elongating barley leaves. The considerable sink strength of the growth zone tissue is also indicated by the accumulation of starch.

The main objective of this study was to compare the enzymology of fructan synthesis in the growth zone of barley leaves raised under normal conditions with that of the more artificial system, where accumulation of huge amounts of fructan is induced in excised mature leaves by continuous illumination (Wagner *et al.*, 1983; Simmen *et al.*, 1993). The results show that the enzymes 1-SST and 6-SFT, which function as key enzymes of fructan synthesis in mature barley leaves, are also highly active in the growth zone of elongating barley leaves. In the case of 6-SFT, the induction was also demonstrated at the mRNA-level, using as a probe the cDNA encoding 6-SFT (Sprenger *et al.*, 1995) (data not shown). The occurrence of similar activities of 1-FFT in the growth zone and in the zone above, where fructans are degraded is consistent with the suggestion that this enzyme plays a role in both fructan synthesis and degradation by modifying the size distribution of the fructan chains. FEH had its maximal activity above the growth zone.

By contrast, invertase activities were minimal in the growth zone and maximal above the growth zone towards the middle of the growing leaf blade. It is unlikely, therefore, that invertases are involved in assimilate import into the growth zone, contrary to the situation in a number of other systems where soluble vacuolar invertase is induced during early growth of newly initiated organs and might hydrolyse a large portion of the imported sucrose (for review see Koch & Nolte, 1995). In the growing barley leaf, invertase may fulfil this function in the distal maturation zone. In the growth zone, however, enzymes of fructan accumulation, such as 1-SST and 6-SFT, appear to replace invertase as the main sucrose-metabolizing enzymes. In this regard, it is of interest that fructan biosynthetic enzymes have been shown to be highly homologous to vacuolar invertases and may have evolved from them by small mutational changes (Sprenger *et al.*, 1995; Lüscher *et al.*, 1996).

We propose, therefore, that in the barley leaf growth zone a change of function has taken place from invertases to fructosyltransferases, with the consequence that part of the imported carbohydrate is deposited in the vacuole in the form of fructan rather than being consumed. We can only speculate about the advantages provided by fructan accumulation in the leaf growth zone. Undoubtedly, the rapidly expanding and differentiating cells in this very delicate tissue are highly dependent on an uninterrupted, continuous supply of carbon and a source of energy, and fructan reserves are ideally suited for this function: fructans can be readily accumulated, particularly at low temperatures (Wagner *et al.*, 1983; Wagner & Wiemken, 1989), where fructan biosynthetic enzymes show remarkably high activities (Wagner & Wiemken, 1986; Simmen *et al.*, 1993), and they can readily be remobilized (Wagner, Wiemken & Matile, 1986). Thus, fructans might buffer metabolism against potentially detrimental temporary shortages of carbon and energy supply, e.g. reduction of assimilate export from source leaves because of unfavourable climatic conditions. This might occur regularly in a temperate grass such as barley, for example in the cold nights of their main growth seasons of autumn and spring. Under such conditions, fructans might also serve as protectants against drought and freezing (Pontis, 1989; Hendry, 1993; Pilon-Smits *et al.*, 1995). These roles of fructan appear to be of particular importance to the cells of the growth zone, since in our experiments the cells mobilize the fructan reserves as soon as they move into the maturation zone, where they might be used for secondary cell wall synthesis, as in tall fescue (Allard & Nelson, 1991).

ACKNOWLEDGEMENTS

We wish to thank Dr M. A. Collinge (Friedrich Miescher-Institute, Basel) for critical reading of the manuscript, Mr Ch. Erdin for technical assistance and Ms E. Roth for typing the manuscript. This work was supported by the Swiss National Science Foundation, Grant 31-42535.94.

REFERENCES

Allard G, Nelson CJ. 1991. Photosynthate partitioning in basal zones of tall fescue leaf blades. *Plant Physiology* **95**: 663–668.

Bradford MM. 1976. A rapid and sensitive method for the quantitation of microgram quantities of protein utilizing the principle of protein-dye binding. *Analytical Biochemistry* **72**: 248–254.

Carpita NC, Housley TL, Hendrix JE. 1991. New features of plant fructan structure revealed by methylation analysis and carbon-13 NMR spectroscopy. *Carbohydrate Research* **217**: 127–136.

Duchateau N, Bortlik K, Simmen U, Wiemken A, Bancal P. 1995. Sucrose:fructan 6-fructosyltransferase (6-SFT), a key enzyme for diverting carbon from sucrose to fructan in barley leaves. *Plant Physiology* **107**: 1249–1255.

Hendry GAF. 1993. Evolutionary origins and natural functions of fructans – a climatological, biogeographic and mechanistic appraisal. *New Phytologist* **123**: 3–14.

Henson CA. 1989. Purification and properties of barley stem fructan exohydrolase. *Journal of Plant Physiology* **134**: 186–191.

Jeong BR, Housley TL. 1992. Purification and characterization of wheat β ($2 \rightarrow 1$) fructan:fructan fructosyl transferase activity. *Plant Physiology* **100**: 199–204.

Koch KE, Nolte KD. 1995. Sugar-modulated expression of genes for sucrose metabolism and their relationship to transport pathways. In: Madore MA, Lucas WJ, eds. *Carbon Partitioning and Source-Sink Interactions in Plants.* Rockville, USA: American Society of Plant Physiologists, 141–155.

Lewis DH. 1993. Nomenclature and diagrammatic representation of oligomeric fructans – a paper for discussion. *New Phytologist* **124**: 583–594.

Lüscher M, Erdin C, Sprenger N, Hochstrasser U, Boller T, Wiemken A. 1996. Inulin synthesis by a combination of purified fructosyl transferases from tubers of *Helianthus tuberosus. FEBS Letters* **385**: 39–42.

Lüscher M, Nelson CJ. 1995. Fructosyltransferase activities in the leaf growth zone of tall fescue. *Plant Physiology* **107**: 1419–1425.

Obenland DM, Simmen U, Boller T, Wiemken A. 1993. Purification and characterization of three soluble invertases from barley (*Hordeum vulgare* L.) leaves. *Plant Physiology* **101**: 1331–1339.

Pilon-Smits EAH, Ebskamp MJM, Paul MJ, Jeuken MJW, Weisbeek PJ, Smeekens SCM. 1995. Improved performance of transgenic fructan-accumulating tobacco under drought stress. *Plant Physiology* **107**: 125–130.

Pontis HG. 1989. Fructans and cold stress. *Journal of Plant Physiology* **134**: 148–150.

Schnyder H. 1993. The role of carbohydrate storage and redistribution in the source-sink relations of wheat and barley during grain filling – a review. *New Phytologist* **123**: 233–245.

Schnyder H, Ehses U, Bestajovsky J, Mehrhoff RR, Kühbauch W. 1988. Fructan in wheat kernels during growth and compartmentation in the endosperm and pericarp. *Journal of Plant Physiology* **132**: 333–338.

Schnyder H, Nelson CJ. 1987. Growth rates and carbohydrate fluxes within the elongation zone of tall fescue leaf blades. *Plant Physiology* **85**: 548–553.

Schnyder H, Nelson CJ. 1989. Growth rates and assimilate partitioning in the elongation zone of tall fescue leaf blades at high and low irradiance. *Plant Physiology* **90**: 1201–1206.

Simmen U, Obenland D, Boller T, Wiemken A. 1993. Fructan synthesis in excised barley leaves. Identification of two sucrose-sucrose fructosyltransferases induced by light and their separation from constitutive invertases. *Plant Physiology* **101**: 459–468.

Spollen WG, Nelson CJ. 1988. Characterization of fructan from mature leaf blades and elongation zones of developing leaf blades of wheat, tall fescue, and timothy. *Plant Physiology* **88**: 1349–1353.

Sprenger N, Bortlik KH, Brandt A, Boller T, Wiemken A. 1995. Purification, cloning and functional expression of sucrose:fructan 6-fructosyltransferase, a key enzyme of fructan synthesis in barley. *Proceedings of the National Academy of Sciences of the USA* **92**: 11652–11656.

Wagner W, Keller F, Wiemken A. 1983. Fructan metabolism in cereals: Induction in leaves and compartmentation in protoplasts and vacuoles. *Zeitschrift für Pflanzenphysiologie* **112**: 359–372.

Wagner W, Wiemken A. 1986. Properties and subcellular localization of fructan hydrolase in the leaves of barley (*Hordeum vulgare* L. cv. Gerbel). *Journal of Plant Physiology* **123**: 429–439.

Wagner W., Wiemken A. 1989. Fructan metabolism in expanded primary leaves of barley (*Hordeum vulgare* L. cv. Gerbel). Change upon ageing and spatial organization along the leaf blade. *Journal of Plant Physiology* **134**: 237–242.

Wagner W, Wiemken A, Matile P. 1986. Regulation of fructan metabolism in leaves of barley (*Hordeum vulgare* L. cv. Gerbel). *Plant Physiology* **81**: 444–447.

Wiemken A, Sprenger N, Boller T. 1995. Fructan – an extension of sucrose by sucrose. In: Pontis HG, Salerno GL, Echeverria EJ, eds. *Sucrose Metabolism, Biochemistry, Physiology and Molecular Biology*. Rockville, USA: American Society of Plant Physiologists, 179–188.

New Phytol. (1997), **136**, 81–88

Rise of fructan exohydrolase activity in stubble of *Lolium perenne* after defoliation is decreased by uniconazole, an inhibitor of the biosynthesis of gibberellins

BY ANNETTE MORVAN, GAËLLE CHALLE, MARIE-PASCALE PRUD'HOMME*, JOËLLE LE SAOS AND JEAN BOUCAUD

Laboratoire de Physiologie et Biochimie Végétales, U.A. INRA, Institut de Recherche en Biologie Appliquée, Université, 14032 Caen Cedex, France

(*Received 24 July 1996*)

SUMMARY

The objectives of this study were (i) to evaluate the relative involvement of fructans from leaf sheaths and from elongating leaf bases to regrowth after defoliation of *Lolium perenne* L. by following fructan exohydrolase (FEH), sucrose-sucrose fructosyl transferase (SST) and invertase (INV) activities and (ii) to examine whether gibberellins could be implicated in regulation of FEH activity. In stubble, 36 % of fructans are located in leaf bases and 64 % in leaf sheaths. During the first phase of regrowth, the depletion of carbohydrates was mainly due to the decline of fructans, which represented 76 % and 50 % of the carbohydrates mobilized from leaf sheaths and elongating leaf bases respectively. Despite a decrease in protein content, FEH activity increased 2·5-fold in leaf sheaths and six-fold in elongating leaf bases, so that the fructan-hydrolysing activity was four times higher in growing leaves than in leaf sheaths, 2 d after defoliation. INV activity also increased, whereas the pattern of SST activity was inversely related to the variations of both hydrolysing activities. SST activity, which is highest in growing leaves, declined approximately threefold in leaf sheaths and elongating leaf bases at the beginning of regrowth.

The increase in activity of FEH in stubble was strongly inhibited by an inhibitor of the biosynthesis of gibberellin, uniconazole. FEH activity was decreased to *c.* 67 %, 45 % and 33 % of the level in nontreated plants 24, 48 and 72 h following defoliation, respectively. The inhibition could be overcome by a subsequent treatment with gibberellic acids (GAs). For the first time, data are provided to support the view that GAs might play a role in the regulation of FEH activity, and the implication of this result is discussed.

Key words: Defoliation, fructan exohydrolase (FEH), gibberellins, *Lolium perenne*, sucrose-sucrose fructosyl transferase (SST).

INTRODUCTION

Perennial ryegrass (*Lolium perenne* L.) is a cool-season grass that accumulates fructans (Chatterton *et al.*, 1989; Gonzalez *et al.*, 1989). Under conditions where photosynthesis exceeds the rate of carbohydrate exports, the fructan concentration increased in leaf sheaths and mature leaf blades (Pollock & Jones, 1979; Housley & Volenec, 1988; Guerrand, Prud'homme & Boucaud, 1996). Interestingly, fructans are also synthesized in bases of expanding leaves, so that the growing leaves might contain a greater concentration of oligosaccharides than sheath or blade tissues (Housley & Volenec, 1988). There, fructans accumulate preferentially in the elongation zone of expanding leaves, which is surrounded by the sheaths of older leaves. Their concentrations decrease after cell elongation ceases and during active deposition of cell wall (Schnyder & Nelson, 1987; Schnyder, Nelson & Spollen, 1988). Defoliation by grazing or cutting reduces the amount of the leaf surface and thereby the supply of photosynthate. Fructans stored in the stem base (or stubble defined as the remaining shoot tissues including both sheaths and bases of expanding leaves) are mobilized and used for the regeneration of new

* To whom correspondence should be addressed.

shoots (Yamamoto & Mino, 1985; Danckwerts & Gordon, 1987; Gonzalez *et al.*, 1989). Examining changes in concentration of non-structural carbohydrates in individual leaf sheaths and bases of elongating leaves during regrowth following defoliation, Volenec (1986) showed that fructan concentrations decline in all sheath tissue as well as in elongating leaf bases. The enzyme responsible for hydrolysis of fructan in plants is fructan exohydrolase (FEH, EC 3.2.1.80), the activity of which increased in *Dactylis*, *Phleum*, *Festuca* and *Lolium* after defoliation (Yamamoto & Mino, 1985; Prud'homme *et al.*, 1992). FEH is located in vacuoles (Frehner, Keller & Wiemken, 1984; Wagner, Wiemken & Matile, 1986; Darwen & John, 1989), and has been partially purified from stem bases of *Dactylis glomerata* (Yamamoto & Mino, 1987), stems and leaf sheaths of *Hordeum vulgare* (Henson, 1989) and shoot tissues of *L. temulentum* and *L. rigidum* (Simpson, Walker & Pollock, 1991; Bonnett & Simpson, 1993; Bonnett & Simpson, 1995). However, relatively little is known about the regulation of FEH activity. The highest level of purity achieved so far has been for FEH from stem base of *D. glomerata* (Yamamoto & Mino, 1989). Activity, amount of FEH and incorporation of [^{14}C]leucine into the enzyme increased in concert after defoliation. Collectively, these results suggest that increase in FEH activity in defoliated stem bases involved *de novo* synthesis of the enzyme. However, FEH activities may also be regulated by metabolite concentrations. Indeed, in many species, sucrose is a strong inhibitor of FEH activity *in vitro* (Simpson & Bonnett, 1993). Moreover, Simpson *et al.* (1991) noted that substantial fructan hydrolysis occurred before FEH activity had increased, and concluded that inhibition of fructan hydrolysis by sucrose was removed when the vacuolar concentration of sucrose declined rapidly at the start of the mobilization period. In the stem base of *D. glomerata*, the increase of FEH activity after defoliation was repressed by exogenous application, not only of sucrose but also of glucose and fructose (Yamamoto & Mino, 1987). To our knowledge, there have been no attempts to explore a possible hormonal regulation of expression of the FEH gene. Gibberellins have been reported to regulate hydrolases involved in the enzymatic degradation and mobilization of endosperm storage macromolecules, namely starch and proteins, in germinating cereal seeds (Bewley & Black, 1985). They induce the production of α-amylase (Varner, 1964; Jacobson, Scandalios & Varner, 1970; Tanaka, Ito & Akazawa, 1970; Baulcombe & Buffard, 1983; Zwar & Hooley, 1986) and cysteine proteinases (Abe, Kondo & Arai, 1987; Koehler & Ho, 1988) and recent studies showed that the gene expression of α-amylase was regulated at the level of transcription by GA_3 (Skriver *et al.*, 1991). Other studies have suggested a relationship between GA3-promoted growth and GA3-augmented invertase activity (Kaufman, Ghosheh & Ikuma, 1968). Recently, GA3 was shown to enhance mRNA concentration of cell-wall invertase in elongating pea shoots, indicating that the expression of the gene could be regulated by GA3 at the transcriptional level (Wu *et al.*, 1993). Uniconazole [S-3307,(E)-1-(4-chlorophenyl)-4,4-dimethyl-2-(1,2,4-triazole-1-yl)-1-penten-3-ol], a triazole-type growth retardant, inhibits the biosynthesis of gibberellins by preventing the conversion of *ent*-kauren to *ent*-kaurenoic acid. Consequently the endogenous levels of gibberellins in treated plants are reduced (Izumi *et al.*, 1985). Therefore, uniconazole was used to study gibberellin effect on α-amylase activity. Mitsunaga & Yamaguchi (1993) reported that application of uniconazole on germinating rice seeds repressed the synthesis of α-amylase and that the repression was overcome by subsequent treatment with GA_3.

The objectives of this study were (i) to evaluate the relative involvement of fructans from leaf sheaths and from elongating leaf bases to regrowth after defoliation by following FEH and sucrose-sucrose fructosyl transferase (SST) activities in these tissues and (ii) to examine whether gibberellins could be implicated in regulation of FEH activity by using uniconazole.

MATERIALS AND METHODS

Plant material and applications

Seeds of *Lolium perenne* L. were germinated and grown for 8 wk in a controlled environment on a nutrient solution as described previously by Gonzalez *et al.* (1989) and Prud'homme *et al.* (1992). After 2 months, plants were defoliated at 4 cm above ground level. The residual material (stubble) was composed of leaf sheaths tightly folded one inside the other and of the bases of elongating leaves. For carbohydrate measurements and enzymatic studies, plants were separated into leaf sheaths and elongating leaf bases, immediately following defoliation and 2, 6 and 28 d later. For analysis of the putative role of gibberellins through the effect of uniconazole on FEH activities during regrowth, a 25 ml solution of uniconazole-P (25 mg) in 10% methanol was sprayed on the stubble of 30 plants immediately after defoliation so that each plant was treated with 833 μg of growth retardant. Control plants were sprayed with 25 ml water containing 10% methanol. Defoliated plants were placed on nutrient solution in the absence or presence of GA_3 (1, 10 or 100 μM), and stubble was sampled 1, 2 and 3 d later in order to measure FEH activities. Uniconazole-P was synthesized at the Sumitomo Chemical Co. in Japan and its purity was 94%, whereas GA_3 was purchased from Sigma Chemical Co. The experiments were carried out with three replicates.

Extraction and purification of water-soluble carbohydrates (WSC)

Samples of leaf sheaths (5 g) and expanding leaves (3 g) were cut into pieces and frozen immediately in liquid nitrogen in which they were stored. Mono- and disaccharides and low-molecular-weight fructans were extracted in 80 % ethanol whilst high-molecular-weight fructans were solubilized in boiling water as described previously (Guerrand *et al.*, 1996). Purified carbohydrates were separated and quantified by high-performance liquid chromatography (HPLC) on a Sugar-PAK column (300 × 6·5 mm, Millipore Waters, USA), eluted with 0·1 mM CaEDTA in water and using mannitol as an internal standard (Guerrand *et al.*, 1996).

Preparation of substrate for FEH

In *Lolium perenne*, FEH activities can be three times higher with native fructans than with 2-1 linked inulin from commercially available dahlia or chicory (Prud'homme *et al.*, 1992). Furthermore, in recent studies of *Lolium rigidum*, Bonnett & Simpson (1993, 1995) found that partially purified fructan-hydrolysing activities did not show the same affinity towards glycosidic linkages. *Lolium* species are assumed to have predominantly 2-6 linked fructans (Sims, Pollock & Horgan, 1992). Consequently, using inulin (2-1 linked fructan) might not be representative of whole fructan hydrolase activities as it would not take into account FEH activities that specifically hydrolyse 2-6 glucosidic linkages. Therefore, FEH activities were assayed with native fructans (i.e. extracted from the same species).

Fructans were extracted from 8-wk old plants of *Lolium perenne* grown under continuous illumination (500 μmol photons m^{-2} s^{-1}) and nutrient solution at 5 °C for 4 d. These conditions induce fructan accumulation (Smouter & Simpson, 1991*a*). Fructans were prepared by a method similar to that described by Bonnett & Simpson (1993); fructans were extracted from stubble first with boiling 80 % ethanol (4 mg g^{-1} f. wt) for 1 h and then with boiling water (5 mg g^{-1} f. wt) for 1 h. Both extracts were combined and the volume was reduced by rotary evaporation. The resulting concentrated extract was depigmented by an overnight contact with polyvinylpolypyrrolidone followed by centrifugation. Fructans contained in the supernatant were then separated from sucrose, glucose and fructose by passage through a gel filtration column (Sephadex G25, 31 cm × 2 cm). Fructans eluted immediately after the void volume were collected; they were free of sucrose as well as of monosaccharides. In order to improve the yield, the subsequent fractions, which contained a mixture of fructans, mono- and disaccharides, were pooled and re-loaded on the Sephadex G25 column. Fructans eluting near the void volume were collected and added to previous ones. By repeating the above procedure six times, it was possible to collect *c.* 90 % of fructans with a degree of polymerization (DP) > 8. Their mean degree of polymerization was estimated after acid hydrolysis to be 28 (Guerrand *et al.*, 1996). The remaining charged material was removed by passage through cation-exchange and anion-exchange resins (Amberlite IR120 and IRA 416, in the hydrogen and formate forms, respectively).

Protein measurement

Protein was measured according to the procedure of Bradford (1976) using a commercial Coomassie blue reagent (Biorad protein dye reagent concentrate). Bovine serum albumin grade V (Sigma) was used as a standard.

Extraction of enzymes

Plant tissue (5 g stubble or leaf sheath, 3 g expanding leaves) was ground at 4 °C at a ratio of 1 ml g^{-1} f. wt in 50 mM citrate-phosphate buffer (pH 5·7) containing 5 mM dithiotreitiol (DTT). After filtration through Miracloth®, the homogenate was centrifuged at 20000 ***g*** for 10 min and the supernatant was adjusted to a ratio of 2 ml g^{-1} f. wt. An aliquot was desalted on Sephadex G50, which also eliminated sucrose from the extract. This is of particular importance, since inhibition of FEH activities by sucrose has been reported for a range of species (Simpson & Bonnett, 1993).

Enzyme assays and product analysis

The assay mixture consisted of 100 μl enzyme extract and 100 μl substrate in 50 mM citrate-phosphate buffer. Triplicate samples were run together with duplicate enzyme blanks (no substrate). After incubation at 30 °C for 2 h (SST, invertase) or 4 h (FEH), 100 μl mannitol (1 g l^{-1}) was added to the assay mixture and the reaction was stopped by boiling for 5 min. The samples were stored at −20 °C until further processing. This method was used for determination of the apparent activities of SST, invertase (INV) and FEH in crude enzyme extracts. Activities of SST and INV were defined as the amount of 1-kestose and fructose, respectively, formed in this assay per unit of time per gram fresh tissue upon incubation in 50 mM sucrose (Guerrand *et al.*, 1996). Activities of FEH were measured as the release of fructose from a solution of high-molecular-weight fructan (8 mg ml^{-1}), extracted from *Lolium perenne*. Samples of 200 μl were passed through a column filled with cation-exchange resin (Dowex 50W, H^+ form), anion-exchange resin (Amberlite IRA 416, formate form) and C18-modified silica (Millipore Waters, USA) to remove charged com-

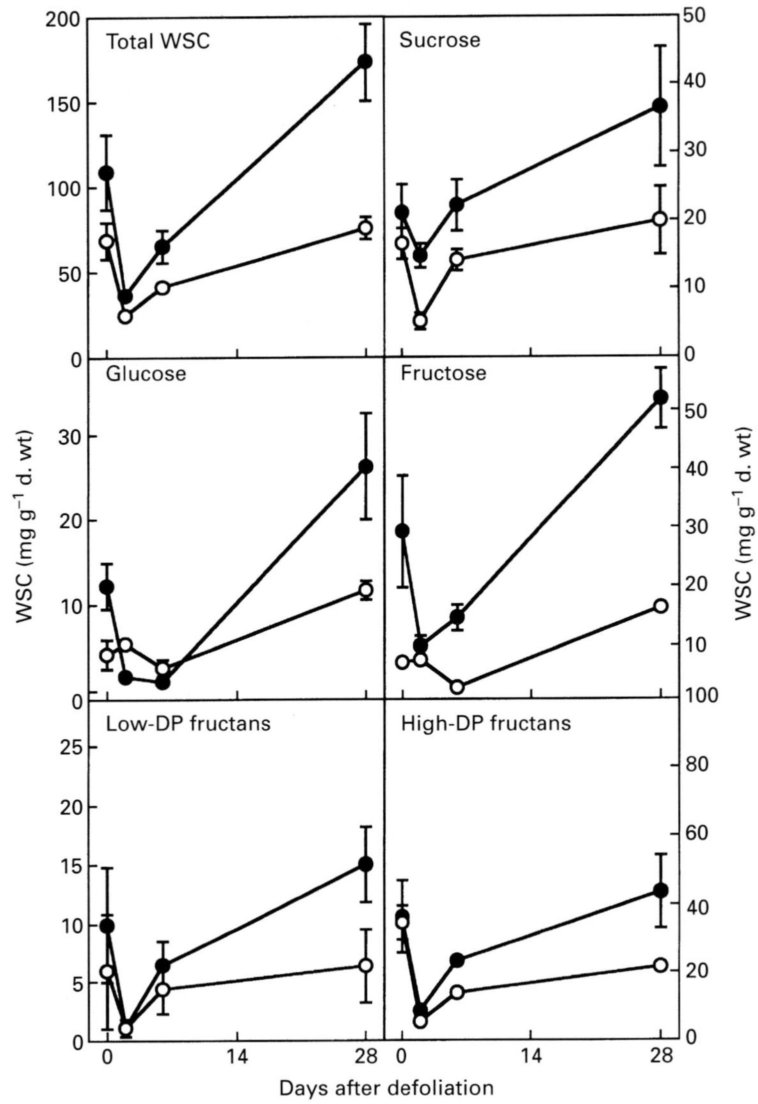

Figure 1. Effect of defoliation on the contents of total WSC, sucrose, glucose, fructose, low-DP fructans and high-DP fructans in sheaths (○) and in elongating leaf bases (●) of perennial ryegrass. Each value is the mean of three replicate determinations ± SE.

pounds, proteins, lipids and pigments (Smouter & Simpson, 1991*b*), as described previously (Guerrand *et al.*, 1996). Carbohydrates in the assay mixtures were separated and quantified by HPLC on the Sugar-PAK column under the conditions defined above for WSC analysis.

RESULTS

Changes in carbohydrate contents in sheaths and in elongating leaf bases

At the onset of the study, fructan and sucrose concentrations were similar in sheaths and in elongating leaf bases whilst concentrations of glucose and fructose were significantly higher in the growing tissue (Fig. 1). In sheaths, high-DP fructans accounted for half, and low-DP fructans for 8·5 %, of total WSC. In elongating leaf bases, high- and low-DP fructans represented 33 % and 9 % of WSC respectively. On a weight basis, stubble tissues of ryegrass consisted of two thirds mature leaf sheaths and one third elongating leaf bases. Therefore, in stubble, 36 % of fructans are located in leaf bases and 64 % in leaf sheaths. As already outlined (Volenec, 1986; Gonzalez *et al.*, 1989), the regrowth period may be divided into two physiological phases. During the first 2 d of regrowth, concentrations of WSC decreased to approximately a third of their initial concentrations in both sheaths and elongating leaf bases. The depletion in carbohydrates was essentially due to the decline in low-DP and high-DP fructan concentrations which represented 76 % and 50 % of the carbohydrates mobilized from sheaths and elongating leaf bases, respectively.

The second phase of the regrowth period is characterized by the restoration of all carbohydrate fractions. Although WSC fractions were restored during regrowth in stubble and reached higher concentrations by the end of the study than at the

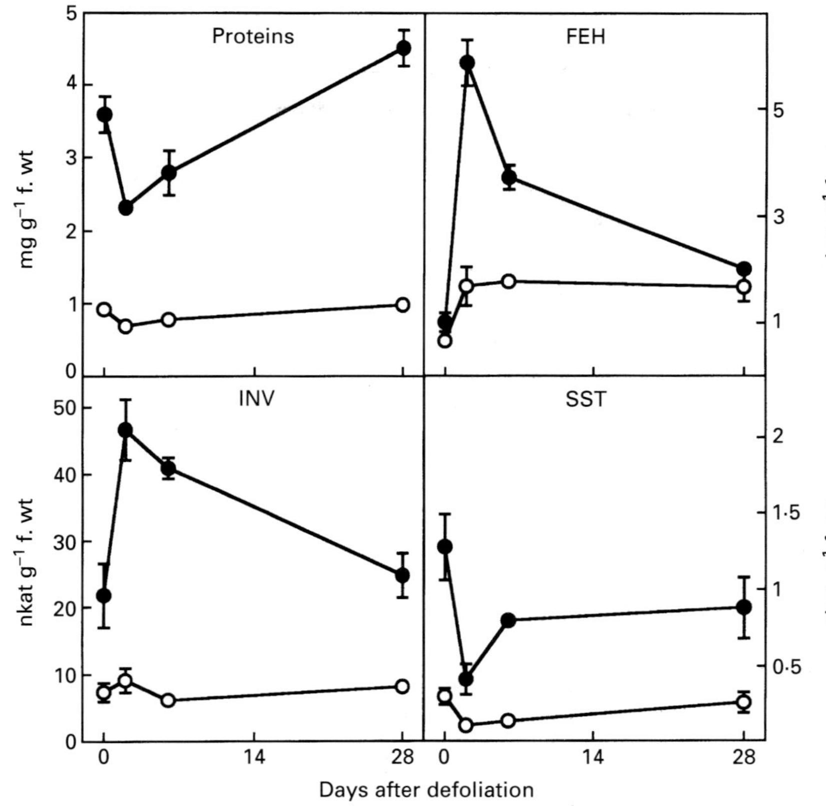

Figure 2. Effect of defoliation on the contents of soluble protein, and on FEH, INV and SST activities in sheaths (○) and in elongating leaf bases (●) of perennial ryegrass. Each value is the mean of three replicate determinations ± SE.

onset (Gonzalez *et al.*, 1989; Prud'homme *et al.*, 1992), this trend was not apparent when only sheaths were examined in the present experiment. Indeed, concentrations of low-DP fructans and sucrose in sheaths were similar on day 28 to initial contents, and high-DP fructans did not reach the concentration observed before defoliation. Elongating leaf bases contributed significantly to the WSC restoration in stubble since they accumulated carbohydrates at a rate of 5·2 mg g^{-1} d. wt per day compared with 1·9 mg g^{-1} d. wt per day in leaf sheaths from days 2–28. A similar pattern occurred for each carbohydrate fraction, and sucrose, glucose, fructose and fructan pools were restored faster in elongating leaf bases than in leaf sheaths.

Changes in enzyme activities of sheaths and elongating leaf bases

In the stubble of ryegrass, changes in soluble protein concentrations showed that early regrowth depended on both carbohydrate mobilization and protein degradation (Ourry, Bigot & Boucaud, 1989). In response to defoliation, soluble protein contents were reduced to 45% in the stubble during the mobilization phase and increased thereafter to the end of the regrowth period. Depletion and restoration of soluble proteins occurred both in sheaths and in elongating leaf bases as reported in Figure 2. Such mobilization was more important in growing tissue than in leaf sheaths since protein concentrations decreased to 35% and 25%, respectively. Despite the decline in soluble protein content, FEH and INV activities increased markedly in elongating leaf bases. During the 2 d following defoliation, FEH activity increased sixfold and INV activity twofold, while fructans and sucrose were being hydrolysed. In leaf sheaths, FEH also increased but only 2·5-fold, remaining at *c.* 1·7 nkat g^{-1} f. wt thereafter. Activity of INV in leaf sheaths did not vary significantly over the regrowth period. Throughout the carbohydrate restoration phase, FEH and INV activities declined gradually in elongating leaf bases so that FEH activity became slightly higher and INV activity similar to activity before defoliation. The pattern of SST activity in either leaf sheaths or elongating leaf bases after cutting was inversely related to the variations in FEH and INV activities during the regrowth period. Activity of SST, which is four times higher in growing leaves than in mature, declined approximately threefold in both tissues of the stubble and increased thereafter until the end of the study.

Effect of uniconazole on FEH activity in stubble

Effects of uniconazole with or without GA_3 on FEH activity in stubble of ryegrass were examined over

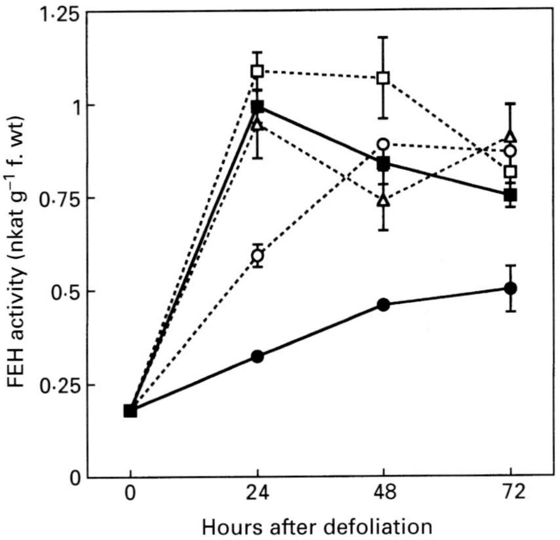

Figure 3. Effect of simultaneous application of uniconazole (883 μg per plant by spraying) and gibberellic acid (1–100 μM added to the nutrient solution) on FEH activity in stubble of perennial ryegrass during the 72 h after defoliation. Control (■), uniconazole (●), uniconazole+1 μM GA_3 (○), uniconazole+10 μM GA_3 (□), uniconazole+100 μM GA_3 (△).

the early regrowth period (Fig. 3). Activity of FEH was decreased by the application of uniconazole to *c.* 67%, 45% and 33% of the level in untreated plants 24, 48 and 72 h following defoliation, respectively. No remarkable changes in the extent of repression were observed at inhibitor concentrations ranging from 166 μg per plant to 333 μg per plant and growth was not affected by uniconazole treatment (data not shown). Interestingly, subsequent applications of 10 μM or 100 μM GA_3 restored FEH activity to the control level. Treatment with a lower concentration of GA_3 (1 μM) partially restored FEH activity, which reached 60% of control values 24 h after cutting. However, 1 and 2 d later the effect of uniconazole was totally overcome by application of 1 μM GA_3.

DISCUSSION

Roles of fructans in elongating leaf bases

In growing leaves, fructan synthesis occurred principally in the elongation zone which is confined to the basal part of the blade surrounded by the sheaths of fully developed leaves (Schnyder *et al.*, 1988; Spollen & Nelson, 1988). Three possible roles have been ascribed to fructans of elongating leaf bases. When either the tip of the elongating leaf or the youngest fully expanded leaf in tall fescue were labelled with $^{14}CO_2$, the photosynthate entered the base of the elongating leaf and *c.* 46% of the ^{14}C accumulated in the cell elongation zone (Allard & Nelson, 1991). Therefore, the elongation zone acts as a strong sink for photosynthate, regardless of the source tissue. Further, almost half of the translocated sucrose was used in fructan deposition (Schnyder & Nelson, 1987). Thus, sucrose which is not involved in biosynthetic processes of the cell appears to be removed from the cytoplasm and sequestered in the vacuole as fructans (Schnyder & Nelson, 1987). This fructan pool might then function to maintain a sucrose gradient between phloem and sink tissues, allowing phloem unloading to proceed. Because of their low mean degree of polymerization, fructans of the elongation zone might also contribute to the osmotic potential necessary for supporting cell expansion (Spollen & Nelson, 1988), and might also serve as a short-term carbohydrate storage since their concentrations decreased distal to the elongation zone (Schnyder *et al.*, 1988) where secondary cell-wall deposition occurred (MacAdam & Nelson, 1987). Moreover, label transfer from fructans to the water-insoluble fraction suggested that fructans were hydrolysed into fructose units which could be used for synthesis of the secondary cell wall (Allard & Nelson, 1991). In the present study, and as already reported once for regrowth of tall fescue (Volenec, 1986), we provide evidence that not only fructans from sheaths but also fructans from elongating leaf bases decreased in concentration in response to defoliation. Further, decline in fructan content was shown to be associated with increase of FEH activity. Surprisingly, increase in growing tissue accounted for *c.* 70% of FEH induction in intact stubble, with only 30% for FEH activity in mature leaf sheaths. Collectively, these results suggested a fourth possible function for fructan of elongating leaf bases. As occurs in the leaf sheaths, they could act as long-term storage carbohydrates, available for early regrowth after defoliation, when elaboration of new tissues is dependent on stored compounds. Therefore, two pools of fructans might be taken into consideration, one located in mature leaf sheaths and the other in elongating leaf bases.

Regulation of fructan exohydrolase activity

Treatment with GA at 62 g ha^{-1} stimulated cool-season regrowth of tropical forage grasses (Whitney, Green & Younge, 1973). Inhibitors of GA biosynthesis are commonly used to change the level of endogenous GAs in plant tissues. If the increase of FEH activity after defoliation results from a synthesis of GAs, then the stubble treated with an inhibitor of GA biosynthesis would be expected to have lower FEH activity. As indicated in Figure 3, this is indeed the case. However, uniconazole used in experiment might not be a specific inhibitor of GAs biosynthesis: according to Izumi & Oshio (1990) it also increases the amounts of ethylene and active forms of cytokinins. The possibility that concentrations of hormonal regulators other than GAs are modified cannot be then discounted, but since the repression of FEH activity caused by the inhibitor could be overcome by a subsequent application of gibberellic acid, we suggest that stubble FEH activity induced by defoliation is mediated, at least partly, through GAs biosynthesis. In vegetative forage grasses, however, and as mentioned above, stubble is composed of both mature and elongating tissues that behave differently throughout the regrowth period. Mature leaf sheaths did not grow, but senesced, and fructans were stored exclusively in newly formed sheaths (Volenec, 1986), accounting for the increase in fructan concentration which occurs in the leaf sheath compartment from the third day of regrowth. Leaves which were elongating before defoliation continued to grow after cutting. Volenec (1986) reported that fructans were first mobilized and further resynthesized in the same growing leaf. The FEH in senescing sheaths might then be regulated in a different manner than that in elongating leaf bases. We hypothesize that control of defoliation-induced FEH activity might involve gibberellins in elongating leaves and rather hormonal signals associated with senescing tissues (i.e. ethylene, abscissic acid and/or polyamines) in mature leaf sheaths. Arguments in favour of this hypothesis are (i) that application of GA_3 to excised leaf sections of *L. temulentum* had no influence on the pattern of enzyme and pigment change associated with the senescence process (Thomas, 1975), (ii) that the leaf elongation rate increased by *c.* 60% in response to defoliation (Volenec, 1986) and (iii) that GAs are known to regulate longitudinal shoot and leaf growth (Metraux, 1987).

Recently, five fructan exohydrolase activities from *L. rigidum* were distinguished (Bonnett & Simpson, 1995). All these activities were markedly inhibited by low concentrations of sucrose. Metabolite regulation of FEH was suggested. Decline of the quantity of sucrose in both mature leaf sheaths and elongating leaf bases following defoliation could act as a regulatory signal to remove inhibition of FEH activity or to derepress expression of FEH genes. If uniconazole only repressed cell expansion, sucrose utilization in regrowth would be low and FEH activity might not increase. Therefore, it could be argued that the repression of FEH rise could be a consequence of the effect of uniconazole on cell expansion rather than of a direct involvement of GAs in FEH activity. Uniconazole had no effect, however, on leaf elongation. But, in order to strengthen the hypothesis of direct implication of GAs in FEH activity, effects of uniconazole on water-soluble carbohydrate pools (sucrose, fructans) are under investigation.

Amylase is another carbohydrate hydrolytic enzyme and its regulation at the molecular level has been extensively studied during seedling development, when endosperm reserves are mobilized to sustain embryo growth. It is of particular interest that both metabolite signals (essentially sucrose) and hormonal signals (GAs and ABA) participate in controlling the total amount of expression of the amylase gene as well as the expression of different amylase genes encoding isozymes with distinct substrate specificities. Moreover, depletion of sucrose has been shown to stimulate synthesis of GAs (Thomas & Rodriguez, 1994). Before establishing such interactions between carbon metabolite and hormonal signals for control of FEH gene expression, FEH enzymes have to be purified to homogeneity in order to raise specific antibodies and make nucleotidic probes.

Distinction between leaf sheaths and elongating leaf bases provides a useful model system to study the reallocation of nutrients from storage compounds to achieve growth of new tissues after defoliation. Fructan exohydrolase(s) should be considered as target enzyme(s) for understanding how plants recover from defoliation and which signals are implicated. Effects of a GAs biosynthesis inhibitor on development of FEH activity in leaf sheaths and elongating leaf bases are under investigation, as is identification and quantification of GAs in these two tissues.

ACKNOWLEDGEMENTS

The authors wish to thank Dr K. Izumi at Sumimoto Chemical Co. for supplying uniconazole. They gratefully acknowledge Dr O. Junttila and Dr R. Simpson for having critically revised the manuscript.

REFERENCES

Abe K, Kondo H, Arai A. 1987. Purification and properties of cysteine proteinase from germinating rice seeds. *Agricultural and Biological Chemistry* **51**: 1509–1514.

Allard G, Nelson CJ. 1991. Photosynthate partitioning in basal zones of all fescue leaf blades. *Plant Physiology* **95**: 663–668.

Baulcombe DC, Buffard D. 1983. Gibberellic-acid-regulated expression of α-amylase and six other genes in wheat aleurone layers. *Planta* **157**: 493–501.

Bewley JD, Black M. 1985. Control of the mobilization of stored reserves. In: *Seeds, Physiology of Development and Germination.* New York: Plenum Press, 305–328.

Bonnett GD, Simpson RJ. 1993. Fructan-hydrolysing activities from *Lolium rigidum* Gaudin. *New Phytologist* **123**: 443–451.

Bonnett GD, Simpson RJ. 1995. Fructan exohydrolase activities from *Lolium rigidum* that hydrolyse β-2,1- and β-2,6- glycosidic linkages at different rates. *New Phytologist* **131**: 199–209.

Bradford MM. 1976. A rapid and sensitive method for the quantitation of microgram quantities of protein utilizing the principle of protein-dye binding. *Analytical Biochemistry* **72**: 248–254.

Chatterton NJ, Harrisson PA, Bennet JH, Asay KH. 1989. Carbohydrates partitioning in 185 accessions in Gramineae grown under warm and cool temperatures. *Journal of Plant Physiology* **134**: 169–179.

Danckwerts JE, Gordon AJ. 1987. Long term partitioning, storage and remobilization of ^{14}C assimilated by *Lolium perenne* (cv. Melle). *Annals of Botany* **59**: 55–56.

Darwen CWE, John P. 1989. Localization of the enzymes of fructan metabolism in vacuoles isolated by a mechanical method from tubers of Jerusalem artichoke (*Helianthus tuberosus* L.) *Plant Physiology* **89**: 658–663.

Frehner M, Keller F, Wiemken A. 1984. Localization of fructan metabolism in the vacuoles isolated from protoplasts of

Jerusalem artichoke tubers. *Journal of Plant Physiology* **116**: 197–208.

Gonzalez B, Boucaud J, Salette J, Langlois J, Duyme M. 1989. Changes in stubble carbohydrate content during regrowth of defoliated perennial ryegrass (*Lolium perenne* L.) on two nitrogen levels. *Grass and Forage Science* **44**: 411–415.

Guerrand D, Prud'homme MP, Boucaud J. 1996. Fructan metabolism in expanding leaves, mature leaf sheath and mature leaf blades of *Lolium perenne* L. Fructan synthesis, fructosyl transferase and invertase activities. *New Phytologist* **134**: 205–214.

Henson CA. 1989. Purification and properties of barley stem fructan exohydrolase. *Journal of Plant Physiology* **134**: 186–191.

Housley TL, Volenec JJ. 1988. Fructan content and synthesis in leaf tissue of *Festuca arundinacea*. *Plant Physiology* **86**: 1247–1251.

Izumi K, Kamiya Y, Sakurai A, Oshio H, Takahashi N. 1985. Studies of sites of action of a new plant growth regardant (E)-1-(4-chlorophenyl)-4,4-dimethyl-2-(1,2,4-triazol-1-yl)-1-pent-en-3-ol (S-3307) and comparative effects of its stereoisomers in a cell-free system from *Cucurbita maxima*. *Plant Cell Physiology* **26**: 821–827.

Izumi K, Oshio H. 1990. Effects of the growth retardant uniconazole-P on the endogenous levels of hormones in rice plants. In: Takahashi N, Phinney BO, MacMilan J, eds. *Gibberellins*, 331–338.

Jacobson JV, Scandalios JG, Varner JE. 1970. Multiple forms of amylase induced by gibberellic acid in isolated barley aleurone layers. *Plant Physiology* **45**: 367–371.

Kaufman PB, Ghosheh N, Ikuma H. 1968. Promotion of growth and invertase activity by gibberellic acid in developing *Avena* internodes. *Plant Physiology* **43**: 29–34.

Koehler SM, Ho THD. 1988. Purification and characterization of gibberellic acid-induced cysteine endoproteases in barley aleurone layers. *Plant Physiology* **87**: 95–103.

MacAdam JW, Nelson CJ. 1987. Specific leaf weight in zones of cell division, elongation and maturation in tall fescue leaf blades. *Annals of Botany* **59**: 369–376.

Metraux JP. 1987. Gibberellins and plant cell elongation. In: Davies PJ, ed. *Plant Hormones and their Role in Plant Growth and Development*. Dordrecht: Nijhoff, 296–317.

Mitsunaga S, Yamaguchi, J. 1993. Induction of α-amylase is repressed by uniconazole, an inhibitor of the biosynthesis of gibberellin, in a dwarf mutant of rice, waito-C. *Plant Cell Physiology* **34**: 243–249.

Ourry A, Bigot J, Boucaud J. 1989. Protein mobilization from stubble and roots, and proteolytic activities during post-clipping regrowth of perennial ryegrass. *Journal of Plant Physiology* **134**: 298–303.

Pollock CJ, Jones T. 1979. Seasonal patterns of fructan metabolism in forage grasses. *New Phytologist* **83**: 9–15.

Prud'homme MP, Gonzalez B, Billard JP, Boucaud J. 1992. Carbohydrate content, fructan and sucrose enzyme activities in roots, stubble and leaves of ryegrass (*Lolium perenne* L.) as affected by source/sink modification after cutting. *Journal of Plant Physiology* **140**: 282–291.

Schnyder H, Nelson CJ. 1987. Growth rates and carbohydrate fluxes within the elongation zone of tall fescue leaf blades. *Plant Physiology* **85**: 548–553.

Schnyder H, Nelson CJ, Spollen WG. 1988. Diurnal growth of tall fescue leaf blades. II. Dry matter partitioning and carbohydrate metabolism in the elongation zone and adjacent expanded tissue. *Plant Physiology* **86**: 1077–1083.

Simpson RJ, Bonnett GD. 1993. Fructan exohydrolase from grasses. *New Phytologist* **123**: 453–469.

Simpson RJ, Walker RP, Pollock CJ. 1991. Fructan exohydrolase activity in leaves of *Lolium temulentum* L. *New Phytologist* **119**: 499–507.

Sims IM, Pollock CJ, Horgan R. 1992. Structural analysis of oligomeric fructans from excised leaves of *Lolium temulentum*. *Phytochemistry* **31**: 2989–2992.

Skriver K, Lok Olsen F, Rogers JC, Mundy J. 1991. Cis-acting DNA elements responsive to gibberellin and its antagonist abscissic acid. *Proceedings of the National Academy of Sciences of the United States of America* **88**: 7266–7270.

Smouter H, Simpson RJ. 1991*a*. Fructan metabolism in leaves of *Lolium rigidum* Gaudin. I. Synthesis of fructan. *New Phytologist* **119**: 509–516.

Smouter H, Simpson RJ. 1991*b*. Fructan metabolism in leaves of *Lolium rigidum* Gaudin. II. Fructosyltransferase, invertase and fructan hydrolase activity. *New Phytologist* **119**: 517–526.

Spollen WG, Nelson CJ. 1988. Characterization of fructan from mature leaf blades and elongation zones of developing leaf blades of wheat, tall fescue, and timothy. *Plant Physiology* **88**: 1349–1353.

Tanaka Y, Ito T, Akazawa T. 1970. Enzymatic mechanism of starch breakdown in germinating rice seds. *Plant Physiology* **46**: 650–654.

Thomas H. 1975. Regulation of alanine aminotransferase in leaves of *Lolium temulentum* during senescence. *Zeitschrift für Pflanzenphysiologie* **74**: 208–218.

Thomas BR, Rodriguez RL. 1994. Metabolite signals regulate gene expression and source/sink relations in cereal seedlings. *Plant Physiology* **106**: 1235–1239.

Volenec JJ. 1986. Nonstructural carbohydrates in stem base components of tall fescue during regrowth. *Crop Science* **26**: 122–127.

Varner JE. 1964. Gibberellic acid controlled synthesis of α-amylase in barley endosperm. *Plant Physiology* **39**: 413–415.

Wagner W, Wiemken A, Matile P. 1986. Regulation of fructan metabolism in leaves of barley (*Hordeum vulgare* L. cv. Gerbel). *Plant Physiology* **81**: 444–447.

Whitney AS, Green RE, Younge OR. 1973. Effects of gibberellic acid and sublethal level of four herbicides on the cool-season regrowth of two tropical forage grasses. *Agronomy Journal* **65**: 473–476.

Wu LL, Mitchell JP, Cohn NS, Kaufman PB. 1993. Gibberellin (GA3) enhances cell wall invertase activity and mRNA levels in elongating dwarf pea (*Pisum sativum*) shoots. *International Journal of Plant Sciences* **154**: 280–289.

Yamamoto S, Mino Y. 1985. Partial purification and properties of phleinase induced in stem base of orchard grass after defoliation. *Plant Physiology* **78**: 591–595.

Yamamoto S, Mino Y. 1987. Effect of sugar level on phleinase induction in stem base of orchardgrass after defoliation. *Physiologia Plantarum* **69**: 456–460.

Yamamoto S, Mino Y. 1989. Mechanism of phleinase induction in the stem base of orchardgrass after defoliation. *Journal of Plant Physiology* **134**: 258–260.

Zwar JA, Hooley R. 1986. Hormonal regulation of α-amylase gene transcription in wild oat (*Avena fatua* L.) aleurone protoplasts. *Plant Physiology* **80**: 459–463.

New Phytol. (1997), **136**, 89–96

Purification and substrate specificity of an extracellular fructanhydrolase from *Lactobacillus paracasei* ssp. *paracasei* P 4134

By M. MÜLLER* and W. SEYFARTH

Centre for Agricultural Landscape and Land Use Research Müncheberg, Institute of Microbial Ecology and Soil Biology, Gutshof 7, D-14641 Paulinenaue, Germany

(*Received 24 July 1996*)

SUMMARY

A novel extracellular fructanhydrolase was isolated from the culture filtrate of *Lactobacillus paracasei* ssp. *paracasei* P 4134 grown on a mineral medium supplemented with fructan extracted from Timothy (*Phleum pratense* L.) as the only carbon source. The enzyme was purified by a combination of ammonium sulphate precipitation, affinity chromatography, preparative isoelectric focusing and anion-exchange chromatography. As a result of these procedures, the specific enzyme activity increased 93-fold, with a final yield of 28·4%. The substrate-specific activities against different fructan types were determined by incubating the enzyme fractions with fructan extracted from Timothy (predominantly β-2,6 fructosyl-fructose linkages), inulin from *Dahlia* tubers (mostly β-2,1 fructosyl-fructose linkages) and sucrose. The purified enzyme catalysed the hydrolysis of β-2,6-linked fructan more rapidly than the β-2,1 linkages of inulin. Additionally, the enzyme showed low ability to hydrolyse sucrose. Fructose was the main product of the degradation of Timothy fructan and inulin, indicating a high exohydrolytic activity of the enzyme. It is proposed that the fructan-degrading enzyme from *L. paracasei* ssp. *paracasei* P 4134 is a β-D-fructan-fructohydrolase (EC 3.2.1.80). The enzyme preparation showed a single protein band in sodium dodecyl sulphate-polyacrylamide gel electrophoresis with a mobility corresponding to molecular weight of *c.* 42 kDa. It was concluded that only one molecular weight of fructan-degrading enzyme exists in *L. paracasei* ssp. *paracasei* P 4134.

Key words: *Phleum pratense* (Timothy), *Lactobacillus paracasei*, fructan, fructanhydrolase, purification.

INTRODUCTION

Successful conservation of temperate grasses by ensilage depends on different factors. The occurrence of lactic acid bacteria and a high concentration of available soluble sugars are of primary importance for the production of high-quality silage (Woolford, 1984). Fructans might be the most abundant fraction (Suzuki, 1993), constituting up to 70% of the total water-soluble carbohydrate. Therefore, the direct fermentation of fructans by lactic acid bacteria (LAB) could make an important contribution to silage preservation. Recent studies in Germany have shown that < 3% of strains of epiphytic LAB isolated from pure stands of orchardgrass (*Dactylis glomerata* L.), Timothy (*Phleum pratense* L.) and perennial ryegrass (*Lolium perenne* L.) were able to ferment fructans (Müller & Lier, 1994). The fructan fermenters were mainly strains of the species *Lactobacillus paracasei* ssp. *paracasei*. There is a number of fructan-degrading enzymes in higher plants and micro-organisms distinguished by the different end-products formed during fructan hydrolysis. These include fructose, inulobiose or levanbiose, di-D-fructose dianhydrides, oligofructans and fructose-only oligomers (Iizuka & Yamamoto, 1979; Vandamme & Derycke, 1983; Ettalibi & Baratti, 1987). These numerous end-products are presumed to be a result of the activities of different enzymes (Avigad, 1982). Uchiyama (1993) classified microbial enzymes according to their affinity for the type of linkage of fructan and for the site of cleavage inside the fructose chain. The β-2,6 hydrolytic enzymes have been separated into three groups. The first group, the unspecific β-D-fructofuranosidases (β-D-fructan-fructohydrolase, EC 3.2.1.80 and β-D-fructofuranoside-fructohydrolase, EC 3.2.1.26) hydrolyse the terminal unsubstituted fructose residue from the fructan chain. Many of these enzymes hydrolyse levan, inulin, sucrose, raffinose and stachyose but there are differences in substrate specificity and affinity (Dacosta & Gibbons, 1968; Xiao, Tanida & Takao, 1989). A considerable number of microbial

* To whom correspondence should be addressed.

fructan-hydrolysing enzymes was classified in this category: enzymes isolated from yeasts (*Candida*, *Saccharomyces*, *Kluyveromyces*, *Debaromyces*), filamentous fungi (*Aspergillus*, *Penicillium*, *Chrysosporium*) and bacteria (*Streptococcus*, *Bacillus*, *Clostridium*, *Bacteroides*). For the most part, the β-2,6 hydrolytic properties of these enzymes have not been tested (Manzoni & Cavazzoni, 1991; Belamri *et al.*, 1994).

Uchiyama (1993) put into a second group a bacterial 2,6-β-fructan 6-levanbiohydrolase (EC 3.2.1.64) isolated from *Pseudomonas* and *Streptomyces* which hydrolysed β-2,6 linked fructan producing levanbiose (Avigad & Zelikson, 1963; Murakami *et al.*, 1990). The third group consists of a 2,6-β-D-fructan fructanohydrolase (EC 3.2.1.65) which split β-2,6 linked fructan chains at random forming short-chain oligomers. This enzyme type was isolated from *Arthrobacter* by Avigad & Bauer (1966). Murakami *et al.* (1992) purified an enzyme from *Bacillus* which hydrolysed levan and phlein to produce levanheptaose as a main product. In this group was a fructofuranosyltransferase which was able to depolymerize fructan to form di-D-fructose dianhydride in cyclical form (Yokota, Enomoto & Tomita, 1991).

Many of these studies were carried out on unpurified or partially purified enzyme preparations. Therefore, it is difficult to determine whether the enzymatic reaction of the degradation of fructans in micro-organisms is catalysed by a single enzyme. Statements about substrate specificity and mode of action of enzymes in unpurified preparations are often questionable.

Recently, we found that inoculation of herbage with a fructan-degrading *Lactobacillus* strain before ensiling could lead to an improvement of silage fermentation (Merry *et al.*, 1995). However, no data regarding either the purification or enzymatic properties of fructanhydrolases from *Lactobacillus* have yet been published. The characterization of a purified fructanhydrolase isolated from *L. paracasei* ssp. *paracasei* will increase the knowledge about bacterial fructan-degrading enzymes and will form the basis for a comparison between that enzyme and other microbial fructanhydrolases.

It was the aim of this study to develop an effective purification procedure and to describe some biochemical properties together with the substrate specificity of the purified enzyme(s) from *L. paracasei* ssp. *paracasei* P 4134.

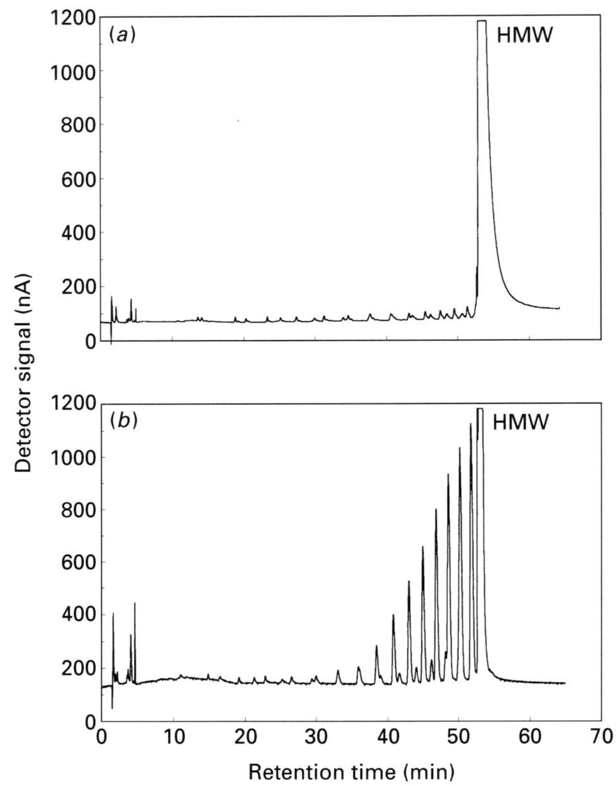

Figure 1. Anion exchange chromatograms of fructan extracted from *Phleum pratense* L. (*a*) and inulin from *Dahlia* tubers (*b*) used as substrates for enzyme assays and for substrate specificity tests. The patterns show the distribution of DP analysed on a CarboPac PA-100 column (Dionex, USA) using pulsed amperometric detection (HPAE/PAD). HMW, high-molecular-weight fructan.

MATERIALS AND METHODS

Strain and growth conditions

Lactobacillus paracasei ssp. *paracasei* P 4134 isolated from pure stand of Timothy (*Phleum pratense* L.) at Paulinenaue in 1991 (Müller & Lier, 1994) was cultivated aerobically for 4 d at 30 °C in MRS medium (Merck, Germany) without glucose. Fructan extracted from Timothy was added to a final concentration of 0·4%, (w/v) as the only carbon source.

Preparation of fructan substrates

Fructan was extracted from Timothy as described by Müller & Lier (1994). The residual content of free fructose, glucose, sucrose, raffinose and mannitol was < 1% and the degree of polymerization (DP) was approx. 90. Further C-sources utilized for substrate specificity tests were: sucrose (Serva, Germany) and inulin from *Dahlia* tubers (DP *c.* 30; Serva, Germany). Figure 1 illustrates the distribution of DP for Timothy fructan and *Dahlia* inulin separated by anion-exchange chromatography.

Enzyme purification procedure

The culture broth of *L. paracasei* ssp. *paracasei* P 4134 (1000 ml) was centrifuged at 11000 ***g*** at 4 °C for 15 min to remove the cells. The supernatant (920 ml) was used as a crude extracellular enzyme solution. The cell material (1·8 g fresh matter) was resuspended in 60 ml of Tris buffer (50 mM pH 7·0) and disrupted by ballistic disintegration with glass beads (diameter 0·25–0·5 mm) in a MM2 disintegrator (Retsch, Germany). The beads were removed by centrifugation (28 000 ***g***, 4 °C, 30 min) and the supernatant (20 ml) was used as a crude cell enzyme solution.

All subsequent operations were carried out at 4 °C. From both crude extracts, proteins precipitated between 30 and 70% saturation with solid ammonium sulphate were redissolved in 100 mM acetate buffer (pH 4·7). The cleared enzyme solution was loaded onto a Concanavalin A Sepharose 4B column (Pharmacia, Sweden) previously equilibrated with 100 mM acetate buffer containing 100 mM NaCl and 1 mM $CaCl_2$ and $MnCl_2$. Proteins not bound to Concanavalin A were eluted in the same buffer and glycosylated proteins that bound were eluted with 500 mM α-D-methylmannopyranoside dissolved in 100 mM acetate buffer. The active fractions were washed, concentrated (see below) and prepared for separation in a preparative isoelectric focusing system with a mini focusing chamber (Rotofor cell, BioRad, USA). The enzyme solution was brought to 10 mM acetate buffer (pH 4·7) containing Bio-Lyte ampholytes (BioRad, USA) in the pH range 3–10. The final concentration of Bio-Lytes used in the Rotofor system depended on the protein concentration in the sample. The procedure was carried out according to the specification of the manufacturers. After the run, 20 enzyme fractions were harvested with the Rotofor vacuum-assisted harvesting system and active fractions were pooled, washed and concentrated. After that, the fructan-degrading proteins were applied to a prepacked anion-exchange column (Econo-Pac Q cartridge 5 ml, BioRad, USA) equilibrated with 100 mM acetate buffer (pH 4·7). The enzyme was eluted with a linear gradient of NaCl (0–500 mM over 20 ml) in the same buffer at a flow rate of 1 ml min^{-1}. Fractions (2 ml) were collected and assayed for enzyme activity. The active fractions were combined, desalted and concentrated for use in subsequent experiments.

All washing and concentration steps were carried out by ultrafiltration in an Amicon stirred-cell concentrator under N_2, using a 30 kDa cut-off membrane (Amicon, USA).

Enzyme terminology and activities

The enzyme fractions were screened for their ability to hydrolyse high-polymer fructans extracted from Timothy, inulin and sucrose. The activity of the enzyme against Timothy fructan was defined as fructanase activity (FRU), that against inulin as inulinase activity (INU) and that against sucrose as invertase activity (INV). The reaction mixture consisted of 1 ml of 1% (w/v) grass fructan, inulin or sucrose solution in 100 mM acetate buffer (pH 4·7) and 200 μl enzyme solution in the same buffer. The mixture was incubated at 37 °C for 1 h and the reaction was stopped by heating at 80 °C for 10 min. Exohydrolytic FRU, INU and INV activities were assayed by measuring the amount of fructose released from grass fructan, inulin and sucrose, respectively. The samples were run together with substrate blanks (without enzymes), enzyme blanks (without substrates) and boiled enzyme blanks (inactivated enzymes plus substrate). Fructose was determined by the hexokinase-phosphoglucose isomerase-glucose 6-phosphate dehydrogenase method using a fructose/glucose kit (Boehringer, Mannheim, Germany). One unit of enzyme activity was defined as the amount of enzyme that produced 1 μmol fructose min^{-1} from one of the three different substrates under the specified conditions. The specific activities were defined as U mg^{-1} of protein. Protein was measured by the method of Lowry *et al.* (1951) using bovine serum albumin as a standard.

Sugar analyses

Fructose and glucose released from fructans were determined as described above. Total fructan content and fructans of lower DP were analysed by isocratic HPLC on an Aminex HPX-42A column (BioRad, USA) at 22 °C with water as an eluent at a flow rate of 0·5 ml min^{-1}. Carbohydrate mass was determined by differential refractometry.

The distribution of DP was analysed by anion-exchange chromatography (HPAE) on a CarboPac PA-100 column (Dionex, USA) at 22 °C and a flow rate of 1 ml min^{-1} using a mixed eluent of 0·1 mol l^{-1} NaOH (A) and 0·5 mol l^{-1} Na-acetate + 0·1 mol l^{-1} NaOH (B) using the gradient 100% B for 5 min, 98% A + 2% B for 5 min, 50% A + 50% B in 49 min, 100% B in 1 min. The separated carbohydrates were detected by pulsed amperometric detection (PAD).

Electrophoresis

The Mini-cell vertical slab electrophoresis system made by Novex (USA) and Novex pre-cast gels were used for denaturating polyacrylamide gel electrophoresis.

The samples were prepared by mixing one part of denaturating Tris-Glycine sample buffer (Novex, USA) and one part of sample. The samples were heated at 85 °C for 2 min. The proteins were separated in a Tris-Glycine gel based on the Laemmli system with a 4% stacking gel and a running gel with a gradient from 4 to 20% acrylamide concentration. Protein bands were detected by staining with 0·1% (w/v) Coomassie brilliant blue R-250 or by silver staining according to Kleinert (1990).

Molecular weight determination

The molecular weight of the purified fructanhydrolase was estimated by sodium dodecyl sulphate-polyacrylamide gel electrophoresis (SDS–PAGE) according to the procedure described above and by using standard proteins as markers (Mark 12®, Novex, USA).

Effects of pH on enzyme activity

The enzyme reaction was done at various pH values in the range from 3·8 to 5·5 in 100 mM acetate buffer. The FRU activity was measured under the standard assay conditions.

RESULTS

Enzyme purification

The extracellular fructan-degrading enzyme from *L. paracasei* ssp. *paracasei* P 4134 was purified 93-fold compared with that of the culture supernatant, with a final yield of 28·4% (Table 1). The greatest increase in purification was achieved by preparative isoelectric focusing. Proteins were separated in 20 focused zones in free solution and harvested into 20 individual test tubes. The enzyme activities against fructan from *P. pratense*, inulin and sucrose were detected in fractions 4–9 (Fig. 2). The isoelectric point of fraction 8 (containing the highest enzyme activity) was pH 4·63. The specific degradation rates of the three substrates were different. Degradation of inulin and sucrose was slower than that of fructan from *P. pratense*. Separation of pooled fractions 4–9 by anion exchange chromatography yielded a single peak with high FRU activity and low INU and INV activity (Fig. 3). The ratio of inulinase activity to fructanase activity (INU/FRU) was *c.* 0·15 and the ratio of invertase to fructanase activity (INV/FRU) *c.* 0·08. No enzyme activity bound to Concanavalin A, suggesting that the enzyme was not a glycoprotein.

Localization of enzyme

The majority of the fructan hydrolase activity against fructan and inulin was detected in the supernatant of the growth medium (crude extracellular enzyme solution). Only a minor part (*c.* 10%) of the enzyme activity was detected in the crude cell enzyme solution (intracellular or cell-wall-bound enzymes).

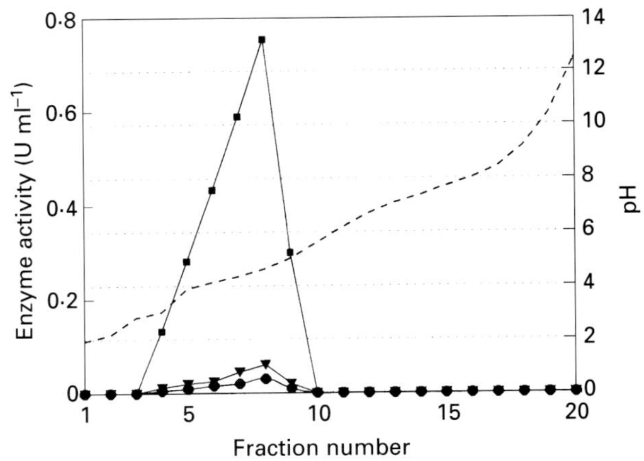

Figure 2. Enzyme activities of the extracellular fructanhydrolase isolated from *Lactobacillus paracasei* ssp. *paracasei* P 4134 against sucrose (●), inulin from *Dahlia* tubers (▼) and fructan extracted from *Phleum pratense* L. (■). Proteins were separated in 20 focused zones along a pH gradient (---) by preparative isoelectric focusing and harvested into 20 individual test tubes.

Estimation of the molecular weight

Examination of the purified enzyme by SDS–PAGE showed a single protein band with a mobility corresponding to molecular weight of 42 kDa (Fig. 4). This suggests that only a single size of fructan-degrading enzyme exists in *L. paracasei* ssp. *paracasei* P 4134.

Effects of pH on enzyme activity

The enzyme activity was assayed over a pH range similar to pH values observed during ensiling, and was most active from pH 4·1 to 4·8.

Substrate specificity of the purified fructanhydrolase

Substrate-specific activities against different fructan types were determined by incubating the purified fructanhydrolase with fructan extracted from *P. pratense*, inulin from *Dahlia* tubers and sucrose. After various incubation times, the reactions were

Table 1. *Purification of extracellular fructanhydrolase isolated from* Lactobacillus paracasei *ssp.* paracasei *P 4134*

Purification step	Total protein content (mg)	Total activity (U)	Specific activity (U mg^{-1})	Yield (%)	Purification (-fold)
Supernatant	4600	8740	1·9	100	1
30–70% $(NH_4)_2SO_4$	2940	7390	2·5	84·6	1·3
Preparative isoelectric focusing	59	6280	106·4	71·9	56
Econo-Pac Q chromatography	14	2480	177·1	28·4	93

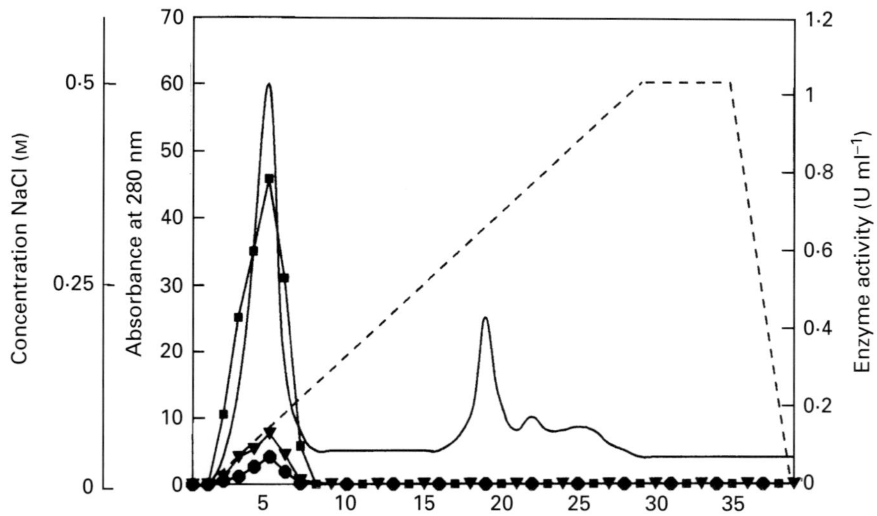

Figure 3. Econo-Pac Q column chromatographic profiles of extracellular fructanhydrolase isolated from *Lactobacillus paracasei* ssp. *paracasei* P 4134 indicated as absorbance at 280 nm (——) and as concentration of NaCl (----). The enzyme activities of the separated fractions against fructan extracted from *Phleum pratense* L. (■), inulin from *Dahlia* tubers (▼) and against sucrose (●) were analysed by determination of revealed fructose.

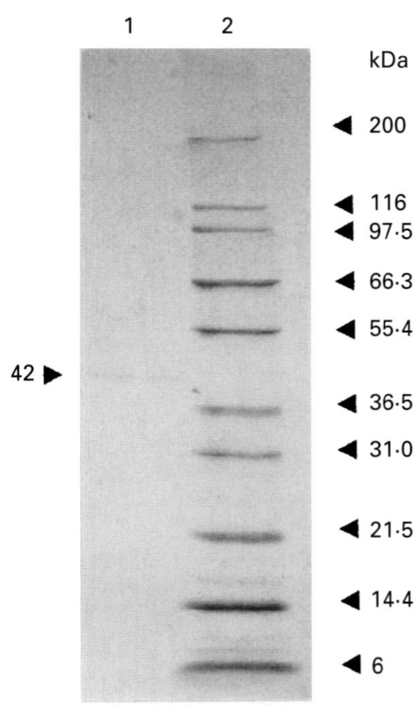

Figure 4. Molecular weight estimation by sodium dodecyl sulphate-polyacrylamide gel electrophoresis of the purified extracellular fructanhydrolase (lane 1) isolated from *Lactobacillus paracasei* ssp. *paracasei* P 4134 after comparison with the molecular weight standard (lane 2). Standard proteins (from top): myosin (200000); β-galactosidase (116000); phosphorylase B (97500); bovine serum albumin (66300); glutamic dehydrogenase (55400); lactate dehydrogenase (36500); carbonic anhydrase (31000); trypsin inhibitor (21500); lysozyme (14400); aprotinin (6000).

stopped and the inactivated fructan solutions were analysed for the distribution of DP of the residual fructans and for fructose accumulation, using anion-exchange chromatography. Figure 5 shows the HPAE/PAD profiles of the fructan from *P. pratense* L. and *Dahlia* inulin after degradation for different times. Fructose was the main product in all phases of the degradation of both fructan and inulin: the fructanhydrolase possessed a high exohydrolytic activity against both substrates. Using chromatographic and enzymatic methods, glucose and sucrose could not be detected during any stage of the reaction against fructan from *P. pratense*. Small amounts of sucrose were produced in the late stages of degradation of inulin. After 60 min, trisaccharides appeared: 1-kestose following breakdown of inulin and 1- and 6-kestose following degradation of fructan from *P. pratense*.

A group of peaks was detected with retention times of 35–50 min, during the degradation of fructan from *P. pratense*. This group of peaks, apparently oligosaccharides, appeared after 20 min reaction time, peaked after 60 min and disappeared at the late stage of the reaction.

The enzyme preparation catalysed the hydrolysis of β-2,6 fructan more rapidly than the β-2,1 linkages of inulin, as indicated in Figure 5. The enzyme showed a low ability to hydrolyse sucrose.

DISCUSSION

The separation of a single protein and the estimation of its functional and structural properties are the key aims in any enzyme purification. In the present study, a fructan-degrading enzyme from *Lactobacillus* was highly purified. It was characterized as an extracellular fructanhydrolase. Intracellular or cell-wall-bound enzyme could only be detected at a very low activity. These findings agreed with reports on inulinases from *Aspergillus niger* (Nakamura *et al.*, 1994), from *Kluyveromyces fragilis* (Gupta *et al.*, 1994) and from *Penicillium*

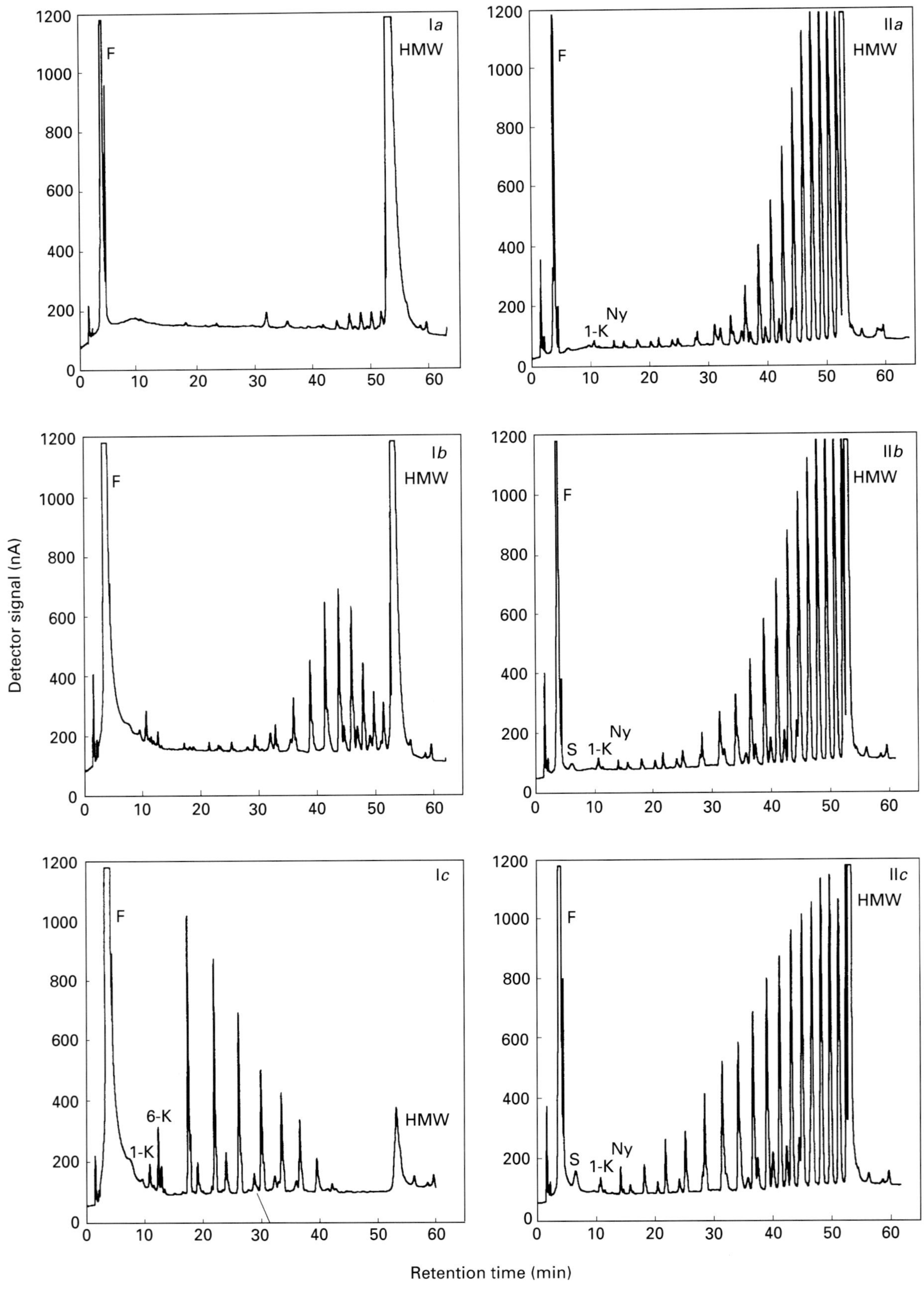

Figure 5. Time-course of degradation of fructan extracted from *Phleum pratense* (I) and inulin from *Dahlia* tubers (II) after incubation with purified extracellular fructanhydrolase isolated from *Lactobacillus paracasei* ssp. *paracasei* P 4134 presented as anion exchange chromatographic profiles after 10 min (*a*), 60 min (*b*) and 120 min (*c*). The patterns show the distribution of DP and the identification of some products analysed on a CarboPac PA-100 column (Dionex, USA) using pulsed amperometric detection (HPAE/PAD). F, fructose; 1-K, 1-kestose; 6-K, 6-kestose; N, nystose; HMW, high-molecular-weight fructan.

trzebinskii (Onodera & Shiomi, 1992), on the β-fructosidase from *Bacillus stearothermophilus* (Belamri *et al.*, 1994) and on two levanases from *Streptomyces* and *Bacillus* (Murakami *et al.*, 1990; Murakami *et al.*, 1992). By contrast, Muramatsu *et al.* (1992) found an intracellular β-fructofuranosidase in *Bifidobacterium* spec. A cell-associated fructan hydrolase of *Streptococcus mutans* was reported by Burne & Penders (1994). Manzoni & Cavazzoni (1992) isolated an extracellular as well as a cell-associated inulinase from *Kluyveromyces cicerisporus*. Although microbial fructanhydrolases could be detected in all cellular locations, most of them are extracellular enzymes.

The present investigation suggests that the fructanhydrolase from *L. paracasei* ssp. *paracasei* P 4134 is not a glycoprotein. In this regard it differs from fructan-degrading enzymes in higher plants, yeasts and fungi. Concanavalin-A-binding capacity is described for fructan-hydrolysing enzyme activities extracted from *Hordeum* sp., *Lolium* sp. and *Cichorium intybus*. The vacuolar enzymes of higher plants are typically glycosylated, whereas extracellular enzymes might or might not be glycosylated (Henson, 1989; Simpson & Bonnett, 1993). The corresponding yeast and fungal extracellular enzymes have been found to contain 7–66 % carbohydrate (Ettalibi & Baratti, 1987; Uhm *et al.*, 1987; Nakamura *et al.*, 1994).

The classification of a purified enzyme requires knowledge about its substrate specificity. Activities against two different fructan types were measured. Meier & Reid (1982) and Chatterton *et al.* (1990) analysed the fructan from *P. pratense* and found that this type was composed mostly of β-2,6 fructan with small proportions of 6-linked β-D-fructosyl units and 1,6-linked branch points. By comparison, inulin from *Dahlia* is considered to have mostly β-2,1 fructosyl linkages (Carpita, Kanabus & Housley, 1989; Bonnett & Simpson, 1993). The purified fructanhydrolase from *L. paracasei* ssp. *paracasei* P 4134 catalysed the hydrolysis of β-2,6 linked fructan more rapidly than the β-2,1 linkages of inulin. Additionally, the enzyme showed a little activity against sucrose. The INU/FRU and INV/FRU ratios were *c.* 0·1; in the same range as found previously for other microbial fructanases (Vandamme & Derycke, 1983; Blatch & Woods, 1993). The focusing and chromatographic purification steps did not change these values, indicating that all the activities were associated with the fructanase.

Hydrolysis of fructans in bacteria, yeasts and fungi are catalysed by both endo- and exohydrolases (Ettalibi & Baratti, 1987; Uchiyama, 1993). In this study, fructose was the main product in all phases of the degradation of both Timothy fructan and inulin; we propose that the fructanhydrolase from *L. paracasei* ssp. *paracasei* P 4134 possessed an exohydrolytic activity which was very high towards β-2,6 linkages and lower against β-2,1 linked inulin. A transient accumulation of a group of peaks was detected during the breakdown of the β-2,6 linked Timothy fructan. This observation leads to different suggestions. (1) An *endo*-like activity with a high specificity against β-2,6 linkages. In a recent paper Müller & Steller (1995) described the occurrence of oligomeric saccharides during fructan degradation by this bacterial strain. These could be fructose-only chains. (2) The accumulation of oligomeric compounds could result from an exo-activity alone because of the different sensitivity of the enzyme to the two types of linkages. The β-2,1 linked branch points of the fructan from *Phleum* could slow hydrolysis and lead to the transient accumulation of branched oligofructans. (3) The accumulation of oligomeric compounds could result from a fructosyl transferase reaction (Ende & Laere, 1996). It is known that fructan hydrolases can also show transferase activity. A depolymerizing inulin fructosyltransferase from *Bacillus circulans* has been described by Kawamura & Uchiyama (1994). Further investigations will be necessary to distinguish between these hypotheses.

The enzyme preparation showed a single protein band on SDS–PAGE. This could represent a single protein with both exo- and endo-type activities and with different activity against different linkage types. On the other hand, the protein band could present two or more proteins of similar size. Most of the authors who have described plant and microbial fructan-depolymerizing enzymes have found several specific enzymes: at least four β-fructofuranosidases in mature vegetative tissues of *Lolium rigidum* (Bonnett & Simpson, 1993); endo- and exoinulinases in *Penicillium* spp. (Onodera & Shiomi, 1992); one invertase, five exoinulinases and three endoinulinases in *Aspergillus ficuum* (Ettalibi & Baratti, 1987) and two forms of extracellular endoinulinase in *Aspergillus niger* (Nakamura *et al.*, 1994). The prevailing opinion is that different end-products are formed by different enzymes (Avigad, 1982).

The fructan-degrading enzyme from *L. paracasei* ssp. *paracasei* P 4134 is not a β-D-fructofuranoside fructohydrolases (EC 3.2.1.26) but is grouped functionally in the β-D-fructan-fructohydrolases (EC 3.2.1.80). However, the classification of microbial invertases (sucrases), inulinases and fructanases (levanases) as separate enzymes is still the subject of discussion since all these enzymes differ only in relative affinity for different fructose-containing molecules (grass fructans, levan, inulin, sucrose, raffinose, etc.) (Uchiyama, 1993).

In future investigations, the use of well defined fructan oligomers and the precise analysis of products will allow the unambiguous determination of the type of linkages this enzyme cleaves and thus permit the exact classification of the enzyme.

ACKNOWLEDGEMENTS

The skilful assistance of Kerstin Umann and Martina Peters is greatly appreciated. Ana L. Winters (Institute of Grassland and Environmental Research, Aberystwyth, UK) is thanked for provision of some sugar standards and Thomas Müller is thanked for critical reading of the manuscript. This study was supported by the Deutsche Forschungsgemeinschaft.

REFERENCES

Avigad G. 1982. Sucrose and other disaccharides. In: Loewus FA, Tanner W, eds. *Plant Carbohydrates I. Intracellular Carbohydrates. Encyclopedia of Plant Physiology New Series vol. 13A.* Berlin, Heidelberg, New York: Springer-Verlag, 217–347.

Avigad G, Bauer S. 1966. Fructan hydrolases. I. Levan-6-fructanohydrolase (levan hydrolase) from *Arthrobacter* sp. In: Neufeld EF, Ginsburg V, eds. *Methods in Enzymology, vol. 8.* New York: Academic Press, 621–628.

Avigad G, Zelikson R. 1963. Cleavage of fructans to levanbiose by a specific hydrolase. *Bulletin Research Council of Israel* **11A4**: 253–257.

Belamri M, Hadj Sassi A, Savart M, Tantaoui-Elaraki A, Cottin P. 1994. Purification and properties of an extracellular inulinase-like β-fructosidase from *Bacillus stearothermophilus. Letters in Applied Microbiology* **19**: 410–413.

Blatch GL, Woods DR. 1993. Molecular characterization of a fructanase produced by *Bacteroides fragilis* BF-1. *Journal of Bacteriology* **175**: 3058–3066.

Bonnett GD, Simpson RJ. 1993. Fructan-hydrolyzing activities from *Lolium rigidum* Gaudin. *New Phytologist* **123**: 443–451.

Burne RA, Penders JEC. 1994. Differential localization of the *Streptococcus* mutans GS-5 fructan hydrolase enzyme, FruA. *FEMS Microbiology Letters* **121**: 243–250.

Carpita NC, Kanabus J, Housley TL. 1989. Linkage structure of fructans and fructan oligomers from *Triticum aestivum* and *Festuca arundinaceae* leaves. *Journal of Plant Physiology* **134**: 162–168.

Chatterton NJ, Harrison PA, Thornley WR, Draper EA. 1990. Oligosaccharides in foliage of *Agropyron, Bromus, Dactylis, Festuca, Lolium* and *Phleum. New Phytologist* **114**: 167–171.

Dacosta T, Gibbons RJ. 1968. Hydrolysis of levan by human plaque streptococci. *Archives of Oral Biology* **13**: 609–617.

Ende WVD, Laere VA. 1996. *In vitro* synthesis of the *in vivo* type chicory root fructans by chicory root SST and FFT. In: Abstracts of the Third International Fructan Conference, 21–24 July, Logan, Utah, USA: 38.

Ettalibi M, Baratti, CJ. 1987. Purification, properties and comparison of invertase, exoinulinases and endoinulinases of *Aspergillus ficuum. Applied Microbiology and Biotechnology* **26**: 13–20.

Gupta AK, Singh DP, Kaur N, Singh R. 1994. Production, purification and immobilisation of inulinase from *Kluyveromyces fragilis. Journal of Chemical Technology and Biotechnology* **59**: 377–385.

Henson CA. 1989. Purification and properties of barley stem fructan exohydrolase. *Journal of Plant Physiology* **134**: 186–191.

Iizuka M, Yamamoto T. 1979. Chemical and physicochemical properties of invertase of *Candida utilis. Agricultural and Biological Chemistry* **43**: 217–223.

Kawamura M, Uchiyama T. 1994. Purification and some properties of cycloinulo-oligosaccharide fructanotransferase from *Bacillus circulans* OKUMZ 31B. *Carbohydrate Research* **260**: 297–304.

Kleinert T. 1990. *Elektrophoretische Methoden in der Proteinanalytik.* Stuttgart, New York: Georg Thieme Verlag, 162–163.

Lowry OH, Rosebrough NJ, Farr AL, Randall RJ. 1951. Protein measurement with the Folin phenol reagent. *Journal of Biological Chemistry* **193**: 265–275.

Manzoni M, Cavazzoni V. 1991. Purification and characterization of the extracellular inulinase from *Kluyveromyces cicerisporus. Lebensmittel-Wissenschaft und-Technologie* **24**: 236–240.

Manzoni M, Cavazzoni V. 1992. Hydrolysis of topinambur (Jerusalem artichoke) fructans by extracellular inulinase of *Kluyveromyces marxianus* var. *bulgaricus. Journal of Chemical Technology and Biotechnology* **54**: 311–315.

Meier H, Reid JSG. 1982. Reserve polysaccharides other than starch in higher plants. In: Loewus FA, Tanner W, eds. *Plant Carbohydrates I. Intracellular Carbohydrates. Encyclopedia of Plant Physiology New Series vol. 13A.* Berlin, Heidelberg, New York: Springer-Verlag, 418–471.

Merry RJ, Winters AL, Thomas PI, Müller M, Müller T. 1995. Degradation of fructans by epiphytic and inoculated lactic acid bacteria and by plant enzymes during ensilage of normal and sterile hybrid ryegrass. *Journal of Applied Bacteriology* **79**: 583–591.

Müller M, Lier D. 1994. Fermentation of fructans by epiphytic lactic acid bacteria. *Journal of Applied Bacteriology* **76**: 406–411.

Müller M, Steller J. 1995. Comparative studies of the degradation of grass fructan and inulin by strains of *Lactobacillus paracasei* subsp. *paracasei* and *Lactobacillus plantarum. Journal of Applied Bacteriology* **78**: 229–236.

Murakami H, Muroi H, Kuramoto T, Tamura Y, Mizutani K, Nakano H, Kitahata S. 1990. Purification and some properties of a levanase from *Streptomyces* sp. No. 7–3. *Agricultural and Biological Chemistry* **54**: 2247–2255.

Murakami H, Kuramoto T, Mizutani K, Nakano H, Kitahata S. 1992. Purification and some properties of a new levanase from *Bacillus* sp. No. 71. *Bioscience, Biotechnology and Biochemistry* **56**: 608–613.

Muramatsu K, Onodera S, Kikuchi M, Shiomi N. 1992. The production of β-fructofuranosidase from *Bifidobacterium* spp. *Bioscience, Biotechnology and Biochemistry* **56**: 1451–1454.

Nakamura T, Nagatomo Y, Hamada S, Nishino Y, Ohta K. 1994. Occurrence of two forms of extracellular endoinulinase from *Aspergillus niger* mutant 817. *Journal of Fermentation and Bioengineering* **78**: 134–139.

Onodera S, Shiomi N. 1992. Purification and substrate affinities of *exo*-inulinase from *Penicillium trzebinskii. Bioscience, Biotechnology and Biochemistry* **56**: 1443–1447.

Simpson RJ, Bonnett GD. 1993. Fructan exohydrolase from grasses. *New Phytologist* **123**: 453–469.

Suzuki M. 1993. Fructans in crop production and preservation. In: Suzuki M, Chatterton NJ, eds. *Science and Technology of Fructans.* Boca Raton, FL, USA: CRC Press, 227–255.

Uchiyama T. 1993. Metabolism in microorganisms Part II. Biosynthesis and degradation of fructans by microbial enzymes other than levansucrase. In: Suzuki M, Chatterton NJ, eds. *Science and Technology of Fructans.* Boca Raton, FL, USA: CRC Press, 169–190.

Uhm RB, Jeon DY, Byun SM, Hong JS, Grootwassink JWD. 1987. Purification and properties of β-fructofuranosidase from *Aspergillus niger. Biochimica et Biophysica Acta* **926**: 119–126.

Vandamme EJ, Derycke DG. 1983. Microbial inulinases: fermentation process, properties and application. *Advances in Applied Microbiology* **29**: 139–176.

Woolford MK. 1984. The silage fermentation. In: *Microbiology Series vol. 14*, New York: Marcel Dekker.

Xiao R, Tanida M, Takao F. 1989. Purification and characteristics of two exoinulinases from *Chrysosporium pannorum. Journal of Fermentation and Bioengineering* **67**: 331–339.

Yokota A, Enomoto K, Tomita F. 1991. Purification and properties of an inulin fructotransferase (depolymerizing) from *Arthrobacter* sp. H65-7. *Journal of Fermentation and Bioengineering* **72**: 262–265.

New Phytol. (1997), **136**, 97–104

Patterns of solute in individual mesophyll, bundle sheath and epidermal cells of barley leaves induced to accumulate carbohydrate

By O. A. KOROLEVA[1]*, J. F. FARRAR[1], A. D. TOMOS[1] and C. J. POLLOCK[2]

[1] *School of Biological Sciences, University of Wales, Bangor, Gwynedd LL57 2UW*
[2] *IGER, Aberystwyth, UK.*

(*Received 24 July 1996*)

SUMMARY

After cooling the roots and shoot apex of barley, there is a reduction in sucrose export from, and an increased total accumulation of carbohydrate and induction of fructan synthesis in, source leaves. We investigated carbohydrate accumulation in individual leaf cells. Using the microsampling technique and microfluorometric enzymatic assay, the amounts of various carbohydrates were measured in mesophyll, bundle sheath and epidermal cells. Epidermal cells take no part in carbohydrate partitioning and have very low concentrations of sugars even when leaf export is reduced. Even without cooling, however, there are remarkable changes in sucrose concentration (up to 200 mM) in mesophyll and bundle sheath cells during the photoperiod. The activity of acid invertase in samples from individual cells was very low, c. 3 μmol sucrose (g f. wt)$^{-1}$ h^{-1}. Interrelationships between these cell types and the control of fructan metabolism are discussed.

Key words: Fructan, sucrose, mesophyll, epidermis, bundle sheath, cooling.

INTRODUCTION

Source leaves of temperate *Gramineae* export photosynthetically-produced sucrose according to demand from other parts of the plant, and accumulate excess carbohydrates. This plant group generally stores sucrose and fructan, rather than starch, in the leaves. Fructan synthesis is relatively insensitive to low temperature, which allows grasses to accumulate reserve carbohydrate during cold seasons, when the conditions allow photosynthesis but restrict growth or transport (Pollock & Jones, 1979). However, if sucrose accumulates in mesophyll cells it might cause down-regulation of photosynthesis (Pollock & Farrar, 1996; Jang & Sheen, 1994). Synthesis of fructan and cellular compartmentation of carbohydrates in leaf tissue could possibly play a key role in avoiding (1) down-regulation of photosynthesis by high sugar concentrations in mesophyll cells and (2) the development of excess osmotic pressure which might lead to a damaging increase in turgor pressure. A way of avoiding overloading leaves with low-molecular-weight carbohydrate is the storage of fructans in non-photosynthesizing parts of the leaf sheaths. Field measurements of the fructan content of mature leaf blades generally yield lower concentrations than those present in sheaths and stems (Pollock & Jones, 1979; Pollock & Cairns, 1991).

Compartmentation in the leaf blade can occur at the tissue or cell level. Sucrose is synthesized in the mesophyll cells and is then transported to the parenchymatous bundle sheath (PBS) cells on the way to the phloem. The synthesis of fructans can occur in mesophyll and bundle sheath cells; it requires high sucrose concentrations (Housley & Pollock, 1985). The paradoxical situation is that the activity of acid invertase in barley leaves is also quite high (Obenland *et al.*, 1993). The reason could be spatial separation of sucrose from acid invertase.

It has been shown that high concentrations of sucrose and fructans occur in the vacuoles of mesophyll cells of barley leaves (Wagner *et al.*, 1983). The carbohydrate metabolism of the parenchymatous bundle sheaths is distinct from that of the mesophyll, at least in terms of starch accumulation (Williams *et al.*, 1989). However, the exact compart-

* To whom correspondence should be addressed.

Abbreviations: DP, degree of polymerization; BSA, bovine serum albumin; G6PDH, glucose-6-phosphate dehydrogenase; HK, hexokinase; PGI, phosphoglucose isomerase; PBS, parenchymatous bundle sheath; SiCSA single-cell sampling and analysis.

mentation of fructan metabolism at the cellular and subcellular level in leaves of temperate grasses is poorly investigated. The epidermis of barley leaves could be an intermediary storage compartment for sugars and contain as much as 5–15% of the leaf's glucose and fructose (Dietz, Hollenbach & Hellwege, 1994). However, in other studies the epidemis was shown to contain extremely low concentrations of sugars, *c.* 2 mM (Fricke, Leigh & Tomos, 1994).

The control of carbon flux from photosynthetic mesophyll cells to the parenchymatous bundle sheath cells and then to the phloem, together with the functional connections between these cells types, needs to be investigated at the cellular level (Altus & Canny, 1985; Williams *et al.*, 1989; Pollock & Cairns, 1991). Mapping of primary carbohydrate metabolism at cell level is the first step in understanding transport to and export via the phloem. The purposes of the work described here were as follows. (1) The exploitation of an experimental system which allows manipulation of carbohydrate accumulation in leaves of intact barley plants. Cooling of sinks will lead to a reduced export of sugars from leaves and therefore to an accumulation of carbohydrates and a synthesis of fructan (Pollock & Cairns, 1991; Smouter & Simpson, 1991; Plum & Farrar, 1996). (2) Investigation of the accumulation of sugars in individual epidermal, mesophyll and bundle sheath cells of barley leaves during the photoperiod. (3) Assay of the activity of vacuolar acid invertase in the same cell types and to compare this with whole-tissue extracts.

MATERIALS AND METHODS

Growth of plants

Plants of barley (*Hordeum vulgare* L. cv. Klaxon) were grown hydroponically under controlled conditions at 20 °C, 16 h light/8 h dark period with a photon flux density of 500 μmol photons m^{-2} s^{-1} and a CO_2 concentration of 350 ppm, in controlled-environmental cabinets (Sanyo PG 660. HQ1, Gallenkamp, UK). Seeds were soaked overnight in water, germinated for 4 d on moist filter paper and then 14 seedlings were transferred to a trough containing 3 l of half-strength Long Ashton nutrient solution (Hewitt, 1966). Solutions were changed twice a week. The third leaf was usually fully expanded at day 14 after transfer. Unless otherwise indicated, all experiments were performed on the middle parts of third leaves, 1–5 d after full expansion. Samples from leaf sheath were taken between 1 and 3 cm below the ligule.

Cooling experiments

Low-temperature conditions were used to induce accumulation of fructan (Smouter & Simpson, 1991). The plant roots, together with the lower part (*c.* 3 cm) of the shoot (where the shoot apex is situated), were submerged in nutrient solution cooled to 10 °C. The rest of the plant was kept at 20 °C. Cooling was continued for 4 d and then the sinks were warmed to 20 °C.

Measurement of carbohydrate content of leaves

For the measurement of non-structural carbohydrate content, samples of *c.* 50 mg (f. wt) were taken from third leaves of three separate plants, weighed and then boiled for 15 min in 90% ethanol. Further extractions were made: one in 5 ml of 90% ethanol at 60 °C for 30 min, and a second in 5 ml of water at 50 °C for 2 h. The total amounts of carbohydrate in the ethanol and water extracts were measured using the phenol-sulphuric acid technique (Dubois *et al.*, 1956).

For HPLC analysis, following Cairns & Pollock (1988), the ethanol extracts were combined, evaporated and redissolved in water. This fraction was deionized using Dowex 50 (H^+) and Dowex 1 (CO_3^{2-}), concentrated, filtered (0·45 μm polysulphene membrane) and analysed on an Aminex HPX 87C column (Biorad), at 85 °C, with water as solvent. The individual fractions were detected by means of a refractive index detector (RID-6A Shimadzu, Japan) and analysed with Biorad Valuchrom integrator software (Biorad). The fractions were quantified using external standards of similar retention time.

Extraction of individual cell sap

Using the single-cell sampling and analysis (SiCSA) technique (Tomos *et al.*, 1994), cell sap was extracted from individual mesophyll, bundle sheath and epidermal cells. A silicone-oil-filled microcapillary was used to obtain cell sap from 'trough' (Fricke *et al.*, 1994) epidermal and underlying cells. To obtain samples from mesophyll or bundle sheath cells, the microcapillary was inserted through the stomatal pore (Fricke *et al.*, 1994). A set of criteria was used to distinguish between mesophyll and bundle sheath cells: position in the leaf, total volume and viscosity of the cell sap entering the microcapillary after puncturing the cell, and the relative amount of chloroplasts in the cell sap (see 'Results' for details). Immediately after extraction, some of the samples were frozen in glass microcapillaries between two phases of silicone oil at −20 °C while others were used for the sugar assay.

Enzymatic microassay of sugar in cell saps

The concentrations of glucose, fructose and sucrose in the single-cell samples were analysed by a microfluorometric enzymatic assay (Tomos *et al.*, 1994). A Leitz MPV Compact 2 Fluorovert microscope photometer fitted with filter block A and Leitz software (Leitz, Wetzlar, Germany) was used to measure fluorescence of 4–5 nl droplets of reaction mixture placed on a microscope slide inside a 4-mm-deep aluminium ring under a 3-mm-thick layer of water-saturated paraffin oil. The reaction mixture contained 68 mM imidazole buffer (pH 7·5), 5·6 mM $MgCl_2$, 5·6 mM ATP, 0·1 % BSA, 37·5 mM NADP and 0·7 U ml^{-1} glucose-6-phosphate dehydrogenase (G6PDH). The initial fluorescence was recorded and the glucose phosphorylation was started by the addition of *c.* 100 pl of hexokinase (HK) (150 U ml^{-1}). The resulting glucose-6-phosphate was subsequently coupled to NADP reduction to NADPH, catalysed by the G6PDH. After 15–20 min the reaction was complete, with the change in NADPH fluorescence of the droplet being proportional to the initial glucose concentration. The second step of the assay was started by the addition of 100 pl of phosphoglucose isomerase (PGI) (230 U ml^{-1}). The changes in NADPH fluorescence after 20–30 min corresponded to the fructose concentration of the droplet. The third step was started with 100 pl of yeast β-fructosidase (invertase) (200 U ml^{-1}). After 40–60 min the reaction was complete and the change of fluorescence at this step was proportional to the sucrose concentration in the droplet.

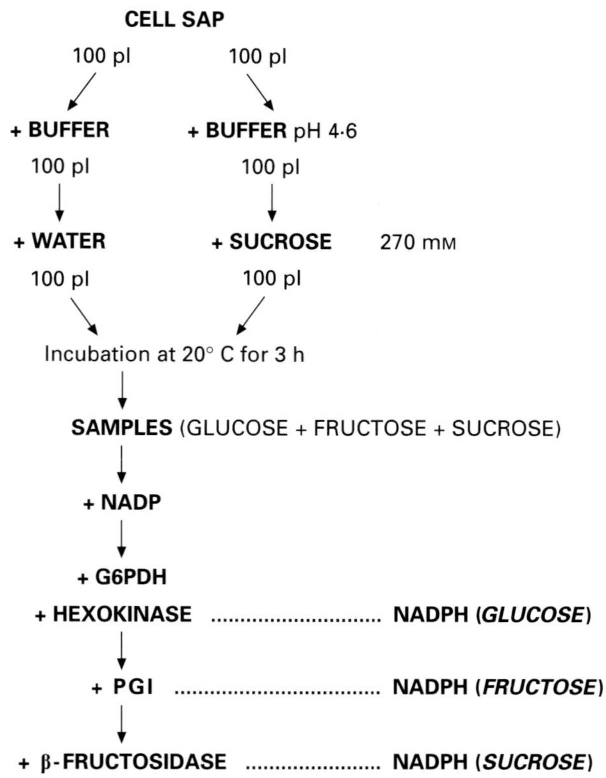

Figure 1. Protocol for determination of acid invertase activity.

Assay of acid invertase activity

For measurements of acid invertase activity, cell sap was obtained from epidermis (three single cell extracts combined), bundle sheath (three single cells) and mesophyll (seven–10 single cells). Then, to a 100 pl droplet (placed under paraffin oil to avoid evaporation) were added immediately equal amounts of 0·32 M citrate buffer (pH 4·6) and 270 mM sucrose solution (or water in the control). The scheme of the assay is shown in Figure 1. After 3 h incubation at room temperature, the concentrations of hexoses and sucrose were measured enzymatically as described above.

For determination of soluble acid invertase activity in bulk leaf tissue, crude extract was prepared from leaves. Leaf tissue (1 g) was frozen with liquid nitrogen, ground, mixed with 1 ml of 0·16 M citrate buffer (pH 4·6) at 4 °C and centrifuged for 2 min at 10000 ***g*** at 4 °C. Then a 5 μl sample of the supernatant was placed under paraffin oil and equal amounts (5 μl each) of 0·32 M citrate buffer (pH 4·6) and 270 mM sucrose (or water in the control) were added. After 3 h incubation at room temperature, samples from these droplets were taken for enzymatic assay of glucose and sucrose concentrations.

Measurement of photosynthesis

The assimilation rate was measured using a clamp-on cuvette system and infra-red gas analyser (CIRAS-PP Systems Herts., UK) at a photon flux density of 500 μmol photons m^{-2} s^{-1}, leaf temperature of 24 °C and CO_2 concentration of 350 ppm.

Most chemicals were purchased from BDH (Poole, Dorset, UK). NADP and coupling enzymes from Boehringer Mannheim (Boehringer, Mannheim, Germany).

RESULTS

Carbohydrate accumulation in leaf tissue

In plants under control conditions (20 °C), total carbohydrate accumulation during the first 11 h of a photoperiod was *c.* 10 mg (g f. wt)$^{-1}$ on a whole leaf basis (Fig. 2). This is equivalent to 5 g m^{-2} of leaf surface. Most of the carbohydrate extracted was 90 % ethanol-soluble (which includes fructan oligosaccharides with DP 5 and less). Only traces of larger fructans (DP > 5) were found in water extracts. The total amount of sugars at the end of the photoperiod increased with the age of the leaf. During dark periods, all sugars accumulated during the light are exported from the leaf blade.

Root cooling increased the accumulation of sugar in the leaves up to more than 20 mg (g f. wt)$^{-1}$ during the photoperiod, with the exception of the first day of cooling, when the accumulation was either indistinguishable from that of, or less than that in, control plants (Fig. 2). During the second and third days of cooling, the content of carbohydrate in

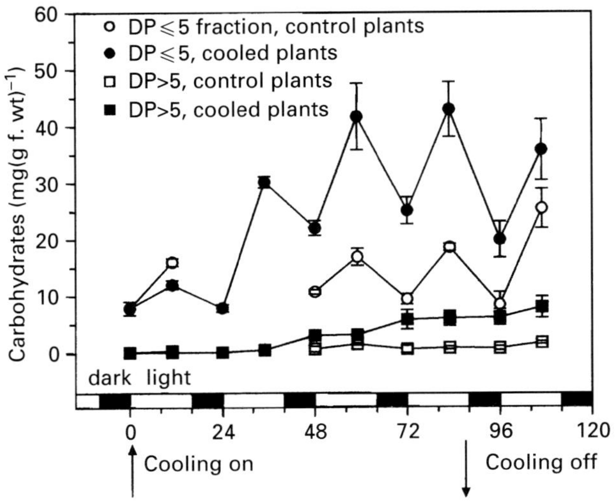

Figure 2. Accumulation of carbohydrates in barley leaf blades and the influence of cooling of the sinks to 10 °C. Each point is the average for three samples from separate plants; bars represent one SD each side of the mean.

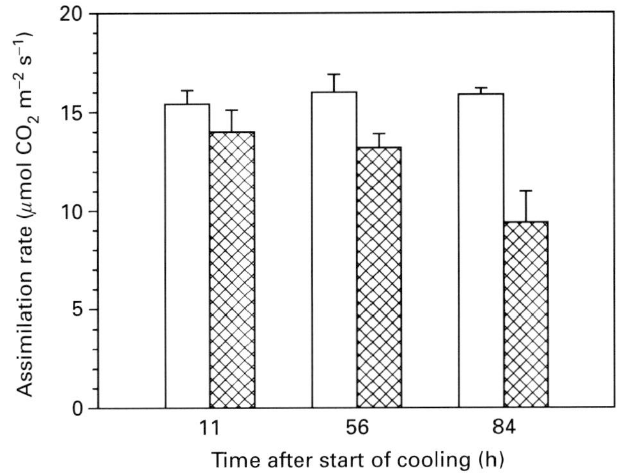

Figure 4. Net assimilation rate of control barley plants (□) and those with sinks cooled to 10 °C (⊠). Each value is the average for three samples from separate plants; bars represent SD.

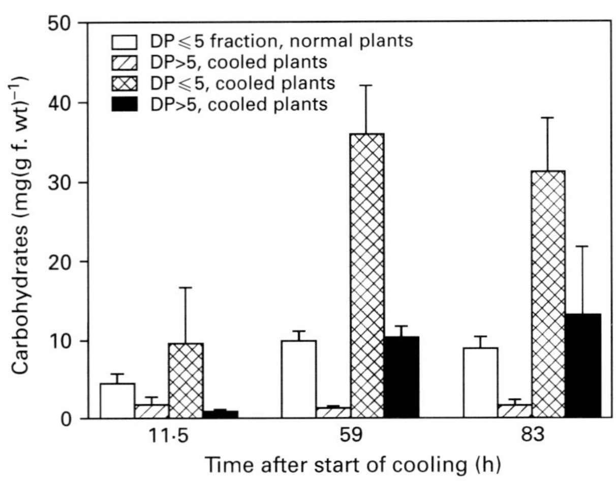

Figure 3. Accumulation of carbohydrates in leaf sheaths of control barley plants and those with sinks cooled to 10 °C. Each value is the average for three samples from separate plants; bars represent SD.

the leaves of cooled plants was at least twice that of the control plants, the difference increasing during the light periods. However, there was a tendency for carbohydrate content to stabilize by the last (4th) day of cooling. During dark periods, the content of ethanol-soluble carbohydrates was still decreasing but the amount of carbohydrates remaining at the end of dark period was increased, showing that cooling reduced export of carbohydrates from leaf blades. A small amount of sugars entered the pool of fructan DP > 5, which therefore increased slightly during the dark periods (Fig. 1). This pool, however, cannot account for the total amount lost from the oligosaccharide (DP ⩽ 5) pool during the night. To check the hypothesis that the export of sugar from the leaf blade was accompanied by storage of fructans in the leaf sheaths under cooling, the carbohydrate content of sheath samples was determined by the phenol-sulphuric acid method. The total accumulation of carbohydrate between the 1st and 3rd day of cooling (both measurements were made 11 h into the light period) in the sheaths was *c.* 33 mg (g f. wt)$^{-1}$ (Fig. 3), and during the next day of cooling the carbohydrate content did not change.

The total amount of carbohydrate accumulated in the leaf sheaths was compared with that released from leaf blades during the period 11–59 h of cooling. Given that for 3rd leaf the fresh weight ratio leaf blade/leaf sheath is 2·5 ($n = 10$ plants, 3 d after full expansion of the leaves), the amount accumulated in the sheath can account for *c.* 13 mg (g f. wt)$^{-1}$ sugars exported from the leaf blade. The actual decrease of carbohydrate content in the leaf blades during the 1st and 2nd dark periods after cooling had begun was 4 and 8 mg (g f. wt)$^{-1}$ respectively (Fig. 2), the sum of which (12 mg (g f. wt)$^{-1}$) is very close to that accumulated in the sheath. However, there might also be some flux of sugars from leaf blade to leaf sheath during the light period, which can be calculated from the difference between net assimilation rate of photosynthesis and accumulation of sugars in the tissue during the light period.

Net assimilation rate per unit area of leaf blade was nearly constant with time in control plants, at *c.* 15 μmol CO_2 m^{-2} s^{-1}, equivalent to a rate of sucrose synthesis of 3 mg sucrose (g f. wt)$^{-1}$ h^{-1} (Fig. 4). However, the photosynthetic rate of plants under cooled conditions declined gradually, and represents at day 4 only 67 % of its value at day 1 of cooling. The approximate rate of sucrose synthesis at day 4 is 2 mg sucrose (g f. wt)$^{-1}$ h^{-1}. Comparing this with the accumulation of sugars (Fig. 2), we conclude that 88 % and 90 % of daily assimilated sugars remain in the leaf blade during the light period on day 2 and 4 cooling respectively. Therefore, the loss of sugars from leaf blades during the light is not high; it might cover the respiratory needs of the leaf.

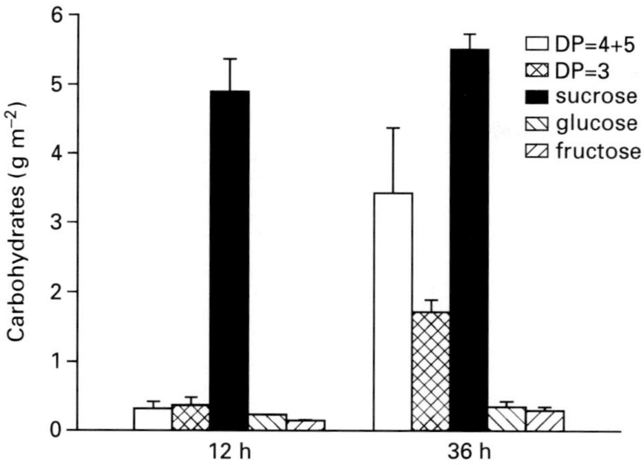

Figure 5. Carbohydrate composition of ethanol-soluble fraction extracted from leaf blades of cooled barley plants. Each value is the average for three samples from separate plants; bars represent SD.

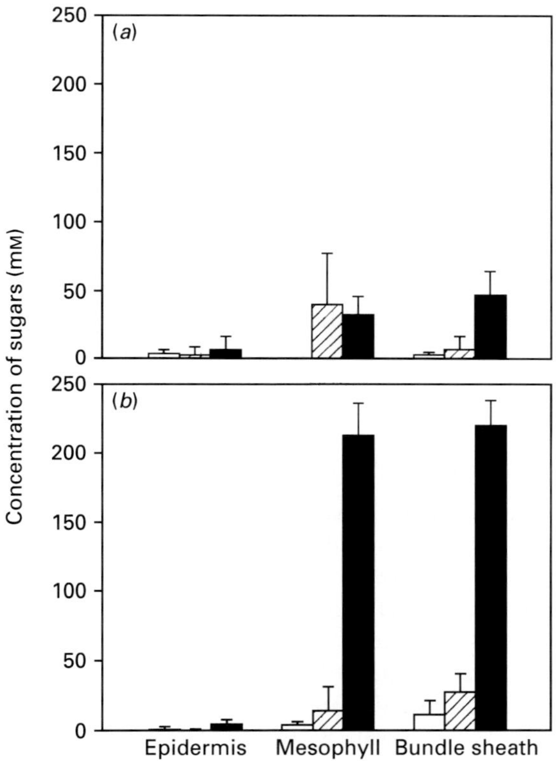

Figure 6. Concentrations of glucose (□), fructose (▨) and sucrose (■) in individual epidermal, mesophyll and bundle sheath cells of control plants: (*a*) at the end of dark period, (*b*) after 10 h of continuous illumination. Each value is average for 4–8 individual cell samples from two separate plants; bars represent SD.

Analysis by HPLC showed that there are both quantitative and qualitative changes in leaf carbohydrates under the cooling conditions (Fig. 5). The concentration of sucrose was similar at the end of the first and second day of cooling, and the content of hexoses was negligible. At the second day, however, content of fructan of DP 3 increases up to six times and that of DP 4+5 up to 10 times the values on the first day. Clearly, synthesis of fructan was induced.

Analysis of individual mesophyll, bundle sheath and epidermal cell saps

Individual cell samples were taken from control plants and analysed. The epidermal cells each yielded comparatively large volume (more than 100 pl) of low-viscosity cell sap. Similar amounts of cell sap were obtained from bundle sheath cells, where the sap was more viscous and sometimes contained a few chloroplasts. By contrast, mesophyll cells gave a very small amount (about 20 pl) of extremely viscous cell sap which contained usually more than 10 chloroplasts from each cell. To avoid blocking of the sampling microcapillary tip by the chloroplasts, only tips with diameters exceeding those of chloroplasts were used to sample from mesophyll cells.

When the samples were taken at the end of dark period (Fig. 6*a*), the total concentration of sugars was *c.* 50–70 mM for mesophyll and bundle sheath cells. In the epidermal cells the sum of the sugar concentration was < 10 mM, in good agreement with the data obtained by Fricke *et al.* (1994) by the same technique and with the results of Dietz *et al.* (1994) obtained using isolated epidermal protoplasts. In both mesophyll and bundle sheath cells, concentration of sucrose was about 30–50 mM at the beginning of the light period, when the fructose concentration was very close to that of sucrose in mesophyll cells.

When the samples were taken after 10 h of continuous illumination (Fig. 6*b*), the sucrose concentration had increased to > 200 mM. The hexose concentrations, especially those of fructose, were higher in mesophyll and bundle sheath cells than in the epidermis, but in most of the samples they were less than 50 mM.

Activity of vacuolar acid invertase in individual cells

After taking cell samples and mixing them with buffer and substrate solution, which took no more than 5 min, the reaction droplets were kept for 3 h at room temperature, then four subsamples of 20 pl were taken from each droplet for the microassay of sugars.

The difference in glucose concentrations between sucrose-loaded and water-loaded control droplets reflected the hydrolysis of sucrose by acid invertase. The concentrations of glucose, fructose and sucrose in the droplets after 3 h of incubation at room temperature, for samples taken from leaf cells of 18-d-old plants are illustrated in Table 1.

The concentration of sucrose given initially (90 mM in the sucrose-loaded droplets) remained practically unchanged. The difference between sucrose-loaded and control droplets measured after commercial yeast β-fructosidase hydrolysis was 91, 108 and 109 mM for epidermal, mesophyll and bundle sheath cells respectively. The sucrose-loaded

Table 1. *Estimation of vacuolar acid invertase activity of cell saps of 18-d-old barley 1 h into the light period*

No.	Sample	Sucrose (mM)	Glucose (mM)
1	Epidermis/water	0	1
2	Epidermis/sucrose	91	3
3	Mesophyll/water	7	8
4	Mesophyll/sucrose	115	12
5	Bundle sheath/water	2	1
6	Bundle sheath/sucrose	111	3

The saps were incubated with water or with sucrose solution to measure glucose release from the sucrose (initial concentration of sucrose in the sucrose-loaded droplets was close to 90 mM). The remaining concentration of sucrose was also measured. (See Fig. 1 for protocol and explanation.)

Table 2. *Estimation of vacuolar acid invertase activity from cell saps of 27-d-old barley*

No.	Sample	Sucrose (mM)	Glucose (mM)
1	Epidermis/water	0	1
2	Epidermis/sucrose	113	0·4
3	Mesophyll/water	191	2
4	Mesophyll/sucrose	260	6
5	Bundle sheath/water	242	0
6	Bundle sheath/sucrose	317	2

The sinks of the plants were cooled from day 19 to day 22 and after that the plants were grown under normal conditions for 4 d. Samples were taken 4 h into light period. (See Table 1 for further details.)

droplets contained glucose (which should appear as a result of sucrose hydrolysis) at a concentration only 2–4 mM higher than in control droplets without sucrose. Therefore, it can be concluded that vacuolar acid invertase activity is very low in all cell extracts. Recalculated, the hydrolysis of 3 mM of sucrose in diluted cell extracts in 3 h corresponds to 3 μmol sucrose hydrolysed (g f. wt)$^{-1}$ h^{-1}.

The same results, very low hydrolysis of sucrose by cell extracts, were obtained when samples were taken 4 h into the light period from 27-d-old plants, 4 d after the cooling experiment was finished and plants had been returned to control conditions (Table 2). Presumably, at this time the leaves are exporting sugars which had been accumulated during the cooling period. Therefore, we might expect increased activity of vacuolar acid invertase. The concentrations of hexoses in cell sap from these plants were again very low, and again the invertase activity in the cell sap was low. The difference between the measured glucose concentrations of sucrose-loaded and control droplets was 0·6–4 mM. The difference between sucrose concentration of sucrose-loaded and control droplets was 113 mM for epidermal, 69 for mesophyll and 75 mM for bundle sheath cells. The difference of less than the ‘expected’ 90 mM, obtained for mesophyll and bundle sheath cells, is probably due to the very high internal concentration of sucrose in the cell saps reaching the upper limit of the capacity of the assay (data not shown) rather than to sucrose hydrolysis, which would result in an equivalent release of hexoses (which was not observed).

Table 3. *Estimation of soluble acid invertase of leaf tissue extract from 20-d-old barley*

No.	Sample	Sucrose (mM)	Glucose (mM)
1	Tissue extract/water	34·4	8·1
2	Tissue extract/sucrose	130·3	19·2

Tissue was extracted with 0·16 M citrate buffer (pH 4·6) and assayed with a assay similar to that described in Tables 1 and 2.

Table 4. *Activity of acid invertase in extracts of whole leaf tissue, and individual epidermal, mesophyll and bundle sheath cells of barley*

Sample	Acid invertase activity (μmol sucrose (g f. wt)$^{-1}$ h^{-1})
Whole tissue	30
Epidermal cells	2
Mesophyll cells	4
Bundle sheath cells	2

Acid invertase activity in tissue extract

Soluble acid invertase activity in crude leaf tissue extract was measured under the same conditions as for individual cell samples (except that the volume of droplet of the reaction mixtures was 15 μl instead 300 pl). The sucrose-loaded droplets contained, after 3 h of incubation, 19·2 mM glucose against 8·1 mM in control droplets (Table 3). Therefore, at least 11 mM, about 10% of the sucrose, was hydrolysed. Obviously, some internal sucrose from the tissue (measured at 34·4 mM) (which can be expected to be about 10% as well and therefore *c.* 4 mM) was hydrolysed, both in control and sucrose-loaded droplets. Given that the tissue sap was diluted six times, we calculated soluble acid invertase activity of 30 μmol sucrose hydrolysed (g f. wt)$^{-1}$ h^{-1}, 10 times higher than vacuolar acid invertase activities measured for individual cells (Table 4).

DISCUSSION

Cooling the roots plus shoot apical meristem of barley plants resulted in a higher total carbohydrate content in leaves, and the induction of fructan synthesis. The assimilation rate and carbohydrate accumulation obtained for the third leaves of barley under sink-cooling are very close to the results reported for the 2nd leaves under the same conditions (Plum & Farrar, 1996). Similar results also were obtained for chilled plants of *Lolium* and *Dactylis* (Pollock & Ruggles, 1976; Pollock *et al.*, 1983; Pollock, 1984; Smouter & Simpson, 1991).

The maximum amounts of sucrose and fructans with DP < 5 were accumulated during days 2 and 3 of cooling. At the fourth day, reduction of assimilation rate to 67% of its value on the first day of cooling resulted in stabilization of carbohydrate content, both in leaf blades and sheaths. Overall, the accumulation rate of carbohydrates in leaf blades under cooling conditions is less than that in excised, illuminated leaf blades (Wagner, Wiemken & Matile, 1986; Cairns & Pollock, 1988; Smouter & Simpson, 1991). Leaf sheaths can take part in carbohydrate compartmentation, accumulating carbohydrates exported from the leaf blade when export to other parts of the plant is reduced. Our results showed that, under cooling, leaf blades and leaf sheaths accumulate carbohydrate to similar concentrations.

In the leaves of non-cooled barley during the photoperiod, mesophyll and bundle sheath cells accumulate sucrose to high concentrations. By contrast, the epidermis contains very low concentrations of sugars and takes no part in carbohydrate accumulation. Our data are in good agreement with those of Wagner *et al.* (1983), who found sucrose and fructans to be deposited in the vacuoles of the mesophyll. The similar sugar composition of mesophyll and parenchymatous bundle sheath cells is consistent with close functional connections of these cells. The PBS cells serve an important role in the rapid removal of assimilate from the apoplast and mesophyll cells (Williams *et al.*, 1989).

Cell samples extracted and measured by SiCSA contain varying proportions of vacuole and cytoplasm (Fricke *et al.*, 1994). In the epidermal cells, the cytoplasmic component will be negligible. Samples from bundle sheath cells are also likely to be predominantly vacuolar (Williams *et al.*, 1989) and the solutes measured are likely to approximate the value of the vacuolar sap. The mesophyll samples, on the other hand, are likely to contain 26% of cytoplasm (Winter, Robinson & Heldt, 1993) and a precise estimation of vacuolar sucrose is not possible. It is thought, however, that the sucrose and fructose values are representative of the vacuole. The comparison of the invertase activity of the whole-leaf extract with that of the samples of individual cells indicates that the enzyme is located outside the protoplasts of the sampled cells. This explains the paradox of acid invertase activity and sucrose accumulation occurring simultaneously in the leaf: the two are spatially separated. Obenland *et al.* (1993) suggested that acid invertase activity also occurs within the vacuoles of epidermal and mesophyll cells, which is contradicted by the evidence given here.

Vacuolar acid invertase activity in mesophyll, bundle sheath and epidermal cells is very low, 2–4 μmol sucrose g f. wt^{-1} h^{-1}, consistent with accumulation of sucrose and fructans. The fact that soluble acid invertase activity in whole tissue is about 30 μmol sucrose g f. wt^{-1} h^{-1}, is in a good agreement with data for mature leaves of wheat and barley (30 and 54 μmol sucrose g f. wt^{-1} h^{-1} respectively (Farrar, unpublished)). The general conclusion is that acid invertase activity is not in protoplasts outside of the veins. Therefore the possible locations for its activity are inside the vascular bundles and in cell walls of the rest of leaf tissue. According to data obtained by Winter *et al.* (1993), the subcellular volumes of primary barley leaf occupied by veins and apoplast are *c.* 10%, and therefore the soluble acid invertase activity could be 10-fold higher in these compartments than its average activity in the tissue.

ACKNOWLEDGEMENTS

We wish to thank Drs P. Richardson, B. Collis and P. Hinde for their technical advice and helpful discussion of the results. The work was supported by a grant from the BBSRC.

REFERENCES

Altus DP, Canny MJ. 1985. Loading of assimilates in wheat leaves. II. The path from chloroplast to vein. *Plant, Cell and Environment* **8**: 275–285.

Cairns AJ, Pollock CJ. 1988. Fructan biosynthesis in excised leaves of *Lolium temulentum* L.I. Chromatographic characterization of oligofructans and their labelling patterns following $^{14}CO_2$ feeding. *New Phytologist* **109**: 399–405.

Dietz KJ, Hollenbach B, Hellwege E. 1994. The epidermis of barley leaves is a dynamic intermediatory storage compartment of carbohydrates, aminoacids and nitrate. *Physiologia Plantarum* **92**: 31–36.

Dubois M, Giles KA, Hamilton JH, Rebus PA, Smith F. 1956. Colorimetric method for the determination of sugars and related substances. *Analytical Chemistry* **28**: 350–356.

Fricke W, Leigh RA, Tomos AD. 1994. Concentrations of inorganic and organic solutes in extracts from individual epidermal, mesophyll and bundle-sheath cells of barley leaves. *Planta* **192**: 310–316.

Hewitt EJ. 1966. *Sand and water culture methods used in the study of plant nutrition. Technical Communication No. 22,* Farnham Royal, UK: Commonwealth Agricultural Bureux.

Housley TL, Pollock CJ. 1985. Photosynthesis and carbohydrate metabolism in detached leaves of *Lolium temulentum* L. *New Phytologist* **99**: 499–507.

Huber SC. 1989. Biochemical mechanism for regulation of sucrose accumulation in leaves during photosynthesis. *Plant Physiology* **91**: 656–662.

Jang JC, Sheen J. 1994. Sugar sensing in higher plants. *Plant Cell* **6**: 1665–1679.

Obenland DM, Simmen U, Boller T, Wiemken A. 1993. Purification and characterisation of three soluble invertases from barley (*Hordeum vulgare* L.) leaves. *Plant Physiology* **101**: 1331–1339.

Plum SA, Farrar JF. 1996. Carbon partitioning in barley following manipulation of source and sink. *Aspects of Applied Biology* **45**: 177–180.

Pollock CJ. 1984. Sucrose accumulation and the initiation of fructan biosynthesis in *Lolium temulentum* L. *New Phytologist* **96**: 527–534.

Pollock CJ, Cairns AJ. 1991. Fructan metabolism in grasses and cereals. *Annual Review of Plant Physiology and Plant Molecular Biology* **42**: 77–101.

Pollock CJ, Cairns AJ, Collis BE, Walker RP. 1989. Direct effects of low temperature upon components of fructan metabolism in leaves of *Lolium temulentum* L. *Journal of Plant Physiology* **134**: 203–208.

Pollock CJ, Farrar JF. 1996. Source: sink relations: the role of sucrose. In: Baker NR, ed. *Photosynthesis and Environment.* Amsterdam: Kluwer, 261–279.

Pollock CJ, Jones T. 1979. Seasonal patterns of fructan metabolism in forage grasses. *New Phytologist* **83**: 8–15.

Pollock CJ, Ruggles PA. 1976. Cold-induced fructan synthesis in leaves of *Dactylis glomerata. Phytochemistry* **15**: 1643–1646.

Pollock CJ, Lloyd ED, Stoddart JL, Thomas H. 1983. Growth, photosynthesis and assimilate partitioning in *Lolium temulentum. Annals of Botany* **55**: 593–596.

Smouter H, Simpson R. 1991. Fructan metabolism in leaves of *Lolium rigidum* Gaudin. I. Synthesis of fructan. *New Phytologist* **119**: 509–516.

Tomos AD, Hinde P, Richardson P, Pritchard J, Fricke W. 1994. Microsampling and measurement of solutes in single cells. In: Harris N, Oparka KJ, eds. *Plant Cell Biology – A Practical Approach.* Oxford: IRL Press, 297–314.

Wagner W, Keller F, Wiemken A. 1983. Fructan metabolism in cereals: induction in leaves and compartmentation in protoplasts and vacuoles. *Zeitschrift für Pflanzenphysiologie* **112**: 359–372.

Wagner W, Wiemken A, Matile P. 1986. Regulation of fructan metabolism in leaves of barley (*Hordeum vulgare* L. cv. Gerbel). *Plant Physiology* **81**: 444–447.

Williams ML, Farrar JF, Pollock CJ. 1989. Cell specialisation within the parenchymatous bundle sheath of barley. *Plant, Cell and Environment* **12**: 909–918.

Winter H, Robinson DC, Heldt HW. 1993. Subcellular volumes and metabolite concentrations in barley leaves. *Planta* **191**: 180–190.

New Phytol. (1997), **136**, 105–113

Fructo-oligosaccharide content and fructosyltransferase activity during growth of onion bulbs

By NORIO SHIOMI*, SHUICHI ONODERA AND HIDEKI SAKAI

Department of Food Science, Faculty of Dairy Science, Rakuno Gakuen University, 582 Bunkyodai, Midorimachi, Ebetsu, Hokkaido 069, Japan

(*Received 24 July 1996*)

SUMMARY

The accumulation of fructo-oligosaccharides and the activities of fructosyltransferase (sucrose:sucrose 1_F-fructosyltransferase (SST), 1_F-fructosyltransferase (1_F-FT) and 6_G-fructosyltransferase (6_G-FT)) in the bulbs of three onion cultivars were investigated from June to September 1993. The total fructo-oligosaccharide content increased from June until August, then decreased in September, except in one cultivar. The levels of neokestose and its related tetrasaccharides (1_F,6_G-di-β-D-fructofuranosyl sucrose and 6_G(1-β-D-fructofuranosyl)$_2$sucrose) were higher than those of 1-kestose and nystose throughout growth. The activities of 6_G-FT, 1_F-FT and SST were high in June and July, then decreased; SST activity was very low in September. The activity ratios of 6_G-FT to 1_F-FT varied between 1·86 and 2·65 over the growth period. Two trisaccharides, three tetrasaccharides and four pentasaccharides were identified, together with a mixture of hexa- and heptasaccharides, all of which were synthesized *in vitro* from 0·1 M sucrose by an enzyme preparation of onion bulbs harvested in August. Octa- and nonasaccharides other than the saccharides formed from sucrose were also synthesized from 0·1 M 1-kestose or 0·1 M neokestose. All the saccharides produced from sucrose, 1-kestose or neokestose by the crude enzyme prepared from onion bulbs were identical to the saccharides occurring naturally in onion bulbs.

Key words: *Allium cepa*, fructo-oligosaccharides, fructosyltransferase.

INTRODUCTION

Fructose-oligomers and fructose-polymers are major reserve carbohydrates in higher plants such as members of the Asteraceae, Liliaceae and Gramineae. Fructose-oligomers (fructo-oligosaccharides (DP 3–*c.* 10)) together with glucose, fructose and sucrose are contained in onion, which is one of the most widely distributed liliaceous plants in the world (Bacon, 1957, 1959; Bose & Shrivastava, 1961; Darbyshire & Henry, 1978). The distribution of fructo-oligosaccharides in onion leaf-bases has been studied, and fructo-oligosaccharide contents were found to be higher in younger leaf-bases (inner) than in older (outer) leaf-bases (Bacon, 1959; Darbyshire & Henry, 1978).

The composition of fructo-oligosaccharide isomers in onion plants was also investigated, and the trisaccharide fraction from onion bulbs was shown to comprise 1-kestose and neokestose (Bacon, 1959; Darbyshire & Henry, 1978).

The tetra-, penta-, and hexasaccharide fractions from onion leaves and bulbs were reported recently to comprise nystose, 6_G(1-β-fructofuranosyl)$_2$sucrose and 1_F,6_G-di-β-fructofuranosylsucrose; 1_F(1-β-fructofuranosyl)$_3$sucrose, 6_G(1-β-fructofuranosyl)$_3$-sucrose, 1_F-β-fructofuranosyl-6_G(1-β-fructofuranosyl)$_2$sucrose and 1_F(1-β-fructofuranosyl)$_2$-6_G-β-fructofuranosyl sucrose; and 1_F(1-β-fructofuranosyl)$_n$-6_G(1-β-fructofuranosyl)$_m$ sucrose ($n+m = 4$), respectively (Shiomi, 1989; Shiomi *et al.*, 1991).

The content, distribution and structure of fructo-oligosaccharides in onion plants were investigated extensively by the researchers cited above, although older studies were carried out using techniques such as paper chromatography (PC), thin-layer chromatography (TLC), and high-performance liquid chromatography (HPLC) with refractive index monitoring. However, little is known about the changes in fructo-oligosaccharide isomers derived from 1-kestose and neokestose, and in the activities of fructosyltransferases (sucrose:sucrose 1_F-

* To whom correspondence should be addressed.

fructosyltransferase (SST, EC 2.4.1.99; Henry & Darbyshire, 1980), 6_G-fructosyltransferase (6_G-FT; Shiomi, 1989) and 1_F-fructosyltransferase (1_F-FT; Henry & Darbyshire, 1980; Shiomi, 1989)) involved in the synthesis of fructo-oligosaccharides in onion bulbs during the growth stage.

Here, we describe the accumulation of fructo-oligosaccharide isomers and the activities of fructosyltransferases (SST, 6_G-FT and 1_F-FT) in bulbs of three onion cultivars during the growth period, and study the synthesis *in vitro* of fructo-oligosaccharide isomers from sucrose, 1-kestose or neokestose by a crude enzyme preparation from onion bulbs.

MATERIALS AND METHODS

Materials

Bulbs of onion (*Allium cepa* L. cv. Sapporo Yellow (SY); cv. Southport White Glove Highsolid (SWGH) and cv. Sweet Spanish Utah Jumbo (SSUT)) were harvested at the experimental farm of the Hokkaido National Agricultural Experiment Station on 25 June, 28 July, 26 August and 27 September 1993, and stored frozen. Standard sugars were prepared as follows. Crystalline 1-kestose (**3a**: 1_F-β-D-fructofuranosylsucrose) and nystose (**4a**: 1_F(1-β-D-fructofuranosyl)$_2$sucrose) were prepared from sucrose by a *Scopulariopsis brevicaulis* enzyme (Takeda *et al.*, 1994). 6-Kestose was synthesized via the transfer action of yeast invertase (grade VII, Sigma Chemicals, St Louis, MO, USA) on sucrose. Neokestose (**3b**: 6_G-β-D-fructofuranosylsucrose) and 1_F(1-β-D-fructofuranosyl)$_m$-6_G (1-β-D-fructofuranosyl)$_n$ sucrose (**4b**: $m = 0$, $n = 2$; **4c**: $m = 1$, $n = 1$; **5a**: $m = 3$, $n = 0$; **5b**: $m = 0$, $n = 3$; **5c**: $m = 2$, $n = 1$; **5d**: $m = 1$, $n = 2$; **6b**: $m = 0$, $n = 4$; **6c**: $m = 3$, $n = 1$; **6d$_1$**: $m = 1$, $n = 3$; **6d$_2$**: $m = 2$, $n = 2$; **8**: $n+m = 6$, **9x**: $n+m \geqq 7$) were obtained from asparagus roots as described in previous papers (Shiomi, Yamada & Izawa, 1976, 1979; Shiomi, 1981). 1_F(1-β-D-Fructofuranosyl)$_m$sucrose(m = 4(**6a**) and 5(**7a**)) were prepared from Jerusalem artichoke tubers.

Quantitative determination of sugars

Total sugars were determined by the anthrone method (Morris, 1948) and reducing sugars were measured by the method of Somogyi and Nelson (Nelson, 1944; Somogyi, 1945).

Analysis of sugars

High-performance anion-exchange chromatography (HPAEC). The synthesized or extracted saccharide mixtures were separated on an HPLC-carbohydrate column (PA1, Carbo Pack, Sunnyvale, CA, USA) with a Dionex Bio LC series HPLC using pulsed amperometric detection (PAD) (Rocklin & Pohl, 1983; Johnson, 1986). The gradient was established by mixing eluent A (150 mM NaOH) with eluent B (500 mM sodium acetate in 150 mM NaOH) in the following two ways: System I: 0–1 min, 25 mM; 1–2 min, 25–50 mM; 2–20 min, 50–200 mM; 20–22 min, 500 mM; 22–30 min, 25 mM. System II: 0–1 min, 5 mM; 1–2 min, 25–50 mM; 2–14 min, 50–500 mM; 14–22 min, 500 mM; 22–30 min, 25 mM; flow rate through the column was 1·0 ml min^{-1}. The applied PAD potentials for E1 (500 ms), E2 (100 ms) and E3 (50 ms) were 0·01, 0·60 and −0·6 V, respectively, and the output range was 1 μC.

Paper and thin-layer chromatography (PC and TLC). Two solvents, *n*-propanol-ethyl acetate-water (7:1:2) (I) and *n*-butanol-acetic acid-water (4:1:2) (II), were used for paper chromatography (filter paper No. 50, Toyo, Tokyo, Japan), and a third solvent, *n*-butanol-isopropanol-water (10:5:4) (III), was employed for thin-layer chromatography (Kieselgel 60F254 pre-coated, Merck, Darmstadt, Germany). After three to five developments, the chromatograms were sprayed with anisidine phosphate solution for detecting sugars (Mukherjee & Srivastava, 1952).

Preparation of enzyme solution

All the operations for enzyme preparation were carried out at 4 °C, and centrifugation was conducted at 10000 ***g*** for 30 min. Onion bulbs (five globes, or 20 globes harvested in June) were peeled, cut into small pieces and mixed thoroughly. The pieces (50 g) were homogenized in 0·05 M phosphate buffer, pH 7·0 (150 ml). The homogenate was filtered through cheesecloth. The residue was homogenized twice more, and the resulting filtrate was centrifuged. The supernatant was made up to 500 ml with the same buffer, 50 ml was centrifuged, dialysed for 1 d against stirred 0·01 M phosphate buffer, pH 6·1, 5 l and was concentrated to 2 ml using a Centriprep 30 (Amicon Beverly, MA, USA). The concentrate was diluted to 50 ml with 0·01 M phosphate buffer then re-concentrated as above, and the process repeated. The concentrate was termed 'supernatant concentrate A' and was assayed for enzyme activities (SST, 1_F-FT, 6_G-FT and invertase).

For studying the synthesis of fructo-oligosaccharides from sucrose, 1-kestose or neokestose, a supernatant concentrate was prepared independently from onion bulbs ('Sapporo Yellow') harvested on 26 August 1993. The five bulbs were peeled and cut into small pieces which were mixed thoroughly. The supernatant (900 ml) obtained from the onion pieces (100 g) was dialysed, concentrated to 2·5 ml by ultrafiltration apparatus (PM-10 membrane, Amicon) and Centriprep30, and termed 'supernatant concentrate B'.

Supernatant concentrates A and B were prepared in duplicate and assayed for enzyme activity.

Measurement of enzyme activity

SST activity. One kat of SST activity was defined as the amount of enzyme which catalysed fructosyl transfer from sucrose to sucrose to produce 1 mole of 1-kestose (isokestose, **3a**) in 1 s under the conditions described below. A mixture (I) of enzyme (0·4 ml), 0·4 M sucrose (0·2 ml), McIlvaine buffer (pH 5·5, 0·2 ml) and toluene (a trace amount) was incubated at 30 °C for 5, 10 and 15 h. After the reaction was stopped by heating in a boiling water bath for 3·5 min, the reaction mixture was diluted, filtered and applied to HPAEC. SST activity was calculated from the amount of 1-kestose.

1_F-FT activity. One kat of 1_F-FT activity was defined as the amount of enzyme which catalysed the fructosyl transfer from 1-kestose to another 1-kestose to synthesis 1 mol of nystose (**4a**) in 1 s under the conditions described below. The enzyme reactions were carried out at 30 °C for 5 and 10 min. A mixture (II) of enzyme (0·4 ml), 0·4 M 1-kestose (0·2 ml) and McIlvaine buffer (pH 5·5, 0·2 ml) was used. The reaction was terminated by heating in a boiling water bath for 3·5 min, and the reaction mixture was diluted, filtered and applied to HPAEC. 1_F-FT activities were calculated from the amount of nystose (**4a**).

6_G-FT activity. One kat of 6_G-FT activity was defined as the amount of enzyme which catalysed the fructosyl transfer from 1-kestose to another 1-kestose to yield 1 mol of 1_F,6_G-di-β-D-fructofuranosylsucrose (**4c**) in 1 s. The activities of 6_G-FT were tested using mixture (II). All the experiments were made similarly to those for 1_F-FT activity, except for calculating 6_G-FT activity from the amount of 1_F,6_G-di-β-D-fructofuranosylsucrose (**4c**).

Invertase activity. One kat of invertase activity was defined as the amount of enzyme which hydrolyses 1 mol sucrose s^{-1}. The experimental conditions were those described for SST measurements. Invertase activity was calculated from the amount of fructose released.

Extraction of sugars

Onion bulbs (five globes, or 20 globes if harvested in June) harvested between June and September were peeled and cut into small pieces which were mixed thoroughly. The pieces (10 g) were homogenized in 70% ethanol (80 ml) containing small amounts of calcium carbonate, and were then immediately boiled in a water bath for 10 min. The residues were extracted five times in the same manner, then extracted with hot water (80 °C, 100 ml). Filtered extracts (25 ml) were concentrated *in vacuo* at 30–33 °C to dryness. The concentrated sugar was added to distilled water (25 ml) and the solution was deionized with Amberlite IRA-410. This process was duplicated.

RESULTS

Contents of saccharides in onion bulbs during the growth period

The neutral sugar extracts from onion bulbs of all three cultivars harvested in September were analysed by HPAEC. As shown in Figure 1, peaks corresponding to glucose, fructose, sucrose and fructo-oligosaccharides isomers (**3a**,1-kestose; **3b**, neokestose and $1_F(1\text{-}\beta\text{-D-fructofuranosyl})_m\text{-}6_G(1\text{-}\beta\text{-D-fructofuranosyl})_n$sucrose (**4a**: $m = 2$, $n = 0$; **4b**: $m = 0$, $n = 2$; **4c**: $m = 1$, $n = 1$; **5a**: $m = 3$, $n = 0$; **5b**: $m = 0$, $n = 3$; **5c**: $m = 2$, $n = 1$; **5d**: $m = 1$, $n = 2$; **6a**: $m = 4$, $n = 0$; **6b**: $m = 0$, $n = 4$; **6c**: $m = 3$, $n = 1$; **6d$_1$**: $m = 1$, $n = 3$; **6d$_2$**: $m = 2$, $n = 2$; **7a**: $m = 5$, $n = 0$; **8**: $n+m = 6$, **9x**: $n+m \geqq 7$)) were detected on the chromatogram of SY onion bulbs. The chromatograms of sugar extracts from three other cultivars were qualitatively similar to that of the extract from SY onion bulbs. The decrease in the abundance of each successive oligosaccharide was

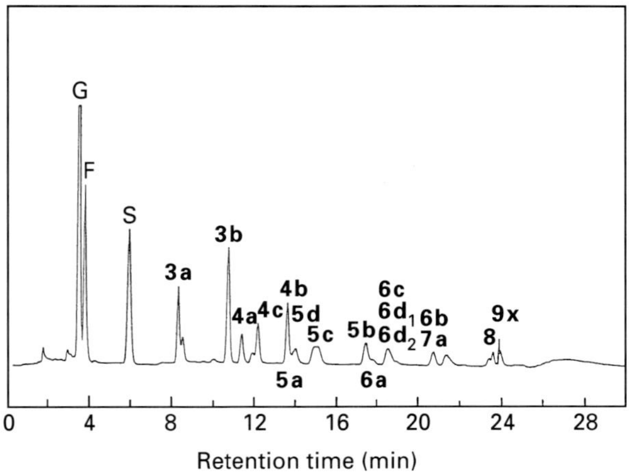

Figure 1. High-performance anion-exchange chromatogram of neutral soluble carbohydrate extracted from onion bulbs in September. G, glucose; F, fructose; S, sucrose; **3a**, 1-kestose; **3b**, neokestose; **4a**, nystose; $1_F(1\text{-}\beta\text{-D-fructofuranosyl})_m\text{-}6_G(1\text{-}\beta\text{-D-fructofuranosyl})_n$sucrose (**4b**: $m = 0$, $n = 2$; **4c**: $m = 1$, $n = 1$; **5a**: $m = 3$, $n = 0$; **5b**: $m = 0$, $n = 3$; **5c**: $m = 2$, $n = 1$; **5d**: $m = 1$, $n = 2$; **6a**: $m = 4$, $n = 0$; **6b**: $m = 0$, $n = 4$; **6c**: $m = 3$, $n = 1$; **6d$_1$**: $m = 1$, $n = 3$; **6d$_2$**: $m = 2$, $n = 2$; **7**: $m+n = 5$; **8**: $m+n = 6$; **9x**: $n+m \geqq 7$). The saccharides were eluted with system I (see text).

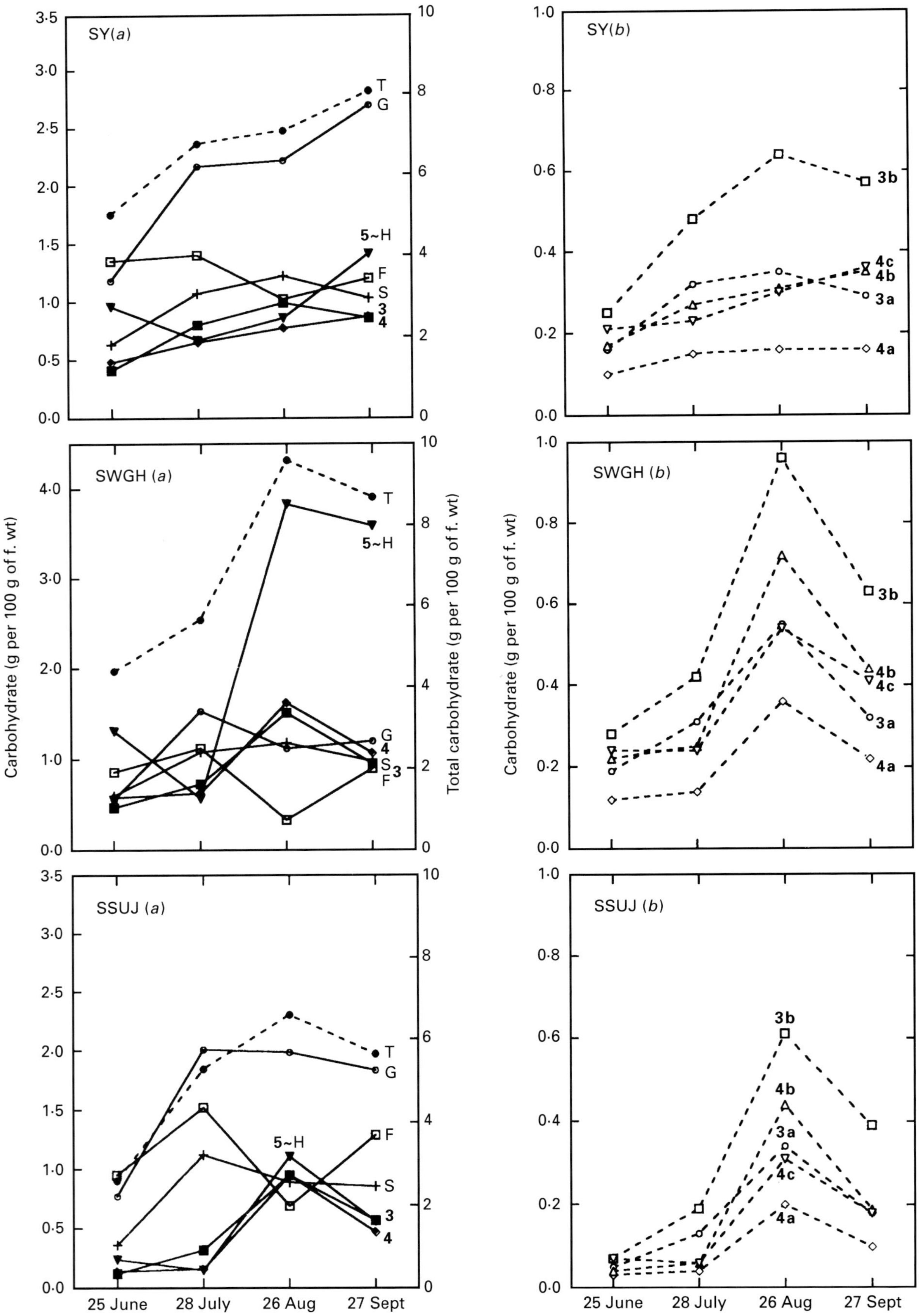

Figure 2. Changes in carbohydrate content of onion bulbs harvested between June and September. (*a*) Saccharides. ●, total sugar (T); ○, glucose (G); □, fructose (F); +, sucrose (S); ■, trisaccharide (DP3); ◆, tetrasaccharide (DP4); ▼, higher oligosaccharides (DP > 4). (*b*) Tri- and tetrasaccharides isomers. ○, 1-kestose (**3a**); □, neokestose (**3b**); ◇, nystose (**4a**); △, **4b**; ▽, **4c**. Cultivars: Sapporo Yellow (SY), Southport White Glove Highsolid (SWGH) and Sweet Spanish Utah Jumbo (SSUJ).

Table 1. *Fructosyltransferase and invertase activities in bulbs of three onion-cultivars harvested between June and September*

	Enzyme activity (nkat g^{-1} f. wt)			
Cultivars	25 June	28 July	26 Aug	27 Sept
1_F-FT				
SY	2·42	2·21	2·16	0·676
SWGH	1·86	2·69	2·59	2·33
SSUJ	1·36	1·48	1·29	0·890
6_G-FT				
SY	5·78	4·65	4·61	1·45
SWGH	4·70	5·50	5·22	4·34
SSUJ	2·82	3·15	2·92	2·07
SST				
SY	0·240	0·252	0·154	0·023
SWGH	0·250	0·256	0·130	0·006
SSUJ	0·151	0·198	0·108	0·007
Invertase				
SY	0·412	0·270	0·100	0·016
SWGH	0·235	0·203	0·043	0·039
SSUJ	0·399	0·447	0·062	0·023

accompanied by increase in the degree of polymerization (DP). The maximum DP (10–15) of fructo-oligosaccharides in onion extracts was shown to vary with cultivar and harvest stage following analysis by HPAEC (elution system II) and PC (solvent system II). The total content of neokestose-series saccharides (**3b**, **4b–c, 5b–d** and **6b–d**) was higher than that of 1-kestose-series saccharides (**3a**, **4a**, **5a**, **6a** and **7a**). The changes in the content of saccharides extracted from bulbs of the three onion cultivars are shown in Figure 2. The total sugar content (f. wt basis) increased in the bulbs of all onion cultivars; SY, SWGH and SSUJ reached 7·08, 9·60 and 6·70% in August, then changed to 8·07, 8·69 and 5·48% in September, respectively.

The increase of glucose and sucrose contents in all of the bulbs was very rapid from June to July, and no level stayed the same or decreased between July and September. From July to September the content of fructose in all of the bulbs was lower than that of glucose.

The contents of trisaccharides increased from June to August then decreased slightly. The tetrasaccharide contents increased from June to September in SY, but declined in September in the other two cultivars. The penta- and higher oligosaccharide levels were low in July, then tended to increase in August.

The changes in the isomers of tri- and tetrasaccharides are shown in Figure 2. All the isomers of tri- and tetrasaccharides except for **4b** in SSUJ increased from June to August and declined in September, but **4b** and **4c** isomers from SY continued to increase until September.

The total content of neokestose derivatives within the tri- and tetrasaccharide isomers (**3b**, **4b** and **4c**) was clearly higher than that of the 1-kestose derivatives (**3a** and **4a**) in all cultivars throughout the growth period.

Fructosyltransferase and invertase activities of onion bulbs

Crude enzyme solutions ('supernatant concentrate A') were used for measuring the activities of SST, 1_F-FT, 6_G-FT and invertase. These enzyme activities are shown in Table 1. The SST activities in the bulbs of the three onion cultivars were highest in July, then decreased. The SST activities in September were the lowest (*c*. 2–9% of the activities in July). The 1_F-FT activities in the bulbs of the three cultivars were highest in June (SY) or July (SWGH and SSUJ) and subsequently decreased very slowly. The 6_G-FT activities in the bulbs of all cultivars changed in parallel to those of the 1_F-FT activities, but were higher (1·9–2·7 times that of 1_F-FT) throughout the growth season. The invertase activities in the bulbs were highest in June (SY and SWGH) or in July (SSUJ), falling rapidly in September.

Enzymatic synthesis of fructo-oligosaccharides from sucrose, 1-kestose or neokestose

The time-course of the *in vitro* synthesis of fructo-oligosaccharides from sucrose, 1-kestose or neokestose was investigated using 'supernatant concentrate B' prepared from onion bulbs (SY) harvested in August. A HPAEC analysis of fructo-oligosaccharides produced from sucrose is shown in Figure 3. 1-Kestose and a trace of neokestose were synthesized from sucrose, liberating glucose after reaction for 15 min. Amounts of 1-kestose and neokestose stabilized after 60 and 120 min respectively. Tetra- (**4a**, **4b** and **4c**) and penta- (**5a**, **5b**, **5c** and **5d**) saccharides were also produced after 60–120 min. Hexa- (**6b**, **6c** and $\mathbf{6d}_{1,2}$) and higher oligosaccharides as well as 1-kestose, neokestose, tetra- and pentasaccharides were formed after 360 min, although the peaks corresponding to saccharides **6b** and **7a** were very small and visible only by amplification (data not shown). The retention times of these oligosaccharides on HPAEC coincided with those of standard saccharides (see 'Materials and Methods'), suggesting that they are identical to the naturally occurring saccharides in onion bulbs.

The synthesis of fructo-oligosaccharides from 1-kestose was measured using 'supernatant concentrate B' diluted tenfold with distilled water. As shown in Figure 4, **4c** and **4a** saccharides were produced from 1-kestose within 5 min, and neokestose was formed within 15 min. After 1200 min reaction, fructo-oligosaccharides including neokestose **4a**, **4b**, **4c**, **5a**, **5b**, **5c**, **5d**, **6a**, **6b**, a mixture of **6c**, $\mathbf{6d}_1$, $\mathbf{6d}_2$ and higher oligosaccharides occurred.

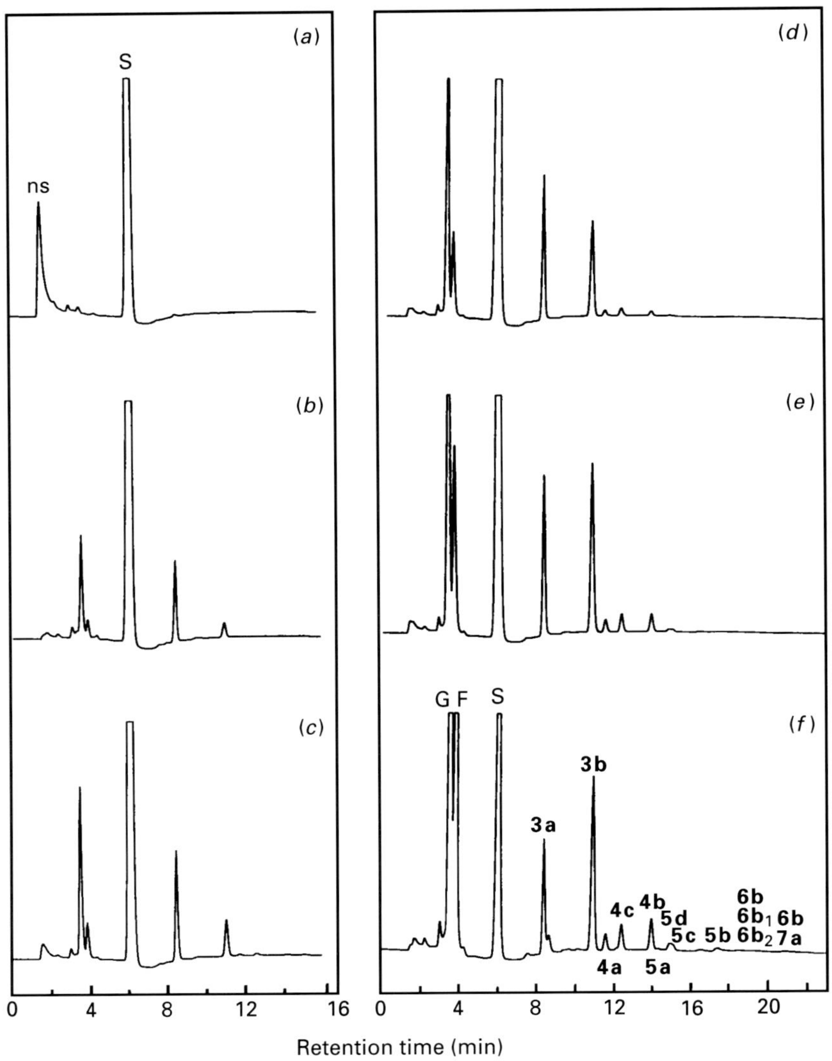

Figure 3. High-performance anion-exchange chromatograms of reaction products from sucrose with 'supernatant concentrate B' from onion bulbs. The enzyme incubation was carried out with 0·1 M sucrose in the mixture for (*a*) 0, (*b*) 15, (*c*) 30, (*d*) 60, (*e*) 120 and (*f*) 360 min (see text for details). The reaction mixture was diluted 100 times with distilled water, and an aliquot (25 μl) was subjected to HPAEC. ns, not saccharide. Markers: G, F, S and **3a–9x** represent the same saccharides as listed in Figure 1. The saccharides were eluted with system I (see text).

The HPAEC traces of the saccharides formed from 1-kestose were very similar to that of the saccharides of onion bulbs (Fig. 1). The formation of fructo-oligosaccharides from neokestrose was also studied, using the same enzyme preparation. As shown in Figure 5, **4c** and **4b** saccharides, together with 1-kestose and sucrose, appeared after 1 h reaction and these saccharides increased in concentration as the reaction time was prolonged. The same saccharides as those synthesized from 1-kestose were detected in the reaction mixture after 720 min. These findings were confirmed by TLC and PC analyses.

DISCUSSION

In the present study, 10 fructo-oligosaccharides corresponding to the saccharides isolated from asparagus roots (Shiomi, Yamada & Izawa, 1976, 1979; Shiomi, 1981) were shown to accumulate in the bulbs of three onion cultivars by HPAEC analysis. These saccharides accumulated in the initial stage of the growth period (June) and increased in August, although the penta- and higher-oligosaccharides tended to decrease in July (Fig. 2). The bulbs (1-cm-diameter) harvested in June were separated into the three parts of the leaf base; outer, middle and inner leaf-bases. The distribution of the saccharides in the leaf-bases of the bulbs in June was also studied. The contents of total sugar and fructo-oligosaccharides were similar among the leaf bases (data not shown). It has been reported that the total fructo-oligosaccharide concentration decreased from the inner leaf-base to the outer leaf-base (Darbyshire & Henry, 1978). Similarly, the oligosaccharide levels were estimated to be lower in the outer leaf-base than in the inner leaf-base of expanding onion bulbs in July. This might explain why the present

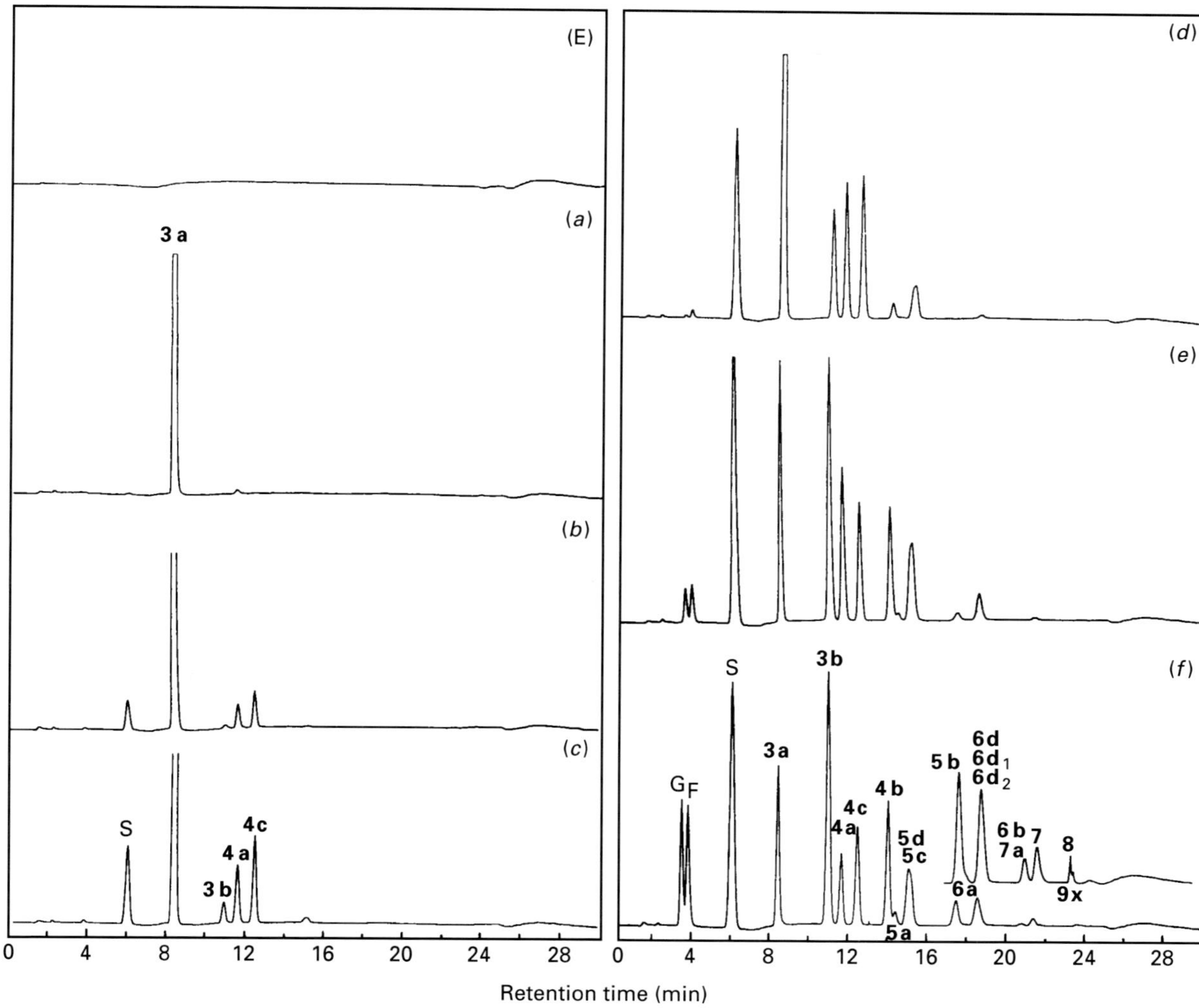

Figure 4. High-performance anion-exchange chromatograms of reaction products from 1-kestose with 'supernatant concentrate B' from onion bulbs. The enzyme incubation was carried out with 0·1 M 1-kestose in the mixture for (*a*) 0, (*b*) 5, (*c*) 15, (*d*) 60, (*e*) 360 and (*f*) 1200 min (see text for details).

'Supernatant concentrate B' was diluted 10 times with distilled water. Sample E contained enzyme preparation, buffer and no substrate. The reaction mixture was diluted 100 times with distilled water, and an aliquot (25 μl) was subjected to HPAEC. ns; not saccharide. Markers: G, F, S and **3a** ~ **9x** represent the same saccharides as in Figure 1. The saccharides were eluted with system I (see text).

fructo-oligosaccharide levels in July were lower than those in June.

SWGH is one of the high solid-type onion cvs. The levels of total sugar and fructo-oligosaccharides in this cv. were higher than those in the other two cvs in August and September. SSUJ, which is a low solid-type of onion cv., had lower levels of total sugar and fructo-oligosaccharides in these months. The ratios of fructo-oligosaccharides to total sugars in SWGH were very high compared with those in SSUJ.

The activities of fructosyltransferase (SST, 1_F-FT and 6_G-FT) involved in the synthesis of fructo-oligosaccharides remained at high levels from June to August, when the onion bulbs expanded rapidly, although the changes in these fructosyltransferase activities did not always parallel those of the fructo-oligosaccharides in the bulbs. The activities of fructosyltransferase in the SWGH bulbs were higher than those in the SSUJ bulbs, similar to the contents of fructo-oligosaccharides in both cultivars.

The SST activity was high in swelling onion bulbs and fell to a very low level in the mature bulbs (Table 1). The levels and changes of these activities (as nkat g^{-1} f. wt) were similar to those in asparagus roots (Shiomi, 1992).

The 1_F-FT activities varied less between June and September (Table 1), a tendency also observed in the activity extracted from asparagus roots (Shiomi, 1992).

The 6_G-FT activities were higher (1·86–2·50 times) than the 1_F-FT activities in onion bulbs. The variation in 6_G-FT activity paralleled that in 1_F-FT activity in the bulbs (Table 1). The higher activity of 6_G-FT throughout the onion growth-period led to a greater accumulation of neokestose and its related saccharides than of 1-kestose and its related saccharides.

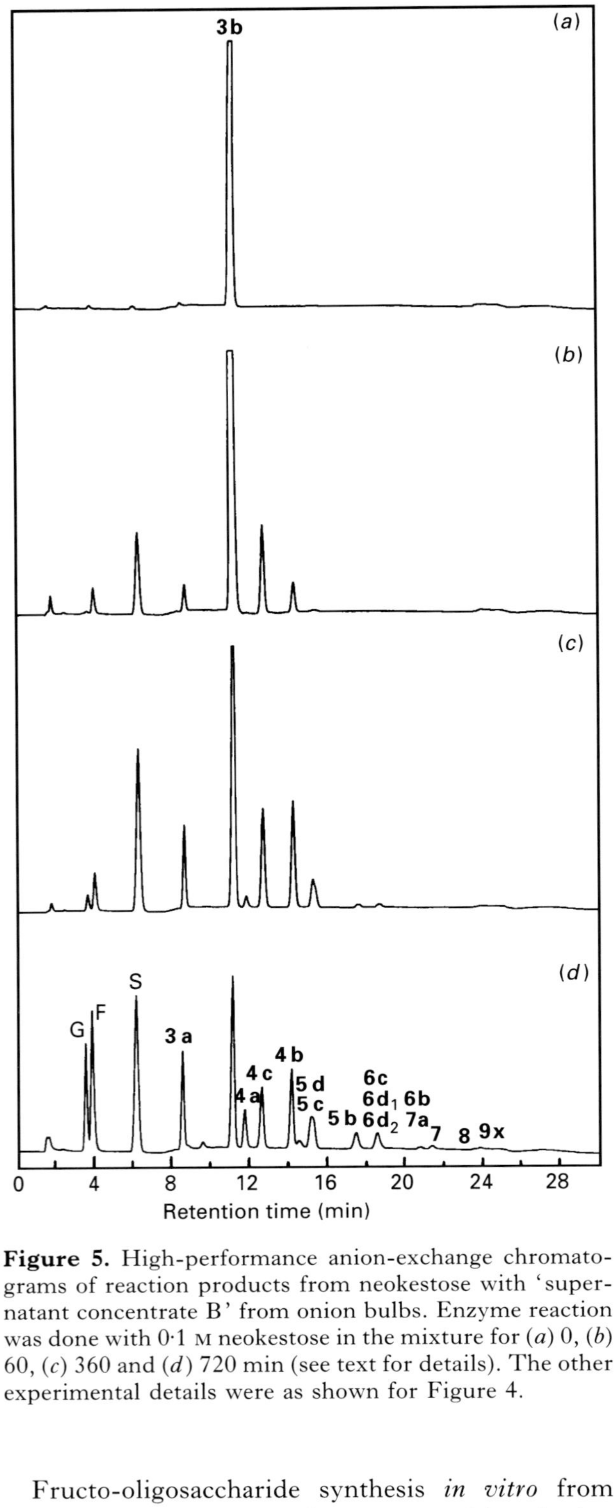

Figure 5. High-performance anion-exchange chromatograms of reaction products from neokestose with 'supernatant concentrate B' from onion bulbs. Enzyme reaction was done with 0·1 M neokestose in the mixture for (*a*) 0, (*b*) 60, (*c*) 360 and (*d*) 720 min (see text for details). The other experimental details were as shown for Figure 4.

Fructo-oligosaccharide synthesis *in vitro* from sucrose, 1-kestose and neokestose was achieved using 'supernatant concentrate B' prepared from onion bulbs harvested in August. Fructo-oligosaccharides corresponding to 1-kestose, neokestose, nystose, **4b**, **4c**, **5a**, **5b**, **5c**, **5d**, **6a**, a mixture of **6c**, $\mathbf{6d_1}$ and $\mathbf{6d_2}$, a mixture of **6b** and **7a**, and higher saccharides were formed from sucrose or 1-kestose (Figs 3, 4) and corresponded closely to the saccharides in onion bulbs (Fig. 1). The retention times of the saccharides synthesized from sucrose or 1-kestose by the onion enzyme preparation were very similar to those synthesized from sucrose or 1-kestose by an asparagus enzyme preparation (Shiomi, 1992). Neokestose was a good substrate for the synthesis of fructo-oligosaccharides by 'supernatant concentrate B'. The activity ratio of 'supernatant concentrate B' for neokestose to 1-kestose as a substrate was 0·21. The saccharides formed from neokestose using 'supernatant concentrate B' were identical to those formed from 1-kestose or sucrose. These findings concerning the action of onion crude enzyme on neokestose will contribute to the clarification of the mechanism of the enzymatic synthesis of fructo-oligosaccharide isomers in onion bulbs.

Based on these results, we propose that the fructo-oligosaccharides in onion bulbs are synthesized *via* 1-kestose from sucrose by the activities of SST, 1_F-FT and 6_G-FT as follows.

SST activity

sucrose + sucrose → 1-kestose + glucose.

1_F-FT activity

1_F(1-β-D-fructofuranosyl)$_m$sucrose
+ 1_F(1-β-D-fructofuranosyl)$_n$sucrose ⇌
1_F(1-β-D-fructofuranosyl)$_{m-1}$sucrose
+ 1_F(1-β-D-fructofuranosyl)$_{n+1}$sucrose.

6_G-FT activity

1_F(1-β-D-fructofuranosyl)$_m$sucrose
+ 1_F(1-β-D-fructofuranosyl)$_n$sucrose ⇌
1_F(1-β-D-fructofuranosyl)$_{m-1}$sucrose
+ 1_F(1-β-D-fructofuranosyl)$_n$
− 6_G-β-D-fructofuranosylsucrose.

4 × neokestose ⇌
1_F,6_G-di-β-D-fructofuranosylsucrose
+ 6_G(1-β-D-fructofuranosyl)$_2$sucrose
+ 2 × sucrose.

The activities of 1_F-FT and 6_G-FT preserved in onion bulbs cannot yet be assigned to a single enzyme.

ACKNOWLEDGEMENTS

We are grateful to Dr Y. Sato (Hokkaido National Agricultural Experiment Station, Sapporo, Japan) for providing onion bulbs.

REFERENCES

Bacon JSD. 1957. The water soluble carbohydrates of onion. *Allium cepa* L. *The Biochemical Journal*: 5p–6p.

Bacon JSD. 1959. The trisaccharide fraction of some monocotyledons. *The Biochemical Journal* **73**: 507–514.

Bose S, Shrivastava AN. 1961. Soluble carbohydrates from onion. (*Allium cepa* L.). *Science and Culture* **27**: 253.

Darbyshire B, Henry RJ. 1978. The distribution of fructans in onions. *New Phytologist* **81**: 29–34.

Johnson DC. 1986. Carbohydrate detection gain potential. *Nature* **321**: 451–452.

Henry RJ. Darbyshire B. 1980. Sucrose:sucrose fructosyltransferase and fructan:fructan fructosyltransferase from *Allium cepa*. *Phytochemistry* **19**: 1017–1020.

Morris DL. 1948. Quantitative determination of carbohydrates with Dreywood's anthrone reagent. *Science* **10**: 254–255.

Mukherjee S, Srivastava HC. 1952. Improved spray reagent for the detection of sugar. *Nature* **169**: 330.

Nelson N. 1944. A photometric adaptation of the Somogyi method for the determination of glucose. *The Journal of Biological Chemistry* **153**: 375–380.

Rocklin RD, Pohl CA. 1983. Determination of carbohydrates by anion exchange chromatography with pulsed amperometric detection. *Journal of Liquid Chromatography* **6**: 1577–1590.

Shiomi N. 1981. Two novel hexasaccharides from the roots of *Asparagus officinalis*. *Phytochemistry* **20**: 2581–2583.

Shiomi N. 1989. Properties of fructosyltransferase involved in the synthesis of fructan in Liliaceous plants. *Journal of Plant Physiology* **134**: 151–155.

Shiomi N. 1992. Content of carbohydrate and activity of fructosyl-transferase and invertase in asparagus roots during fructo-oligosaccharide- and fructo-polysaccharide accumulating season. *New Phytologist* **122**: 421–432.

Shiomi N, Onodera S, Chatterton NJ, Harrison PA. 1991. Separation of fructooligosaccharide isomers by anion exchange chromatography. *Agricultural and Biological Chemistry* **55**: 1427–1428.

Shiomi N, Yamada J, Azawa M. 1976. Isolation and identification of fructo-oligosaccharides in roots of *Asparagus* (*Asparagus officinalis* L.). *Agricultural and Biological Chemistry* **40**, 567–575.

Shiomi N, Yamada J, Izawa M. 1979. A novel pentasaccharide in the roots of *Asparagus* (*Asparagus officinalis* L.). *Agricultural and Biological Chemistry* **43**, 1375–1377.

Somogyi M. 1945. Determination of blood sugar. *The Journal of Biological Chemistry* **160**: 69–73.

Takeda H, Sato K, Kinoshita S, Sasaki H. 1994. Production of 1-kestose by *Scopulariopsis brevicaulis*. *Journal of Fermentation and Bioengineering* **79**, 242–246.

New Phytol. (1997), **136**, 115–122

Fructan storage in tubers of Jerusalem artichoke: characterization of sink strength

By SVEN SCHUBERT* and ROBERT FEUERLE

Institute of Plant Nutrition (330), University of Hohenheim, Fruwirthstr. 20, D-70599 Stuttgart, Germany

(*Received 24 July 1996*)

SUMMARY

Possible limitations of sink strength for fructan storage in tubers of Jerusalem artichoke (*Helianthus tuberosus* L. cv. Waldspindel) were studied in 2-yr field experiments. Treatments included the removing of side branches and the breaking of apical dominance in unbranched plants. Data from both treatments and the comparison of results obtained for 2 yr, which differed in temperature and total irradiance, indicated that fructan storage in tubers during the storage phase was not limited by supply of assimilate (sucrose). Moreover, fructan concentrations were very constant despite large variations in tuber yields and yield components (e.g. number of tubers per plant, tuber length). However, stimulation of cell expansion and diameter of tubers increased fructan concentrations, probably owing to a larger vacuolar volume in the storage cells. We concluded that fructan storage in tubers of Jerusalem artichoke is not source-limited as far as carbon import into tubers is concerned. However, a strong shoot control, most probably regulated via phytohormones, limits sink strength. Fructan concentrations in tubers might not only be increased by sink activity (fructan-synthesizing enzymes) but also by sink capacity (vacuolar volume). In this respect the shoot might play a decisive role not only as a temporary storage pool for carbohydrates but also as a source of native growth-regulators which control sink capacity.

Key words: Cell expansion, fructan, source–sink relations, tuber growth, *Helianthus tuberosus* L. (Jerusalem artichoke).

INTRODUCTION

Source–sink relations determine economically relevant crop yield. Different developmental stages of a crop, genetic (particularly species) differences, and environmental factors interact in a dynamic way to modulate the prevailing influence of either source- or sink-limiting factors on crop yield. The source–sink relations in wheat have been well studied. In this crop, the vegetative phase and flowering stage are source-limited (Waters, Martin & Lee, 1984; Stoy & Bertholdsson, 1986), whereas grain filling is sink-limited for carbon (Michael & Beringer, 1980; Mengel, Friedrich & Judel, 1985; Jenner, Ugalde & Aspinall, 1991).

The conclusion that growth of the storage organ is sink-limited does not apply to all crops. In potato, tuber yield is apparently limited by photosynthesis (Engels & Marschner, 1987). However, source limitation might not be generalized for all vegetative storage organs. An example is Jerusalem artichoke. Fructan (inulin) storage in tubers of this crop appears to be sink-limited (Meijer & Mathijssen, 1991; Denoroy, 1993; Schubert & Feuerle, 1993). A major reason that photosynthate supply does not limit fructan storage in tubers of this crop is that storage of fructan in shoots is reversible. Shoots of Jerusalem artichoke can act as additional sinks during periods of abundant photosynthesis and can supply carbohydrates during phases when photosynthesis alone cannot meet the assimilate demand of the 'irreversible' storage organs (Incoll & Neales, 1970; Meijer & Mathijssen, 1991).

To characterize the productivity of Jerusalem artichoke, we applied the concept of sink strength (Ho, 1988), the ability of a sink organ to import assimilate. Potential sink strength can differ from actual sink strength in that carbohydrates from the shoot are not supplied in sufficient quantities to meet the demand of the sink organ. Assuming that sucrose concentrations are well controlled in the storage organ (Koops & Jonker, 1994; Geigenberger *et al.*, 1995) and that sucrose is the major long-distance transport metabolite (Edelman & Popov, 1962; Soja, Haunold & Praznik, 1989), sucrose concentrations in tubers will provide reliable information as to whether at any time the storage process is source- or sink-limited. After manipulation of sink strength or

* To whom correspondence should be addressed.

source capacity, a decrease in sucrose concentrations in the sink organ will indicate source limitation, whereas no change is expected when sink limitation predominates.

Apparent sink strength can easily be quantified as the rate of d. wt increase of tubers. It differs from actual sink strength because it does not account for respired carbohydrates (Ho, 1988). Apparent sink strength is the product of sink capacity and sink activity. For fructan storage, sink capacity is ultimately the vacuolar volume of tubers, because it is well established that fructans are synthesized and stored in vacuoles (Keller, Frehner & Wiemken, 1984; Pollock, 1986; Darwen & John, 1989). The vacuolar volume is determined by cell number and particularly by cell size. These parameters are intimately related to the number of tubers, individual tuber length, and individual tuber diameter.

Regarding tuber growth as a dynamic process, cell division and subsequent cell expansion are processes that characterize sink activity and create the sink capacity of fully expanded cells. Further determinants of sink activity in tubers of Jerusalem artichoke are enzyme activities of SST (sucrose:sucrose fructosyl transferase) and FFT (fructan:fructan fructosyl transferase), which provide the driving force for fructan storage in vacuoles (Martinoia, 1992). Therefore, we quantified sink activity as the rate of fructan deposition in the tubers.

Using these parameters, we studied the effect of removing side branches of shoots, breaking of apical dominance, as well as differences in temperature and in total irradiance between two years, on source–sink relations in unbranched Jerusalem artichoke plants under field conditions. The rationale of these experiments was to test whether, during the storage phase, fructan accumulation in tubers of Jerusalem artichoke is source- or sink-limited.

MATERIALS AND METHODS

Field experiments

Two field experiments using Jerusalem artichoke (*Helianthus tuberosus* L. cv. Waldspindel) were made at Hohenheim, in 1993 and 1994. The soil of the experimental site is an orthic luvisol derived from loess. In order to guarantee ample nutrient supply, the soil was analysed for N_{min} (39 and 25 kg N ha^{-1} in 1993 and 1994, respectively), P (157 mg kg^{-1}), K (315 mg kg^{-1}), Mg (160 mg kg^{-1}), and pH (6·3) and was subsequently fertilized with 60 kg N ha^{-1} as Nitrochalk (74 % NH_4NO_3 + 36 % $CaCO_3$) in spring 1993. Further N and other nutrients were applied in the form of a multi-element fertilizer (Blaukorn®: N/P/K/Mg = 12/6/14/1 + B + Zn) at a rate of 50 kg N ha^{-1} (1993) and 107 kg N ha^{-1} (1994). Mutual shading of plants was minimized by extremely low planting density (one plant m^{-2}). Therefore, nutrient supply far exceeded nutrient demand by the plants.

To avoid drought stress, the water balance of the soil was estimated by taking into account data provided by a meteorological station on water consumption of a sugarbeet stand nearby, and current precipitation. Random soil samples confirmed the efficacy of this procedure. Water deficits were compensated by drip irrigation twice in 1993 (19 and 21 l m^{-2} 62 and 133 d after planting, respectively) and once in 1994 (29 l m^{-2} 96 d after planting).

In both years, tubers of Jerusalem artichoke were planted on 6 April. Because all growth factors except temperature and total irradiance were optimized, differences in the growth and developmental pattern of plants between the years might be explained primarily in terms of one or both these factors. During the vegetative period in 1993 (6 April–5 November), average temperature was 12·5 °C with a standard deviation of 5·8. Minimum and maximum temperatures were 8·4 and 17·0 °C, respectively. Average temperature (14·7 °C with a standard deviation of 5·1) and temperature extremes (minimum: 9·6 °C, maximum: 19·1 °C) were higher in 1994. Monthly total irradiance paralleled temperature data. Therefore, no distinction can be made between these two factors on yield components. In the following, differences in temperature referred to always include differences in total irradiance.

In both years, plants were grown without side branches, which were removed as soon as they appeared. Data for these plants (control plants) were used for comparison between years. In 1993, half the plants were decapitated to terminate shoot growth 95 d after planting (14 d before the first harvest). This treatment broke apical dominance without reducing the photosynthetically active leaf area. In 1994, side branches were not removed from half the plants. Treated plants were randomized in the plot, and guard rows that surrounded the plot avoided position effects.

Harvests were taken at regular intervals after tuber initiation, which was defined as the beginning of a macroscopically visible swelling of the stolon apex. For each harvest data, tubers of five plants per treatment were dug out of the soil, washed with tap water and separated from stolons for sucrose and fructan analyses. Subsequently, the number of tubers and individual tuber weight were determined. Individual tuber diameter and individual tuber length of the medium-sized tubers were measured after removal of the largest and smallest 25 %.

Sucrose and fructan analyses

For the extraction of glucose, fructose, sucrose and fructan (inulin), a representative aliquot of medium-

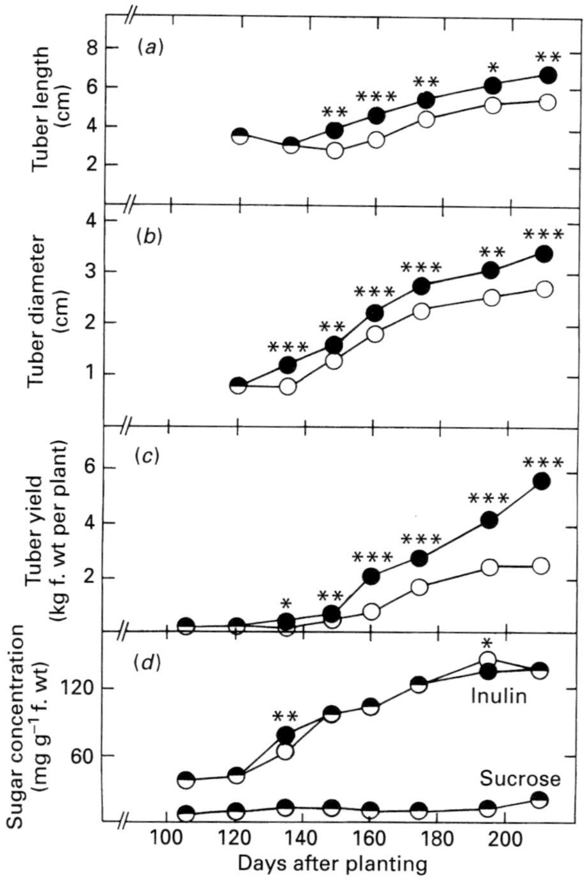

Figure 1. Effect of side branches on tubers of Jerusalem artichoke. (*a*) Individual tuber length. (*b*) Individual tuber diameter. (*c*) Tuber yield per plant. (*d*) Sugar concentrations. Branched plants (●); unbranched plants from which side branches were removed as soon as they appeared (○). Significant differences between treatments at the *$P = 5\%$, **$P = 1\%$, or ***$P = 0{\cdot}1\%$ level, respectively.

sized tubers of each plant was cut into small pieces and homogenized in boiling water to inactivate hydrolases. We also investigated sugar concentrations in the smallest and largest 25 % of tubers but could not find a significant difference between these and the medium-sized tubers. Extraction was completed by incubation of the homogenate at 60 °C for 30 min, during which samples were gently shaken. In the filtrate fructans were hydrolysed, with some modifications, according to Klaushofer & Wolfslehner (1983). Sugar extract (100 μl) was pipetted into a disposable 4 ml cuvette together with 200 μl of 100 mM NaOH, 30 μl of 0·5 M H_2SO_4, 200 μl of HCl/citrate buffer pH 4 (Merck), 1370 μl redistilled water and 100 μl of 1:100 diluted inulase solution (Novozyme 230 SP, 3000 U g^{-1}, Novo-Industri, Copenhagen). Cuvettes were sealed with Parafilm® and incubated at 25 °C for 3 h. After adjustment of pH to 7·6, fructose and glucose were measured enzymatically by means of test kits (Boehringer Mannheim, Germany). Total fructans were calculated as the difference in fructose and glucose before and after hydrolysis. Sucrose was also analysed enzymatically by means of test kits (Boehringer Mannheim, Germany) which use β-fructosidase to hydrolyse sucrose; it is possible therefore that some of the glucose moieties originating from fructans are also detected in this test, overestimating sucrose concentrations in the samples. However, analyses using ion chromatography and the low sucrose concentrations found in the tuber tissues indicate that this error was negligible.

Determination of the cell number in various tissues

Cell counts were carried out with eight replicates per tissue for two tubers per plant. Tissues were sampled in the region of largest diameter, fixed in 5 % (v/v) formaldehyde, 5 % (v/v) propionic acid, and 90 % (v/v) ethanol, and prepared for cutting of 4–8 μm slices by microtome as described by Hermanns & Schulz (1981). Cell walls were stained with haematoxylene dye. The microscopic picture was recorded by a video-camera (Sony CCD) and computerized. The number of cells in cortex, extension zone, storage tissue and pith was counted at a magnification of ×160 on a 1·25 mm × 1·69 mm section.

Statistical treatment

Significant effects of treatments were calculated using Student's *t*-test. Standard error (SE) was calculated for the comparison between years and is indicated in the figures as error bars. Where the value of SE does not exceed the size of symbols used in the diagrams it is not shown.

RESULTS

Continual removal of side branches as they appeared throughout the experiment reduced the total leaf area of Jerusalem artichoke plants by 78 % at the end of shoot growth (not shown). Despite this drastic treatment, there was no significant effect on the number of tubers per plant (not shown), indicating that the stimulus for tuber initiation was not impaired by the removal of side branches. In contrast, length (Fig. 1*a*) and diameter (Fig. 1*b*) of individual tubers were significantly decreased. Whereas tuber length was clearly affected during fast tuber growth only (150 d after planting), the effect on individual tuber diameter was clear 1 wk earlier. However, the transition from slow tuber growth to fast tuber growth (*c*. 140 d after planting) was not influenced by the removal of side branches (Fig. 1*c*). Instead, the difference in tuber weight at the end of the experiment was mainly due to decreased tuber growth rates between 150 and 160 as well as 190 and 210 d after planting (not shown).

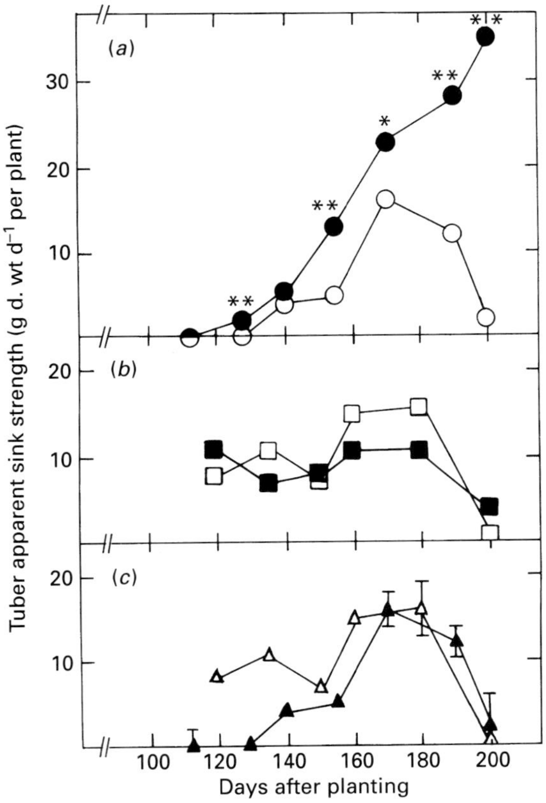

Figure 2. Effect of side branches (*a*), apical dominance (*b*), and different years (*c*) on the apparent sink strength (d. wt accumulation) of tubers of Jerusalem artichoke. ○, Unbranched plants; ●, branched plants; □, apical dominance; ■, no apical dominance, △, 1993; ▲, 1994. Significant differences between treatments at $*P = 5\%$, $**P = 1\%$ level, respectively. Vertical bars, where they exceed the symbols, indicate the SE.

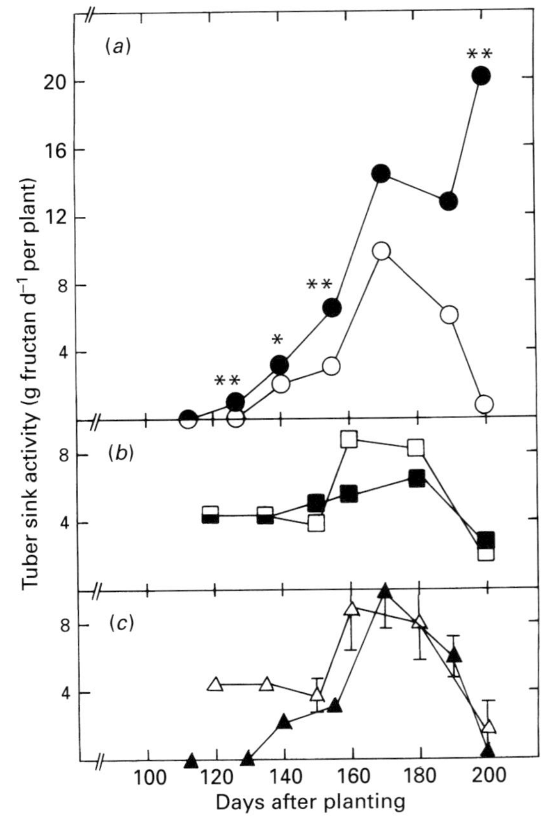

Figure 3. Effect of side branches (*a*), apical dominance (*b*), and different years (*c*) on the sink activity (fructan accumulation) of tubers of Jerusalem artichoke. ○, Unbranched plants; ●, branched plants; □, apical dominance; ■, no apical dominance; △, 1993; ▲, 1994. Significant differences between treatments at $*P = 5\%$, $**P = 1\%$ level, respectively. Vertical bars, where they exceed the symbols, indicate the SE.

The decrease in tuber yield by more than 50% in unbranched plants (Fig. 1*c*) was apparently not caused by limitation of assimilate supply to tubers because sucrose concentration was not reduced (Fig. 1*d*). Therefore, it is not surprising that fructan concentration in tubers did not differ consistently between the two treatments (Fig. 1*d*). Apparent sink strength during the whole vegetation period was significantly decreased by removing side branches (Fig. 2*a*). The effect was particularly spectacular 180 d after planting. During this time, apparent sink strength of branched plants further increased whilst a decline was observed in tubers of unbranched plants. An identical result was obtained for sink activity (Fig. 3*a*).

Breaking the apical dominance of unbranched plants at the end of shoot growth increased the number of tubers per plant, indicating a control of the shoot apex on tuber initiation (Fig. 4*a*). At the same time, individual tuber diameter was decreased to such an extent (Fig. 4*b*) that the net result was an unchanged tuber yield (not shown). Although sucrose and fructan concentrations remained unchanged during most phases of the experiment, there was a sharp decrease of fructan concentration at the second harvest as a result of removing the shoot apex (Fig. 4*c*). Because this phenomenon coincided with a drop in radial extension growth of tubers, we quantified the effect of apical dominance on cell expansion in the various tuber tissues. The increase in cell number per area in the extension zone, in contrast with other tissues, indicates inhibited cell expansion in tubers of plants without apical dominance at the first harvest (Table 1). There was no significant effect of apical dominance on apparent sink strength or sink activity (Figs 2*b*, 3*b*).

A comparison was made between tuber yields and yield components of the 1993 and 1994 field experiments. The temperature sums of the years differed (Fig. 5). Whereas the first 3 wk in 1993 were warmer, temperatures in the succeeding period did not differ. Higher temperatures were recorded during the period between 100 and 140 d after planting in 1994, making the difference in temperature sums disappear. However, lower temperatures were again recorded in 1994 140 d after planting.

Higher temperatures during early spring of 1994 almost doubled tuber yields at the end of the experiment (Fig. 6*d*). This result was caused by both

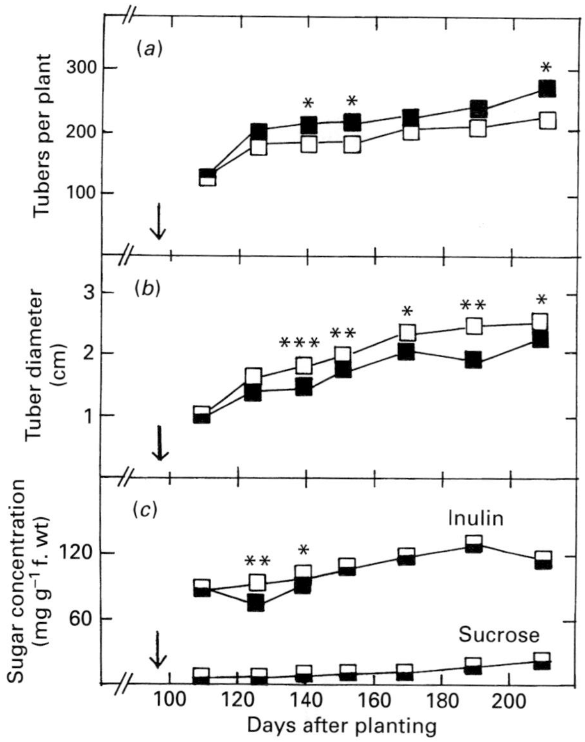

Figure 4. Effect of apical dominance on tubers of Jerusalem artichoke. (*a*) The number of tubers per plant. (*b*) Individual tuber diameter. (*c*) Sugar concentrations. Side branches of control plants with apex (□, apical dominance) and of plants without apex (■, no apical dominance) were removed as soon as they appeared. Shoot apices were removed as indicated by arrows. Significant differences between treatments at $*P = 5\%$, $**P = 1\%$, or $***P = 0{\cdot}1\%$ level, respectively.

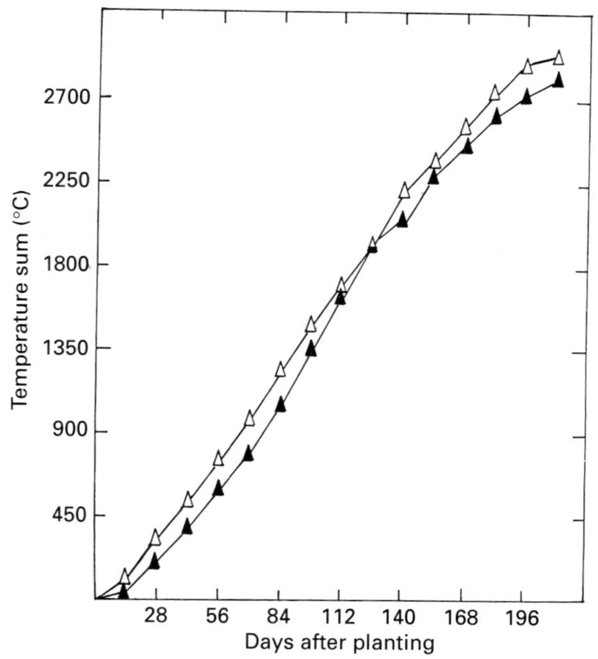

Figure 5. Temperature sums during two different years of field experiments with Jerusalem artichoke. 1993 (△), 1994 (▲). Vertical bars of SE do not exceed the symbols.

the number of tubers per plant (Fig. 6*a*) and by the individual tuber length (Fig. 1*b*). Higher temperatures in early 1994 made the phase of slow tuber growth disappear due to stimulation of individual tuber length (Fig. 6*b*) and diameter (Fig. 6*c*). Lower temperatures in 1993 during the storage phase (160 d after planting) apparently had no inhibiting effect on tuber diameter (Fig. 6*c*).

Despite large differences in tuber yield per plant between the two years there were no differences in sucrose concentrations between years. An increase of sucrose concentrations was observed at the end of both experiments (Fig. 7). Large differences in fructan concentrations between years were recorded, with higher values for 1994 100–140 d after planting and for 1993 at 180 d after planting. Interestingly, fructan concentrations as a function of time very much resembled the development of individual tuber diameter during the two years (compare Figs 6*c* and 7). Apparent sink strength and sink activity were positively influenced by high temperature 110–160 d after planting, but no effect was found during later stages (Figs 2*c*, 3*c*).

DISCUSSION

Source or sink limitation?

Tuber yields of Jerusalem artichoke grown at suboptimal plant density in the field were drastically reduced when side branches were removed (Fig. 1*c*) or when plants were grown at lower temperature and/or lower total irradiance (Fig. 6*d*). Removal of side branches decreased leaf area by 78%, but diminished tuber yield by only 50%. It is unlikely that this discrepancy is due to mutual shading of side

Table 1. *Effect of apical dominance on the number of cells per unit area (1·25 mm × 1·69 mm) in various tissues of tubers of Jerusalem artichoke at the first harvest*

Treatment	Cortex	Extension zone	Storage tissue	Pith
Apical dominance	60·4 ± 3·4	30·7 ± 1·1	17·9 ± 0·6	31·5 ± 1·0
No apical dominance	56·8 ± 3·6	34·3 ± 0·6*	18·8 ± 0·6	28·3 ± 1·6

Plants were treated as described in Figure 1. Significant difference between treatments at $*P = 1\%$ level.

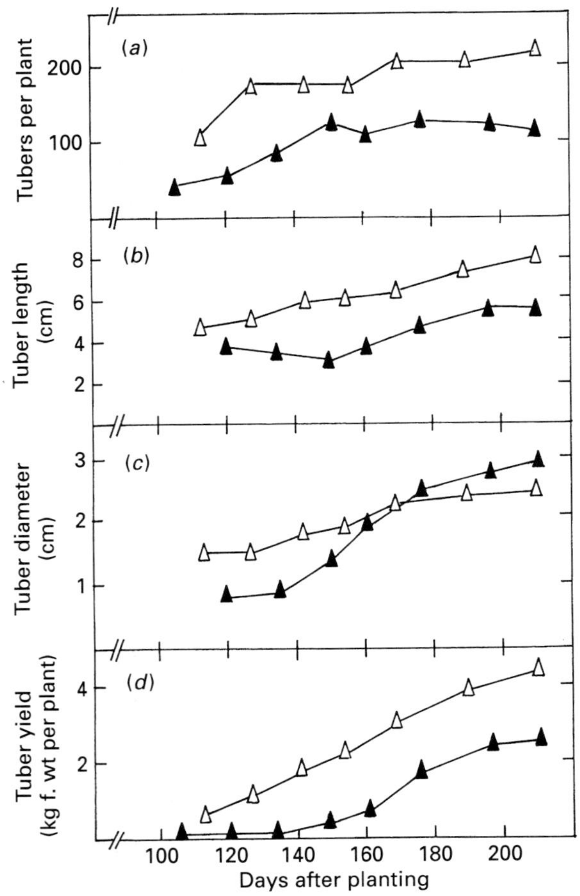

Figure 6. Yield components and tuber yields of Jerusalem artichoke in two different years of field experiments. (*a*) Number of tubers per plant. (*b*) Individual tuber length. (*c*) Individual tuber diameter. (*d*) Tuber yield per plant. 1993 (△), 1994 (▲). Vertical bars of SE do not exceed the symbols.

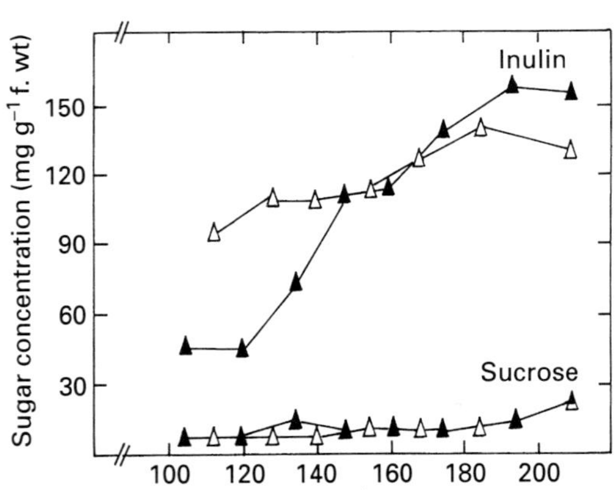

Figure 7. Sugar concentrations in tubers of Jerusalem artichoke in two different years of field experiments. 1993 (△), 1994 (▲). Vertical bars of SE do not exceed the symbols.

branches, and consequent decrease of total light interception by the canopy, because optimal plant density is assumed to be four plants m^{-2} or even higher (Meijer & Mathijssen, 1991; Pejin *et al.*, 1993), at least four times higher than that in our experiment, yet the discrepancy between leaf area reduction and tuber yield amounted to less than a factor of two. Furthermore, it has been shown that shading of plants might be compensated by a longer duration of photosynthesis accompanied by delayed senescence of individual leaves (Schubert & Feuerle, 1993).

Differences between potential and apparent sink strength might occur when the amounts of assimilates delivered by current photosynthesis and/or remobilization of carbohydrates from temporary storage organs do not meet the demand of the sink organs. Sucrose is the major long-distance transport metabolite in Jerusalem artichoke (Edelman & Popov, 1962; Soja *et al.*, 1989); its concentration in tubers was not affected by the presence of side branches, by apical dominance (Figs 1*d*, 4*c*) nor by temperature effects over two different years (Fig. 7). Differences in tuber yields can therefore not be attributed to variable availability of assimilate for growth or storage processes. A third argument against source limitation is the drastic decrease of apparent sink strength in unbranched plants in contrast to branched plants. The latter even showed increased apparent sink strength towards the end of the vegetation period (Fig. 2*a*). If source limitation were responsible for these large differences, differences in sugar, particularly in sucrose concentration, should be expected. However, this was not the case (Fig. 1*d*).

Consequently, fructan storage in tubers of Jerusalem artichoke was sink-limited, in agreement with earlier results (Meijer & Mathijssen, 1991; Denoroy, 1993; Schubert & Feuerle, 1993). Sink limitation was particularly the case at the beginning of the storage phase when apparent sink strength was very low (Fig. 2*a*). Jerusalem artichoke might differ from potato in this respect (Engels & Marschner, 1987), because it disposes of a large temporary storage pool of carbohydrates in the shoot (Incoll & Neales, 1970; Meijer & Mathijssen, 1991). An increase in sucrose concentrations in tubers at the end of the vegetative period (Figs 1*d*, 4*c*, 7) suggests that the fine regulation of sucrose concentrations by the concerted activities of sucrose phosphate synthase and sucrose synthase, as demonstrated for potato tubers (Geigenberger *et al.*, 1995), might be disturbed during this phase.

Sink strength

Apparent sink strength and sink activity were significantly influenced by the removal of side branches (Figs 2*a*, 3*a*) and by temperature (and/or total irradiation; Figs 2*c*, 3*c*). Breaking of apical dominance transiently decreased fructan concentration in tubers (Fig. 4*c*) without affecting sink activity (Fig. 3*b*), so we conclude that this treatment decreased sink capacity. As discussed above, in all three cases, limitation by assimilate supply by the shoot can be safely excluded. Therefore, the question arises: which internal factors are responsible for the variation in sink capacity and sink activity?

Decapitating unbranched plants eliminated an important source of auxin for tubers (Sachs, 1993; Cline, 1994). For potato tubers of similar physiological age, a correlation has been found between tuber growth rate and auxin concentration (Marschner, Sattelmacher & Bangerth, 1984). Auxins play an essential role in the control of cell expansion (Kutschera & Schopfer, 1986; Fry, 1989). Because cell expansion is required for the establishment of a large vacuolar volume, the transient effect of the breaking of apical dominance might be explained by decreased cell expansion in the extension zone (Table 1) that reduced the total vacuolar volume and thus diminished fructan concentrations (Fig. 4*d*). The transient nature of this phenomenon possibly implies that new auxin sources (e.g. root tips) might be able to compensate for the lack of auxins from the shoot.

Removing of the side branches might involve the deletion of auxin and gibberellin sources, because not only individual tuber diameter (Fig. 1*b*) but also individual tuber length decreased (Fig. 1*a*). In potato, gibberellins increase stolon growth (Railton & Wareing, 1973; Mares, Marschner & Krauss, 1981; Vreugdenhill & Struik, 1989); therefore it is possible that the increased longitudinal growth of Jerusalem artichoke tubers was triggered by gibberellins (Fig. 1*a*).

The positive influence of high temperature on tuberization (Fig. 6*a*–*c*) agrees with earlier findings that tuberization depends strongly on cumulative temperature (Spitters, Lootsma & van de Waart, 1989). The parallel between tuber diameter (Fig. 6*c*) and fructan concentrations in tubers (Fig. 7) points to the importance of radial extension growth in providing a large vacuolar storage volume for fructans that are preferentially stored in peripheral tuber tissues (Mazza, 1985).

CONCLUSIONS

The results have shown that fructan storage in tubers of Jerusalem artichoke during the storage phase is not limited by assimilate supply from the shoot, but rather that sink capacity controls fructan storage in vacuoles. Internal and external factors influence sink capacity considerably, probably via native growth regulators (Kosegarten & Mengel, 1995). It may thus be concluded that the shoot of Jerusalem artichoke not only plays an important role as a temporary storage pool for carbohydrates but also as a decisive source of native growth regulators that determine sink capacity via cell division and, particularly, cell expansion. This hypothesis deserves further experimental treatment.

ACKNOWLEDGEMENTS

The authors are grateful to Professor Dr R. Stoesser and Dr S. F. Anvari for the opportunity to carry out cell counts by microscopic imaging techniques.

REFERENCES

Cline MC. 1994. The role of hormones in apical dominance. New approaches to an old problem in plant development. *Physiologia Plantarum* **90**: 230–237.

Darwen CWE, John P. 1989. Localization of the enzymes of fructan metabolism in vacuoles isolated by a mechanical method from tubers of Jerusalem artichoke (*Helianthus tuberosus* L.). *Plant Physiology* **89**: 658–663.

Denoroy P. 1993. Jerusalem artichoke productivity modelling. In: Fuchs A, ed. *Inulin and Inulin-Containing Crops*. Amsterdam: Elsevier Science Publishers B.V., 45–50.

Edelman J, Popov K. 1962. Metabolism of $^{14}CO_2$ by illuminated shoot of *Helianthus tuberosus* L. with special reference to fructose polymers. *Compte Rendu de l'Académie bulgare des Sciences* **15**: 627–630.

Engels C, Marschner, H. 1987. Effects of reducing leaf area and tuber number on the growth rates of tubers on individual potato plants. *Potato Research* **30**: 117–186.

Fry SC. 1989. Cellulases, hemicellulases and auxin-stimulated growth: a possible relationship. *Physiologia Plantarum* **75**: 532–536.

Geigenberger P, Krause K-P, Hill LM, Reimholz R, MacRae E, Quick P, Sonnewald U, Stitt M. 1995. The regulation of sucrose synthesis in leaves and tubers of potato plants. In: Pontis HG, Salerno GL, Echeverria EJ, eds. *Sucrose Metabolism, Biochemistry, Physiology and Molecular Biology*. Proceedings of the International Symposium on Sucrose Metabolism. Rockville, MD, USA: American Society of Plant Physiology, 14–24.

Hermanns W, Schulz LC. 1981. Enzyme histochemical demonstration of certain cell populations in synovium using plastic-embedded whole rat joints. In: Deicher H, Schulz LC, eds. *Arthritis Models and Mechanisms*. Berlin: Springer-Verlag, 137–143.

Ho LC. 1988. Metabolism and compartmentation of imported sugars in sink organs in relation to sink strength. *Annual Review of Plant Physiology and Plant Molecular Biology* **39**: 355–378.

Incoll LD, Neales TF. 1970. The stem as a temporary sink before tuberization in *Helianthus tuberosus* L. *Journal of Experimental Botany* **21**: 469–476.

Jenner CF, Ugalde, TD, Aspinall D. 1991. The physiology of starch and protein deposition in the endosperm of wheat. *Australian Journal of Plant Physiology* **18**: 211–226.

Keller F, Frehner M, Wiemken A. 1988. Sucrose synthase, a cytosolic enzyme in protoplasts of Jerusalem artichoke tubers (*Helianthus tuberosus* L.). *Plant Physiology* **88**: 239–241.

Klaushofer H, Wolfslehner L. 1983. Bestimmung des vergärbaren Zuckers in Zichorien- und Topinamburknollen. *Zuckerindustrie* **108**: 552–554.

Koops AJ, Jonker HH. 1994. Purification and characterization of the enzymes of fructan biosynthesis in tubers of *Helianthus tuberosus* 'Colombia'. I. Fructan:fructan fructosyl transferase. *Journal of Experimental Botany* **45**: 1623–1631.

Kosegarten H, Mengel K. 1995. Carbohydrate metabolism and partitioning in crop production. In: Singh BB, Mengel K, eds. *Plant Physiology and Biochemistry*. New Delhi, India: Panima Publishing Corporation, 1–66.

Kutschera U, Schopfer P. 1986. *In-vivo* measurement of cell-wall extensibility in maize coleoptiles: effects of auxin and abscisic acid. *Planta* **169**: 437–442.

Mares DJ, Marschner H, Krauss A. 1981. Effect of gibberellic acid on growth and carbohydrate metabolism of developing tubers of potato (*Solanum tuberosum*). *Physiologia Plantarum* **52**: 267–274.

Marschner H, Sattelmacher B, Bangerth F. 1984. Growth rate of potato tubers and endogenous contents of indolylacetic acid and abscisic acid. *Physiologia Plantarum* **60**: 16–20.

Martinoia E. Transport processes in vacuoles of higher plants. *Botanica Acta* **105**: 232–245.

Mazza G. 1985. Distribution of sugars, dry matter and protein in Jerusalem artichoke tubers. *Canadian Institute of Food Science and Technology Journal* **18**: 263–265.

Meijer WJM, Mathijssen WJM. 1991. The relation between flower initiation and sink strength of stems and tubers of Jerusalem artichoke. *Netherlands Journal of Agricultural Science* **39**: 123–135.

Mengel K, Friedrich B, Judel GK. 1985. Effect of light intensity on the concentrations of phytohormones in developing wheat grains. *Journal of Plant Physiology* **120**: 255–266.

Michael G, Beringer H. 1980. The role of hormones in yield formation. In: *Physiological Aspects of Crop Productivity*. Proceedings of the 15th Colloquium of the International Potash Institute, Bern, 85–115.

Pejin D, Jakovlevic J, Razmovski R, Berenji J. 1993. Experiences in cultivation, processing and application of Jerusalem artichoke (*Helianthus tuberosus* L.) in Yugoslavia. In: Fuchs A, ed. *Inulin and Inulin-Containing Crops*. Amsterdam: Elsevier Science Publishers B.V., 51–56.

Pollock CJ. 1986. Fructans and the metabolism of sucrose in vascular plants. *New Phytologist* **104**: 1–24.

Railton ID, Wareing PF. 1973. Effects of daylength on endogenous gibberellins in leaves of *Solanum andigena*. Changes in levels of free acidic gibberellin-like substances. *Physiologia Plantarum* **28**: 88–94.

Sachs T. 1993. The role of auxin in the polar organization of apical meristems. *Australian Journal of Plant Physiology* **20**: 541–553.

Schubert S, Feuerle R. 1993. Inulinspeicherung in Topinamburknollen – Analyse der begrenzenden Faktoren. *Mitteilungen der Gesellschaft für Pflanzenbauwissenschaften* **6**: 133–136.

Soja G, Haunold E, Praznik W. 1989. Translocation of ^{14}C-assimilates in Jerusalem artichoke (*Helianthus tuberosus* L.). *Journal of Plant Physiology* **134**: 218–223.

Spitters CJT, Lootsma M, van de Waart M. 1989. The contrasting growth pattern of early and late varieties in *Helianthus tuberosus*. In: *Proceedings of the Jerusalem Artichoke EEC Workshop 1987, Madrid*. Report EUR 11855. CEC Luxembourg, 37–43.

Stoy V, Bertholdsson N-O. 1986. Studies on grain filling in cereals. In Olsson G, ed. *Svalöf 1886–1986. Research and Results in Plant Breeding*. Stockholm: Lts förlag, 232–240.

Vreugdenhill D, Strik PC. 1989. An integrated view of the hormonal regulation of tuber formation in potato. *Physiologia Plantarum* **75**: 525–531.

Waters SP, Martin P, Lee BT. 1984. The influence of sucrose and abscisic acid on the determination of grain number in wheat. *Journal of Experimental Botany* **35**: 829–840.

New Phytol. (1997), **136**, 123–129

Fructan metabolism in two species of *Bromus* subjected to chilling and water stress

By ANDREA F. PUEBLA, GRACIELA L. SALERNO AND HORACIO G. PONTIS*

Centro de Investigaciones Biológicas, FIBA and PROBIOP-CONICET, Casilla de Correo 1348, Mar del Plata, 7600, Argentina

(Received 24 July 1996)

SUMMARY

Two species of *Bromus* adapted to different climatic conditions were studied to determine their responses to chilling and water stress. Enzymes of sucrose and fructan metabolism as well as fructoside content were measured in roots and shoots. The results indicate that *Bromus pictus* (adapted to a cold desert area) has a constitutive fructan synthesis, whilst *B. auleticus* (adapted to a warmer northern climate) only accumulated fructan under cold stress. Microscopic observations of roots exposed to fluorescent probes ascertained cell viability and showed that the apex of roots in the northern species were more affected by the water-deficit treatment than the southern species. The presence of fructans in *B. pictus* and their absence in *B. auleticus* support the hypothesis that fructans play a role in plant tolerance to environmental stress.

Key words: Fructan, sucrose, *Bromus*, cold stress, water stress.

INTRODUCTION

Among the abiotic stresses, water deficit and cold constitute major hazards to agriculture throughout the world. Plants native to Patagonia are subjected to drought, chilling and freezing temperatures during their life cycle. Among native Patagonian Gramineae, the genus *Bromus* occurs also in a more temperate region of Argentina. Thus, *Bromus* plants might provide an appropriate system for a comparative study of the water-deficit and chilling tolerance of plants. Environmental stresses elicit a complex of responses in plants at the cellular, physiological and developmental levels (Levitt, 1980). One response, an alteration in the level of sucrose and sucrose-derived oligosaccharides, is a consequence of water and cold stress (Levitt, 1980; Calderón & Pontis, 1985; Hendry, 1987; Pressman *et al.*, 1989; Hendry, 1993; Spollen & Nelson, 1994; Bachmann *et al.*, 1994). Frutans, sucrose-derived oligosaccharides, accumulate when grasses from temperate and cool climatic areas are exposed to chilling temperatures (Eagles, 1967; Pontis & del Campillo, 1985; Chatterton *et al.*, 1989; Suzuki, 1989; Livingston III, Knievel & Gildow, 1994), an accumulation that is preceded by an increase in the concentration of sucrose (Pontis, 1970; Chandorkar & Collins, 1974). The increase in fructan content might be connected with acclimation to low temperatures (Pontis, 1989; Tognetti, Calderón & Pontis, 1989; Santoiani *et al.*, 1993), it might be a consequence of the low demand for photosynthates at chilling temperatures (Pollock, 1984), or both might occur simultaneously. Tognetti *et al.* (1989) and Santoiani *et al.* (1993) studied fructan metabolism in young leaves and roots of wheat at chilling temperatures, whilst Spollen & Nelson (1994) investigated their relationship to water deficit in growing leaves of tall fescue. It is unclear how fructan metabolism might be involved in osmotic adjustment during water deficit, and it has also been argued whether there is a relationship between fructan accumulation and the tolerance of plants to low temperatures (Pollock & Cairns, 1991). Species of *Bromus* growing in North America accumulate fructan (Chatterton *et al.*, 1989, 1993) and therefore it was thought that the use of *Bromus* species growing in southern latitudes and under very different climatic conditions might constitute a very good system for evaluating changes in sucrose and fructan metabolism under stress conditions.

This paper reports the comparison of enzyme activities related to sucrose and fructan synthesis, the amounts of fructose-containing sugars, and the microscopic observations of drought injury, in *B.*

* To whom correspondence should be addressed.
E-mail: pontis@fibamp.org.ar

pictus and *B. auleticus* (Patagonian and central Argentinian species), during exposure to chilling temperatures and water stress.

MATERIALS AND METHODS

Plant material

Two *Bromus* species native to different regions of Argentina were used. Seeds of *Bromus auleticus* (northern species) were kindly provided by Ing. O. Royo Pallarés of INTA Mercedes whilst seeds of *Bromus pictus* (Patagonian species) were supplied by Dr A. Soriano of IFEVA, Facultad de Agronomía, Universidad de Buenos Aires. Plants were grown on vermiculite in a chamber at 25 °C, 200 μmol m^{-2} s^{-1} photosynthetic photon flux density (PPFD), a 14/10 h day/night regime, watered every other day and fertilized weekly with Hoagland's solution. After 90 d, plants were divided into three groups: (1) plants which were transferred to a chamber at 4 °C, with all other conditions unchanged; (2) plants which were watered with polyethylene glycol (PEG) 8000 solution at increasing concentrations (5, 10 and 20 %) and (3) control plants which were maintained under the original conditions. At harvest times samples were collected, weighed, frozen in liquid nitrogen and kept at −80 °C until further use.

Preparation of protein extracts

Samples of roots and shoots were powdered in liquid nitrogen with a mortar and pestle. Homogenates were prepared by extracting the powder in a buffer containing 100 mM HEPES (*N*-[2-hydroxyethyl]-piperazine-*N′*-[2-ethanesulphonic acid]), pH 7·5, 20 mM β-mercaptoethanol, 10 mM, $MgCl_2$, 2 mM EDTA, 0·5 mM PMSF, 2 % ethylene glycol (v/v) and 20 % glycerol (v/v). Extracts were centrifuged at 20000 ***g*** for 30 min. The clarified supernatant was either used immediately for sucrose-sucrose fructosyl transferase (SST, EC 2.4.1.99) assays or was filtered through Sephadex G-50 columns and used to assay other enzymes.

Enzyme assays

SST activity was assayed by incubating the extracts for 4–6 h at 200 mM [^{14}C]sucrose (250 Bq μmol^{-1}) at pH 5·2. The solutions were spotted on Whatman No. 3 paper and chromatographed in *n*-butanol: pyridine: water (6:4:3 (v/v), for 30 h at 20 °C. The trisaccharide zone was cut out and counted in a liquid scintillation counter. Sucrose-phosphate synthase (SPS, EC 2.4.1.14) and sucrose synthase (SS, EC 2.4.1.13) were assayed by determining the sucrose formed after incubation in 100 mM HEPES pH 7·5, 20 mM $MgCl_2$, 10 mM UDPglc and in the presence of 10 mM fructose-6P, 50 mM glucose-6P and 10 mM arbutine (SPS), and 10 mM fructose (SS). The sucrose formed was determined by the thiobarbituric acid (TBA) method according to Percheron (1962). Acid invertase (INV, β-D-fructofuranosidase, EC 3.2.1.26) activity was assayed in 100 mM acetate buffer, pH 4·5 in 20 mM sucrose, and the reducing sugars produced were measured by the method of Somogyi & Nelson (Spiro, 1966). Protein content in the extracts was determined by the method of Bradford (1976). The results presented are the mean of at least three independent experiments.

Extraction and determination of sugars

Carbohydrates were extracted in boiling alkaline water as previously reported (Tognetti *et al.*, 1990), except that dry tissue was used. Bound fructose was determined by the TBA method. Carbohydrates were analysed by paper chromatography (PC) as previously reported (Santoiani *et al.*, 1993) or by thin-layer chromatography (TLC) using Schleicher & Schuell F-1500 'ready foils' developed three times in 1-butanol: 2-propanol: water (3:12:4). The position of sucrose and fructans was ascertained by urea-phosphoric acid reagent (Wise *et al.*, 1955), using *Helianthus tuberosus* fructans as standards.

Solute potential and relative water content

Solute potentials (ψ_s) were obtained for PEG 8000 solutions at different percentages with a Wescor C-52 thermocouple psychrometer. PEG 8000 was used at three different concentrations: 5 % (−0·02 MPa), 10 % (−0·05 MPa) and 20 % (−0·71 MPa). Relative water content (RWC) of the first expanded leaves was calculated as:

$$\mathrm{RWC} = \frac{\text{fresh wt} - \text{dry wt}}{\text{turgid wt} - \text{dry wt}}.$$

Turgid weight is the weight after incubation of the leaf in an atmosphere saturated with water until a constant weight is attained. The d. wt of the leaves was taken after drying plant material in an oven at 85 °C to constant weight.

Determination of drought injury

Roots of control and water-deficit plants were viewed under fluorescence microscopy. Roots were prepared using propidium iodide (100 μg ml^{-1}) as a fluorescent probe to ascertain cell viability. A fluorescent nucleus indicates that the cell is dead.

RESULTS AND DISCUSSION

Osmotic potential and RWC

Plants were exposed to gradual increases in concentrations of PEG 8000. Leone *et al.* (1994) showed in cultured potato cells that a direct exposure to 20 %

Table 1. *Relative water content (RWC) of primary expanded leaves of control* Bromus auleticus *and* Bromus pictus, *and plants exposed to cold temperatures and water stress*

		Control			Chilling			Drought		
Species	Day...	2	4	7	2	4	7	2	4	7
B. auleticus		95±1	95±2	96±1	92±2	95±2	97±1	93±2	96±1	77±11
B. pictus		96±3	95±3	94±4	98±2	97±1	99±1	92±2	96±1	82±3

Fresh weight (f. wt), turgid weight (t. wt) and dry weight (d. wt) were measured.

$$\text{RWC} = \frac{\text{f. wt} - \text{d. wt}}{\text{t. wt} - \text{d. wt}\,\%} \pm \text{SD}.$$

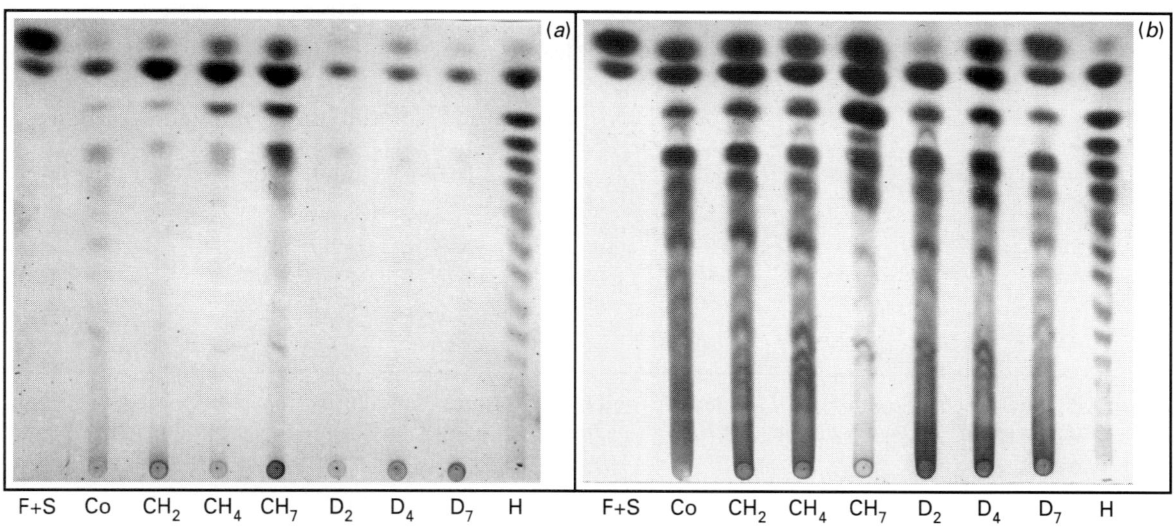

Figure 1. Thin-layer chromatography of fructo-oligosaccharides from shoots of *Bromus auleticus* (*a*) and *Bromus pictus* (*b*) at different stages of chilling and water stress. Each lane contained the water soluble carbohydrates equivalent to 6 mg f. wt. Sucrose and fructose (lane 1) and fructans of *Helianthus tuberosus* (lane 9) were used as standards. Co, Control; Ch, chilling; D, drought; 2, 4, 7 d of treatments; H, *Helianthus tuberosus*; F+S, fructose+sucrose.

PEG 8000 produced severe cell injury, a dramatic modification in protein patterns which might lead to cellular death. To avoid detrimental effects, Leone *et al.* (1994) demonstrate that a slow increase in PEG concentration triggers an adaptive response avoiding cellular death at the maximum PEG concentration. Table 1 shows RWC of control plants and those submitted to stress conditions. The water status of plants was not modified by low temperature treatment but was altered in drought-treated plants at the highest concentration of PEG. The low RWC found in drought-stressed plants suggests that the induced water deficit was sufficiently low to change the cellular water status of the leaves in both species.

The carbohydrate composition of leaves and roots

TLC of water extracts of roots and shoots of the two *Bromus* species indicated a clear and outstanding difference in their carbohydrate composition. Shoots of *B. auleticus* grown under control conditions (25 °C and normal irrigation) contained only traces of fructans (Fig. 1*a*). On the other hand shoots of *B. pictus* contained a complete series of fructans to at least DP > 15 (Fig. 1*b*). The large difference found between total carbon status between the two species in the control plants indicates that before the treatment was imposed there already was a considerable difference in accumulation of sucrose and fructans between the two species. The possible reasons are complex and might involve the balance in the native maximal photosynthesis rate, overall sink demand and relative growth rates of each species. During the first 4 d *B. pictus* plants subjected to chilling temperatures (4 °C) exhibited no difference in fructan patterns from control plants. In some experiments, an increase in sucrose and trisaccharide contents appeared at the seventh day of exposure to low temperatures (Fig. 1*b*). However, even if the pattern on the seventh day appeared modified, the total fructoside content was unchanged. *B. auleticus* presented a distinct modification of the level of fructans detected by TLC. Sucrose and the trisaccharide increased with time at the cold temperature. Water deficit did not elicit a change in the fructan patterns in either *B. auleticus* or in *B. pictus*.

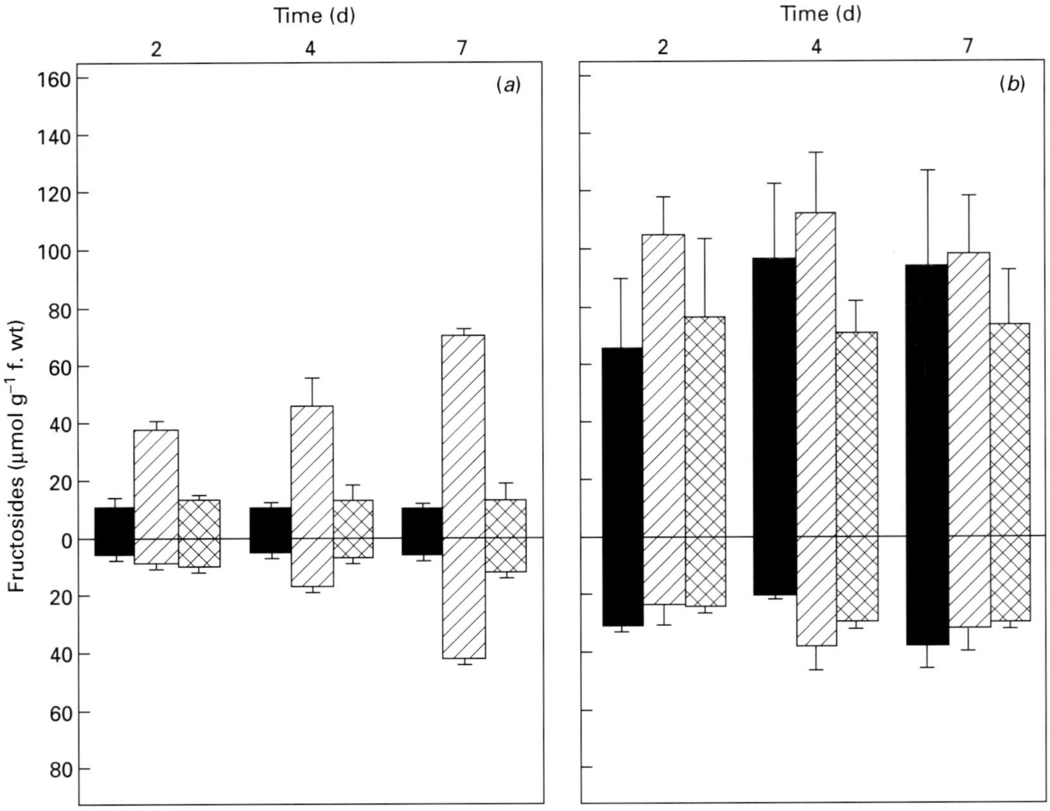

Figure 2. Total fructoside contents of shoots (top) and roots (bottom) of two *Bromus* species exposed to chilling and water stress. (*a*) *Bromus auleticus*; (*b*) *Bromus pictus*. ■, Control; ▨, chilling; ⊠, drought.

Table 2. *Activities of enzymes of sugar metabolism from shoots of* Bromus *species exposed to cold temperatures and drought*

Species/ enzyme	Day...	Control				Chilling				Drought		
		1	2	4	7	1	2	4	7	2	4	7
B auleticus												
SS		2·0	2·3	2·1	1·5	2·5	1·8	1·9	1·5	2·6	1·8	4·0
SPS		1·1	0·8	1·4	1·0	1·2	0·9	0·9	1·1	1·9	1·8	2·8
INV		9·7	16·0	11·1	9·3	13·4	15·1	14·2	7·2	9·6	17·9	21·2
B. pictus												
SS		1·0	0·5	0·6	0·9	0·6	0·9	0·6	0·4	0·5	0·4	0·5
SPS		1·4	0·9	0·6	1·1	0·5	0·9	0·6	0·7	0·7	0·5	1·0
INV		5·2	4·1	5·5	6·7	3·4	5·8	3·5	5·0	2·5	1·8	5·0

SS, sucrose synthase; SPS, sucrose phosphate synthase; INV, acid invertase. Results are expressed in nmol of product per g of f. wt per hour of incubation at 30 °C: SS and SPS, sucrose; INV, reducing sugars.

Nevertheless, a slight increase in the amount of fructose was observed in *B. pictus* (Fig. 1*b*). Analysis of root carbohydrates showed the same picture (data not shown).

Total fructoside content of shoots and roots of both species is shown in Figure 2. The concentration of fructosides reached by the northern species was similar to that in both control and treated plants of the southern species. It should be noted that the higher standard deviation observed for the plants of Patagonian origin suggests a greater variation in the natural plant population from which seeds were obtained.

Enzyme determinations

As fructan synthesis is related to sucrose concentration, the enzymes associated with sucrose metabolism were measured. The results (Table 2) show few if any modifications to SS, SPS and INV activities by the treatments. There were no differences in the amount of SPS and SS, as measured by Western blots using rice anti-SPS and anti-SS antibodies obtained by Salerno *et al.* (1989) and Crespi *et al.* (1991) (data not shown). However, the greater activity of INV and SS found in *B. auleticus* than in *B. pictus* might explain the lower con-

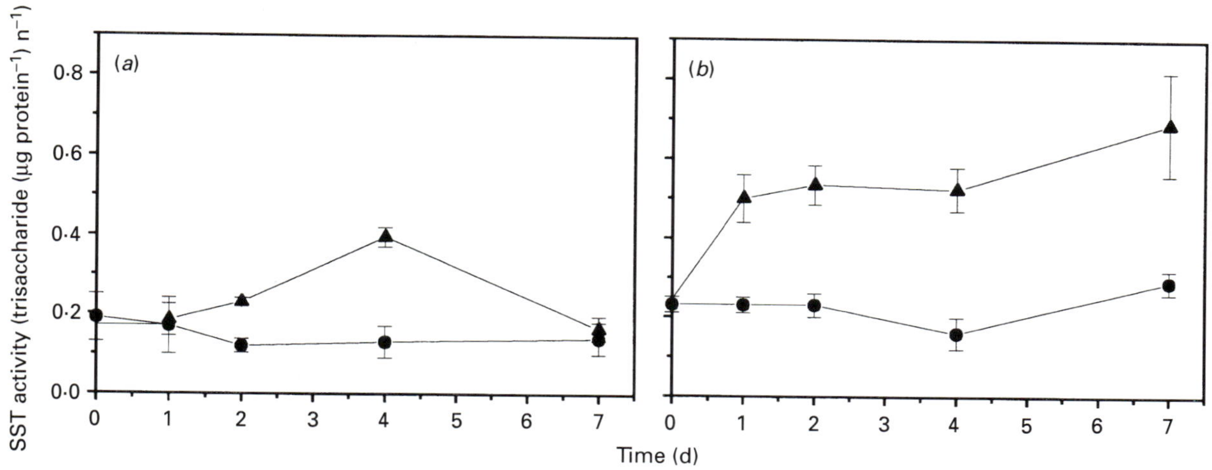

Figure 3. Activity SST from shoots of *Bromus auleticus* (*a*) and *Bromus pictus* (*b*) during exposure to chilling temperatures. ●, control; ▲, chilling.

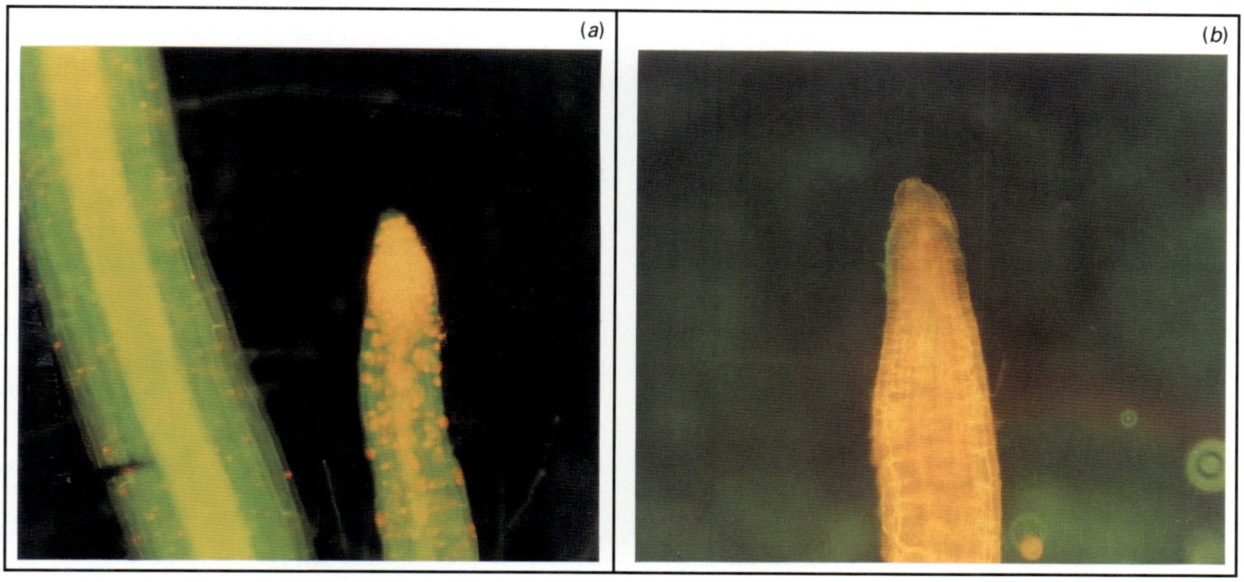

Figure 4. Drought injury to root apices of *Bromus auleticus* and *Bromus pictus* observed by fluorescence microscopy. Roots exposed to PEG 8000 solutions were incubated in propidium iodide solution (100 µg ml^{-1}). Fluorescent nucleated cells were dead. (*a*) *Bromus auleticus* and (*b*) *Bromus pictus* apex after water deficit treatment showing mainly dead and intact cells respectively.

centration of sucrose found in the northern species. Stitt (1988) showed that in order to see a modification of SPS activity, measurements had to be made with limiting substrate conditions. However, under low temperature, growth is severely reduced. The data indicate the possibility of an excess of photosynthates and a diminished demand, as suggested by Pollock (1986). When subjected to drought, cell growth is reduced much more than photosynthesis (Salisbury & Ross, 1992).

Variation in SST activities in *B. auleticus* and *B. pictus* exposed to chilling temperatures is presented in Fig. 3. *B. pictus* had a higher enzyme activity throughout the period of exposure to low temperatures. The situation might appear to conflict with the apparent depolymerization of fructans found at the seventh day of exposure to cold temperatures. However, it should be noticed that during chilling there is an increase in the sucrose concentration in *B. pictus* which might explain activity of SST being higher than in *B. auleticus*. In this species an increase in activity corresponded to the initiation of fructan synthesis. Similar results were observed by Santoiani *et al.* (1993) when wheat roots were subjected to chilling temperatures.

There were no changes in SST activity under drought conditions. Spollen & Nelson (1994) found a depolymerization of fructans under water deficit in tall fescue. Some sugar extracts of *B. pictus* showed similar results. In future it will be important to measure fructan exohydrolase activities under water stress.

Microscopic observation of root injury induced by drought

We used fluorescent probes (Krishan, 1975; Jones & Senft, 1985) to check the presence of possible injury to roots resulting from drought. Microscopic observations of roots of the two *Bromus* species studied showed that the roots were differently affected by drought. In *B. pictus*, only 50% of the apical cells were dead; by contrast, in *B. auleticus* up to 95% of the apical cells were dead, as evidenced by a fluorescent nucleus (Fig. 4). These observations suggest that *B. pictus* has adapted to water deficit better than *B. auleticus*, a species not normally exposed to drought. These observations might be related to the fact that in *B. pictus*, the Patagonian species, fructans were present under all treatments, whilst in *B. auleticus* they were not. Some authors have shown that various carbohydrates protect membrane stability at low water activities (Crowe *et al.*, 1984, 1988; Crowe, Hoekstra & Crowe, 1992). Trehalose, for example, is very effective. It would be enlightening to study the possible roles of individual fructans in protecting the plant organelles and membranes during water stress.

Bromus is a good system for studying sucrose and fructan metabolism in plants exposed to water and cold stresses. The use of closely related species of *Bromus*, native to extremely different climatic areas, allowed us to compare sugar metabolisms and to suggest that their metabolic differences have implications for tolerance. Further studies are in progress.

ACKNOWLEDGEMENTS

The authors are grateful to: Lic. M. Barrio for helping with the microscopic observations; Dr J. Casal and Ing Agr. J. Chimenti for lending the thermocouple psychrometer; Dr J. Tognetti for criticism and helpful discussion; and for the technical assistance of Mrs C. Fernández and Ms Carmen Rodríguez. G.L.S. and H.G.P. are Career Investigators of the Consejo Nacional de Investigaciones Científicas y Técnicas (CONICET). A.F.P. is a fellow of CONICET.

REFERENCES

Bachmann M, Matile P, Keller F. 1994. Metabolism of raffinose family oligosaccharides in leaves of *Ajuga reptans* L. cold acclimation, translocation, and sink to source transition: discovery of chain elongation enzyme. *Plant Physiology* **105**: 1335–1345.

Bradford MM. 1976. A rapid and sensitive method for the detection of microgram quantities of protein utilizing the principle of protein-dye binding. *Analytical Biochemistry* **72**: 248–254.

Calderón P, Pontis HG. 1985. Increase of sucrose synthase activity in wheat plants after a chilling shock. *Plant Science* **42**: 173–176.

Chandorkar K, Collins FW. 1974. The mechanism of *de novo* synthesis of fructo-oligosaccharides in leaf disks of certain Asteraceae III. *Canadian Journal of Botany* **52**: 181–188.

Chatterton NJ, Thornley WR, Harrison PA, Bennett JH. 1989. Fructosyltransferase and invertase activities in leaf extracts of six temperate grasses grown in warm and cool temperatures. *Journal of Plant Physiology* **135**: 301–305.

Chatterton NJ, Harrison PA, Thornley WR, Bennett JH. 1993. Structure of fructan oligomers in cheatgrass (*Bromus tectorum* L.). *New Phytologist* **124**: 389–396.

Crespi MD, Zabaleta EJ, Pontis HG, Salerno GL. 1991. Sucrose synthase expression during cold acclimation in wheat. *Plant Physiology* **96**: 887–891.

Crowe JH, Crowe LM, Carpenter JF, Rudolph AS, Aurell Wistrom C, Spargo BJ, Anchordoguy TJ. 1988. Interactions of sugars with membranes. *Biochimica et Biophysica Acta* **947**: 367–384.

Crowe JH, Hoekstra FA, Crowe LM. 1992. Anhydrobiosis. *Annual Reviews in Plant Physiology* **54**: 579–599.

Crowe LM, Mouradian JH, Jackson SA, Womersley C. 1984. Effects of carbohydrates on membrane stability at lower water activities. *Biochimica et Biophysica Acta* **769**: 141–150.

Eagles CF. 1967. The effect of temperature on vegetative growth in climatic races of *Dactylis glomerata* in controlled environment. *Annals of Botany* **31**: 645–651.

Hendry GAF. 1987. The ecological significance of fructan in a contemporary flora. *New Phytologist* **106**: 201–216.

Hendry GAF. 1993. Evolutionary origins and functions of fructan – a climatological, biogeographic and mechanistic appraisal. *New Phytologist* **123**: 3–14.

Jones KH, Senft JA. 1985. An improved method to determine cell viability by simultaneous staining with fluorescein diacetate–propidium iodide. *Journal of Histochemistry and Cytochemistry* **33**: 77–79.

Krishan A. 1976. Rapid flow cytofluorometric analysis of mammalian cell cycle by propidium iodide staining. *Journal of Cell Biology* **66**: 188–193.

Leone A, Costa A, Tucci M, Grillo S. 1994. Adaptation versus shock response to polyethylene glycol-induced low potential in cultured potato cells. *Physiologia Plantarum* **92**: 21–30.

Levitt J. 1980. *Responses of plants to environmental stresses, 2nd edn, vol. 1.* New York: Academic Press, 13.

Livingston III D, Knievel DP, Gildow FE. 1994. Fructan synthesis in oat. I. Oligomer accumulation in stems during cold hardening and their *in vitro* synthesis in a crude enzyme extract. *New Phytologist* **127**: 27–36.

Percheron F. 1962. Colorimetric determination of fructose and fructofuranosides by the thiobarbituric acid reaction. *Comptes Rendus de l'Académie des Sciences, Paris* **255**: 2521–2522.

Pollock CJ. 1984. Sucrose accumulation and the initiation of fructan biosynthesis in *Lolium temulentum* L. *New Phytologist* **96**: 527–534.

Pollock CJ. 1986. Environmental effects on sucrose and fructan metabolism. In: Randal DD, Miles CD, Nelson CJ, Blevins DG, Miernik JA, eds. *Current Topics in Plant Biochemistry and Physiology, vol. 5.* Columbia, MO, USA: University of Missouri, 32–46.

Pollock CJ, Cairns AJ. 1991. Fructan metabolism in grasses and cereals. *Annual Review of Plant Physiology and Plant Molecular Biology* **42**: 77–101.

Pontis HG. 1970. The role of sucrose and fructosylsucrose in fructosan metabolism. *Physiologia Plantarum* **23**: 1089–1100.

Pontis HG. 1989. Fructans and cold stress. *Journal of Plant Physiology* **134**: 148–150.

Pontis HG, del Campillo E. 1985. Fructans. In: Dey PM, Dixon RA, eds. *Biochemistry of storage carbohydrates in green plants.* London: Academic Press, 205–227.

Pressman E, Schaffer AA, Compton D, Zamski E. 1989. The effect of low temperature and drought on the carbohydrate content of asparagus. *Journal of Plant Physiology* **134**: 209–213.

Salerno GL, Ianiro JL, Tognetti JA, Crespi MD, Pontis HG. 1989. Differential induction of sucrose metabolizing enzymes in wheat (*Triticum aestivum* cv. San Agustín) leaf sections. *Journal of Plant Physiology* **134**: 214–217.

Salisbury FB, Ross CW. 1992. *Plant physiology*. Belmont, CA, USA: Wadsworth Publishing Company.

Santoiani CS, Tognetti JA, Pontis HG, Salerno GL. 1993. Sucrose and fructan metabolism in wheat roots at chilling temperatures. *Physiologia Plantarum* **87**: 84–88.

Spiro RG. 1966. Analysis of sugars found in glycoproteins. *Methods in Enzymology* **8**: 3–26.

Spollen WG, Nelson CJ. 1994. Response of fructan to water deficit in growing leaves of tall fescue. *Plant Physiology* **106**: 329–336.

Stitt M, Wilke I, Feil R, Heldt HW. 1988. Coarse control of sucrose-phosphate synthase in leaves: alterations of the kinetic properties in response to the rate of photosynthesis and the accumulation of sucrose. *Planta* **174**: 217–230.

Suzuki M. 1989. Fructans in forage grasses with varying degrees of cold hardiness. *Journal of Plant Physiology* **134**: 224–231.

Tognetti JA, Calderón P, Pontis HG. 1989. Fructan metabolism: reversal of cold acclimation. *Journal of Plant Physiology* **134**: 232–236.

Tognetti JA, Salerno GL, Crespi MD, Pontis HG. 1990. Sucrose and fructan metabolism of different wheat cultivars at chilling temperatures. *Physiologia Plantarum* **78**: 554–559.

Wise CS, Dimler RJ, Davis HA, Rist CE. 1955. Determination of easily hydrolyzable fructose units in dextran preparation. *Analytical Chemistry* **27**: 33–36.

New Phytol. (1997), **136**, 131–135

Effects of short-term phosphorus deficiency on carbohydrate storage in sink and source leaves of barley (*Hordeum vulgare*)

BY C. WANG AND J.-E. TILLBERG*

Department of Botany, Stockholm University, S-106 91 Stockholm, Sweden

(*Received 24 July 1996*)

SUMMARY

Barley (*Hordeum vulgare* L. cv. Agneta) was cultivated hydroponically under continuous light, constant temperature and relative humidity. The relative growth rate (RGR) was kept at 25% by an exponential supply of complete medium for 2 wk. The cultures were then transferred to phosphorus-free media and the amounts of total P, fructans, starch, sucrose, glucose and fructose in sink and source leaves were measured at 0, 12, 24, 48, 72, 120 and 156 h. The capacity of net fructan synthesis (i.e. total activity of sucrose:sucrose 1-fructosyltransferase (1-SST) and sucrose:fructan 6-fructosyltransferase (6-SFT)), the activity of fructan exohydrolase (FEH), as well as the activity of acid invertase were also measured in the two types of leaves. Omission of P from the growth medium caused a decrease in the P content in both sink and source leaves, whereas the RGR was not affected during the 156 h of the experiment. The fructan concentration increased in low-P plants whilst there was little or no effect on the concentrations of starch, sucrose, glucose and fructose. There was a slight increase in the capacity of net fructan synthesis in both sink and source leaves at 156 h after the onset of P-starvation. There was no change in the activity of FEH or acid invertase. The results indicate that in barley leaves the mechanism for carbon partitioning into fructans is more sensitive towards low P conditions than the mechanism for carbon partitioning into starch.

Key words: Fructan, *Hordeum vulgare*, phosphorus deficiency, sink and source leaves, starch, sucrose.

INTRODUCTION

Besides carbon, the quantitatively most important mineral elements needed by plants are nitrogen and phosphorus. For fructan-accumulating plants like barley, long-term deficiency of either N or P induces an increase in fructan accumulation in addition to that of other carbohydrates in the leaves (Archbold, 1938; Russell, 1938). Short-term effects of N deficiency on carbon partitioning and on the regulation of fructan metabolism in barley leaves has been recently investigated (Wang & Tillberg, 1996). Short-term effects of P-starvation on fructan metabolism in leaves has not yet been reported.

Assimilated carbon in a leaf is either stored as starch in the chloroplast or exchanged with cytosolic P for triosephosphates (TP) in the cytosol, providing the substrate for sucrose synthesis. The pool sizes of P and TP are thus important in regulating carbon flow across the chloroplast membrane and through sucrose and starch synthesis (Walker & Robinson, 1978; Champigny, 1985). Sucrose metabolism is closely related to fructan metabolism in leaves of fructan-producing plants (Pollock & Cairns, 1991). We are therefore interested in studying how P starvation affects the regulation of carbon partitioning between starch, sucrose and fructan.

According to a new model for fructan biosynthesis (Duchateau *et al.*, 1995; Wiemken, Sprenger & Boller, 1995), three enzymes, sucrose:sucrose 1-fructosyltransferase (1-SST), sucrose:fructan 6-fructosyltransferase (6-SFT) and fructan:fructan 1-fructosyltransferase (1-FFT) are necessary to synthesize all of the fructan species known to occur in barley. 1-SST and 6-SFT use sucrose exclusively as the fructosyl donor for the net synthesis of fructans.

The present study examines the effects of a short-term nutritional P deficiency on carbon partitioning in sink and source leaves of barley grown hydroponically under continuous light at 20 °C. The effect of P deficiency on fructan metabolism under these conditions was analysed by assaying the total activity of 1-SST and 6-SFT, the activity of invertase and the activity of fructan exohydrolase (FEH) in leaf extracts.

* To whom correspondence should be addressed.

MATERIALS AND METHODS

Seeds of barley (*Hordeum vulgare* L. cv. Agneta) were germinated in moist vermiculite at 27 °C for 3 or 4 d, then seedlings were transferred to slits in black Styrofoam® sheets floated on 7 l of nutrient solution in a climate chamber under continuous light (350 μmol m^{-2} s^{-1}), r.h. 65 %, at 20 °C. Exponentially increasing nutrient concentrations were given daily to keep the plants at a relative growth rate (RGR) of 25 % (Ericsson, Larsson & Tillberg, 1982; Thorsteinsson *et al.*, 1987). After preculture on complete medium for 2 wk, the plants were transferred to P-free but otherwise complete media. Leaves at two developmental stages were analysed. The youngest expanding leaf (without sheath) was collected as a sink leaf. The third leaf (without sheath) from the tip of the plant, fully expanded, was collected as a source leaf. The term 'sink leaf' used in this paper does not mean that the expanding leaves behave only as a sink for carbohydrates. The sink part of the leaf is mainly represented by the meristematic zone whilst the tip becomes photosynthetically active as soon as it appears above the whorl of sheaths and is thus able to transfer newly synthesized sucrose to the basal part of the leaf. For each experiment, sink and source leaves from 15 plants were separately sampled at the beginning (0 h) of the starvation period, then at 12, 24, 48, 72, 120 and 156 h after transfer to P-free media. The samples were ground immediately in a mortar in liquid nitrogen, part of the ground sample was lyophilized, and the dried plant material was powdered and stored in a desiccator at 4 °C for carbohydrate and P analysis. A separate aliquot of the sample ground in liquid nitrogen was used for determination of enzymatic activity.

Methods for quantifying carbohydrates (including fructan, starch, sucrose, glucose and fructose) and for determining fructan-metabolizing enzyme activities (including total activity of 1-SST and 6-SFT, invertase and FEH) were as previously described (Wang & Tillberg, 1996).

The total content of P was measured with a modified Fiske & Subbarow (1925) method. Dried material (1 g) from each sample was wet-digested in 10 ml of HNO_3:$HClO_4$ (7:3 (v/v)), and the digested sample was then diluted with distilled water to 20 ml. An aliquot of 0·2 ml of the diluted sample was mixed with 3 ml of a reagent made of 4·41 g of $(NH_4)_6Mo_7O_{24}.4\,H_2O$ in 44 ml of 1·6 *N* H_2SO_4 and 2 ml of 0·5 g of $SnCl_2$, 2 g of ascorbic acid in 16·5 ml of conc. H_2SO_4. The P content was determined spectrophotometrically at 680 nm 90 min after mixing. KH_2PO_4 was used as a standard.

The data presented are means of replicates from three different experiments. For comparing the results on carbohydrate concentrations and enzymatic activities, analysis of variance was applied with $P \leqslant 0{\cdot}05$.

RESULTS

Omission of P from the growth medium caused a decrease in the total P concentration in both source and sink leaves (Fig. 1*a*). The P in the source leaves decreased somewhat faster and to a lower concentration than in sink leaves after 100 h of starvation. In spite of the decrease in P, the relative growth rate (RGR) did not change during P-starvation (Fig. 1*b*).

Fructans accumulated in both sink and source leaves at low P concentrations though the increase was much more pronounced in source leaves (Fig. 2*a*, *b*). The increase was first detectable after 24 h in P-depleted growth medium. This accumulation was apparently not caused by a corresponding decrease in the carbon utilization rate (measured as RGR) (Fig. 1*b*). Starch accumulation did not increase in either sink or source leaves of low-P plants (Fig. 2*c*, *d*).

The content of sucrose, a pivotal substance in carbohydrate metabolism, was not affected by low P in either sink or source leaves (Fig. 3*a*, *b*). The sucrose concentration was consistently about 50 % higher in source than in sink leaves.

Glucose and fructose concentrations were not affected by low P. The glucose concentrations were somewhat higher in the sink leaves than in the source leaves. The fructose concentrations were lower than those of glucose and were the same in both sink and source leaves (Fig. 3*a*–*f*)

Analysis of the activities of fructan-metabolizing enzymes during P-starvation showed that although

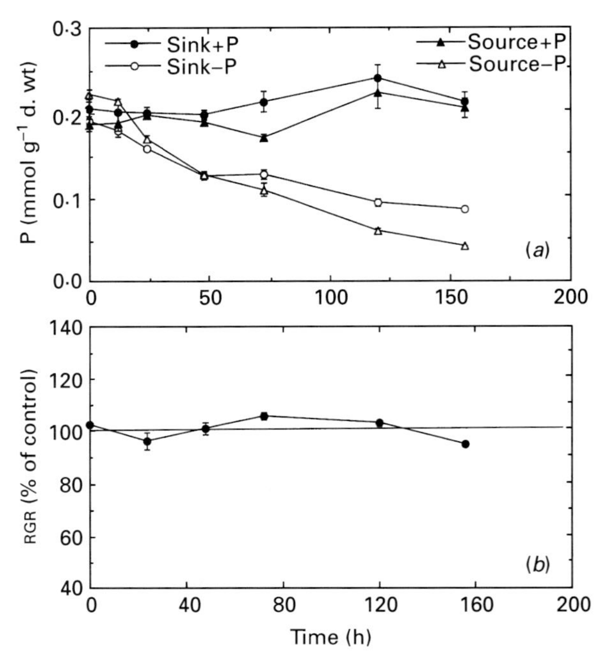

Figure 1. Changes over time in total phosphorus content in leaves (*a*) and relative growth rate (RGR) (*b*). Optimally growing barley plants were transferred at 0 h from a complete medium to a P-free but otherwise complete medium. The RGR is expressed as a percentage of the control value (25 % at 0 h). The values were obtained from 15 plants for each sampling. Error bars denote SE ($n = 3$).

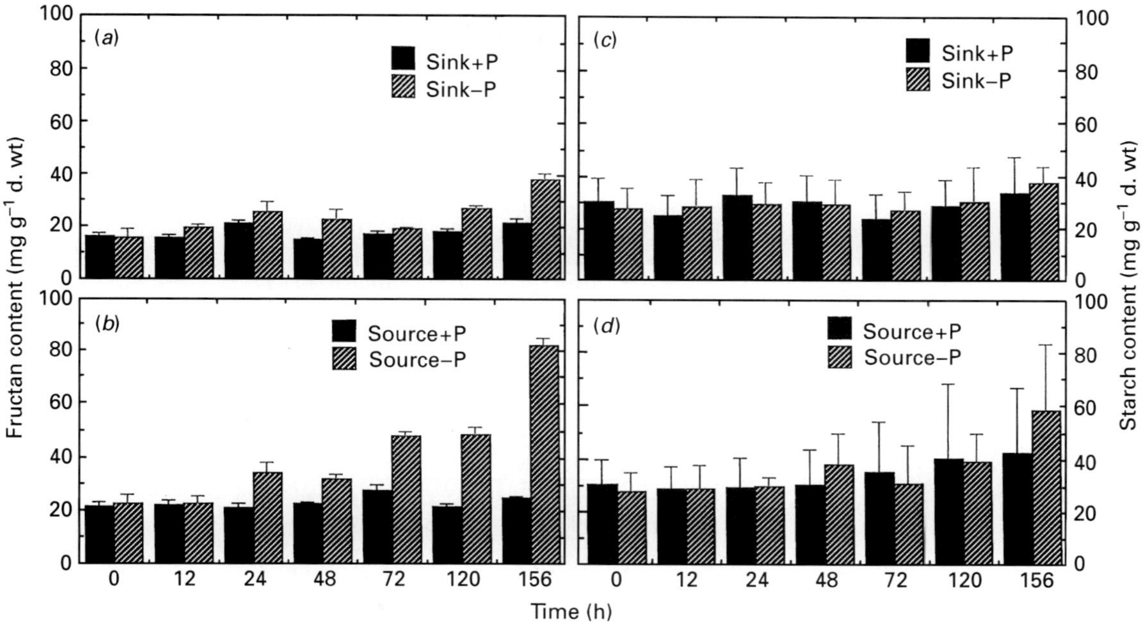

Figure 2. Fructan (*a*, *b*) and starch (*c*, *d*) content in sink and source leaves in controls and in P-starved barley plants. Sampling began at the start (0 h) of the starvation period. Error bars denote SE ($n = 3$).

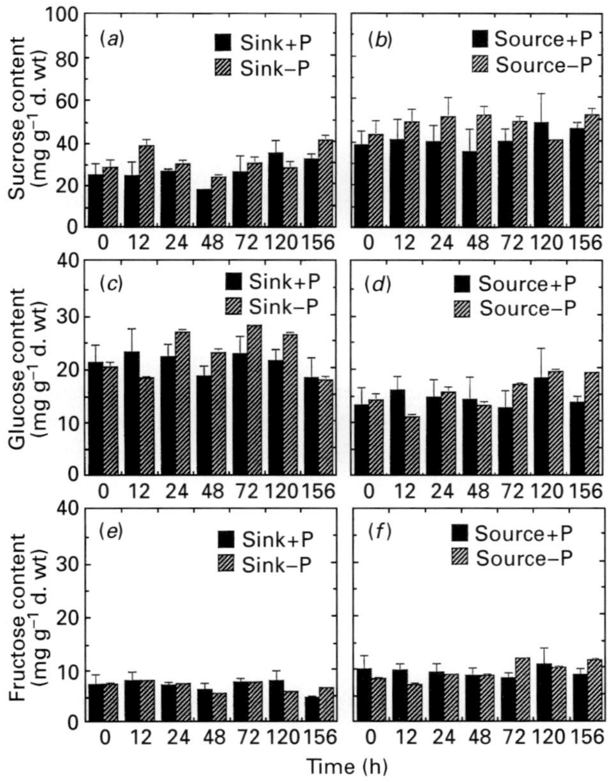

Figure 3. Sucrose content in (*a*) sink and (*b*) source leaves in controls and P-starved barley plants. Glucose content in (*c*) sink and (*d*) source leaves in controls and P-starved barley plants. Fructose content in (*e*) sink and (*f*) source leaves in controls and P-starved barley plants. Error bars denote SE ($n = 3$).

the increase in fructan was already detectable after 24 h of starvation and the increase was large during the starvation period, the total activity of 1-SST and 6-SFT did not change significantly until 120 h after the onset of P deficiency. A slight increase in total activity of 1-SST and 6-SFT was observed in both sink and source leaves at the end of the starvation period (156 h) (Fig. 4*a*, *b*). Acid invertase activity and FEH activity were not affected by P-starvation (Fig. 4*c*–*f*).

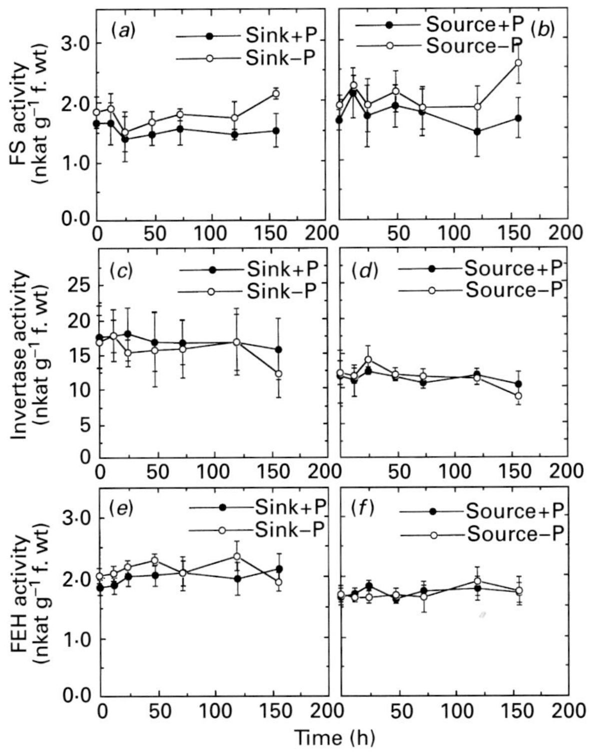

Figure 4. Enzyme activities in sink and source leaves from barley plants. (*a*) Activity of fructan synthetase (FS) (including 1-SST and 6-SFT) in samples prepared from sink leaves and (*b*) from source leaves; (*c*) invertase activity in samples prepared from sink leaves and (*d*) from source leaves; (*e*) activity of fructan exohydrolase (FEH) in samples prepared from sink leaves and (*f*) from source leaves. Error bars denote SE ($n = 3$).

DISCUSSION

There have been many reports on the increase in starch accumulation during P depletion in leaves that use either starch or fructans as the major storage carbohydrate (Lehnert, Vieweg & de Fekete, 1979; Foyer & Spencer, 1986; Thorsteinsson, Tillberg & Tillberg, 1987; Sicher & Kremer, 1988). The present experiment shows that in barley an increase in the accumulation of fructans rather than of starch was an early response to P-starvation (Fig. 2). The increase in frucants might not be due to a reduction of assimilate demand for growth processes since the RGR did not change (Fig. 1*b*), and a significant accumulation of starch did not occur simultaneously with the increase in fructan over the 156 h studied. In addition, it has been observed that photosynthetic activity in leaves of barley plants increased during the first days of P-starvation, although activity decreased after prolonged treatment with low-P growth medium (Thorsteinsson, 1989). The stimulation of fructan accumulation might thus be caused by different sensitivities of photosynthesis and carbon utilization to initial P-starvation. The present results are different from those of studies of nitrogen deficiency, in which nitrogen omission induced a fast decline in RGR (Wang, unpublished), and starch in the leaves increased parallel to fructan over time (Wang & Tillberg, 1996).

Fructan metabolism is closely related to sucrose metabolism, since sucrose is the substrate for fructan synthesis and fructan can be regarded as an extension of the sucrose storage pool (Pollock, 1986). The pool size of P is very important in regulating sucrose synthesis in the cytosol because two of the enzymes involved in sucrose synthesis, cytoplasmic fructose 1,6-bisphosphatase (Cyt FBP′ase, EC 3.1.3.11) and sucrose phosphate synthase (SPS, EC 2.3.1.14) are sensitive, directly or indirectly, to the local concentration of inorganic phosphate and phosphorylated intermediates (Stitt, Huber & Kerr, 1987; Huber, 1986; Huber *et al.*, 1991). Phosphorus inhibits SPS and Cyt FBP′, therefore low levels of P are associated with higher activity of the sucrose synthesis pathway. The constant concentrations of both sucrose and starch (Fig. 3*a*, *b*) in the present experiment imply that the increase in fructan accumulation in P-starvation functions to maintain the cytoplasmic sucrose concentration below that required to stimulate additional starch accumulation in the chloroplast. There is no evidence yet for the direct involvement of P in the pathway of fructan synthesis. The accumulation of fructan rather than starch as the early response to P-starvation might thus be a way to save P for metabolic processes in the cytosol and for exchange with triosephosphates across the chloroplast envelope.

Phosphorus is a controlling factor in photosynthate partitioning but the buffering role through the ‘non-metabolic’ vacuole P pool is slow (Rebeille,

Bligny & Douce, 1982; Woodrow *et al.*, 1984; Sicher & Kremer, 1988). The differences in fructan accumulation between source and sink leaves could thus be caused by differences in short-term changes in the cytosolic P pool. Photosynthesis in sink leaves that have not fully expanded might not be as high as in source leaves. The amount of triosephosphate used for exchange with P over the chloroplast envelope in the sink leaves could thus be less than in the source leaves, i.e. the extent of P stress in the source leaves might be greater than in the sink leaves. This might be the reason for the much greater increase of fructan in the source leaves than in the sink leaves (Fig. 2*a*, *b*).

Concentrations of hexose did not change during the experiment (Fig. 3*c–f*). This result supports the suggestion that glucose released by fructan synthesis can be phosphorylated in the cytosol and used to form sucrose (Housley & Pollock, 1993).

The changes in fructan concentrations did not coincide with significant changes in fructan-metabolizing enzyme activities during the short-term P-starvation experiment, but the total activity of 1-SST and 6-SFT increased by the end of the starvation period (Fig. 4). This indicates that fructan accumulation during the first 120 h of starvation is mainly regulated by substrate concentration. The regulation process might be very rapid in the attached barley leaves because an increase in sucrose concentration could not be monitored. Fructan synthesis enzymes appear to be more important in the regulation after 120 h of starvation.

In conclusion, increases in fructan in both sink and source barley leaves were observed during the short-term P-starvation experiment. Relative growth rate, and concentrations of starch, sucrose and hexoses did not change during the study. These results indicate that the mechanism for carbon-partitioning into fructan is more sensitive towards low-P conditions than is the mechanism for carbon-partitioning into starch in leaves of barley, a fructan-producing plant.

ACKNOWLEDGEMENTS

The authors thank Dr Maria Greger for helpful advice and instructions regarding P analysis.

REFERENCES

Archbold HK. 1938. Physiological studies in plant nutrition. VII. The role of fructosans in the carbohydrate metabolism of barley plant. Part. 2. Seasonal changes in the carbohydrates with a note on the effect of nitrogen deficiency. *Annals of Botany* **2**: 403–436.

Champigny ML. 1985. Regulation of photosynthetic carbon assimilation at the cellular level: a review. *Photosynthesis Research* **6**: 273–286.

Duchateau N, Bortlik K, Simmen U, Wiemken A, Bancal P. 1995. Sucrose:fructan 6-fructosyltransferase, a key enzyme for diverting carbon from sucrose to fructan in barley leaves. *Plant Physiology* **107**: 1249–1255.

Ericsson T, Larsson C.-M., Tillberg E. 1982. Growth responses of *Lemna* to different levels of nitrogen limitation. *Zeitschrift für Pflanzenphysiologie* **104**: 331–340.

Fiske CH, Subbarow Y. 1925. The colorimetric determination of phosphorus. *Journal of Biological Chemistry* **66**: 375–400.

Foyer C, Spencer C. 1986. The relationship between phosphate status and photosynthesis in leaves. Effects of intracellular orthophosphate distribution photosynthesis and assimilate partitioning. *Planta* **167**: 369–375.

Housley TL, Pollock CJ. 1993. The metabolism of fructan in higher plants. In: Suzuki M, Chatterton NJ, eds. *Science and Technology of Fructans*. Boca Raton, FL: CRC Press, 191–226.

Huber JL, Hite DRC, Outlaw WH, Jr, Huber SC. 1991. Inactivation of highly activated spinach leaf sucrose–phosphate synthase by dephosphorylation. *Plant Physiology* **95**: 291–297.

Huber SC. 1986. Fructose 2,6-bisphosphate as a regulatory metabolite in plants. *Annual Review of Plant Physiology* **37**: 233–246.

Lehnert H-J, Vieweg GH, de Fekete MR. 1979. Starch accumulation in leaves from maize and *Zebrina pendula* Schinzl. dependent on nitrogen and phosphorus deficiency. *Zeitschrift für Pflanzenphysiologie* **92**: 162–169.

Pollock CJ. 1986. Environmental effects on sucrose and fructan metabolism. *Current Topics in Plant Biochemistry and Physiology* **5**: 32–46.

Pollock CJ, Cairns AJ. 1991. Fructan metabolism in grasses and cereals. *Annual Review of Plant Physiology and Plant Molecular Biology* **42**: 77–101.

Rebeille F, Bligny R, Douce R. 1982. Regulation of P uptake by *Acer pseudoplatanus* cells. *Archives of Biochemistry and Biophysics* **219**: 371–378.

Russell RS. 1938. Physiological studies in plant nutrition. IX. The effect of mineral deficiency on the fructosan metabolism of the barley plant. *Annals of Botany* **2**: 865–882.

Sicher RC, Kremer DF. 1988. Effects of phosphate deficiency on assimilate partitioning in barley seedlings. *Plant Science* **57**: 9–17.

Stitt M, Huber SC, Kerr P. 1987. Control of photosynthetic sucrose formation. In: Hatch MD, Boardman NK, eds. *The Biochemistry of Plants, vol. 10*. New York: Academic Press, 327–409.

Thorsteinsson B. 1989. *The effects of nutrient limitation on photosynthesis, respiration and levels of carbohydrates and hormones in plants*. Ph.D. thesis, University of Stockholm, Sweden.

Thorsteinsson B, Tillberg J-E, Tillberg E. 1987. Carbohydrate partitioning, photosynthesis and growth in *Lemna gibba* G3. II. Effects of phosphorus limitation. *Physiologia Plantarum* **71**: 271–276.

Walker DA, Robinson SP. 1978. Chloroplast and cell. A contemporary view of photosynthetic carbon assimilation. *Berichte der Deutschen Botanischen Gesellschaft* **91**: 513–526.

Wang C, Tillberg J-E. 1996. Effects of nitrogen deficiency on accumulation of fructan and fructan metabolizing enzyme activities in sink and source leaves of barley (*Hordeum vulgare*). *Physiologia Plantarum* **97**: 339–345.

Wiemken A, Sprenger, N, Boller T. 1995. Frucan – an extension of sucrose by sucrose. In Pontis HG, Salerno GL, Echeverria EJ, eds. *Sucrose Metabolism, Biochemistry, Physiology and Molecular Biology – Current Topics in Plant Physiology, An American Society of Plant Physiologists Series*. **14**: 179–189.

Woodrow IE, Ellis JR, Jellings A, Foyer CH. 1984. Compartmentation and fluxes of inorganic phosphate in photosynthetic cells. *Planta* **161**: 525–530.

New Phytol. (1997), **136**, 137–144

Accumulation of fructans following oxygen deficiency stress in related plant species with different flooding tolerances

By GERD ALBRECHT[1]*, SOPHIA BIEMELT[1] AND SABINE BAUMGARTNER[2]

[1] *Institute of Biology, AG-Botany, Humboldt University, Späthstr., 80/81, D-12437 Berlin, Germany*

[2] *IFA-Tulln, Centre of Analytical Chemistry, Konrad Lorenz Str. 20, A-3430 Tulln, Austria*

(*Received 24 July 1996*)

SUMMARY

In the present work we compare responses to hypoxia of the carbohydrate content of related plant species, which grow naturally on sites prone to flooding (*Senecio aquaticus* Hill., *Myosotis palustris* (L.) Lehm. Rchb.) with plants from habitats with only a low risk of oxygen shortage (*Senecio jacobaea* L., *Myosotis arvensis* (L.) Hill.). Whilst the amounts of glucose, fructose and sucrose changed only slightly following hypoxia and peaked at a maximum approximately double that of the aerated control specimens, the fructan content increased fivefold to tenfold. In nearly all the plants studied, fructan became the main polysaccharide reserve. For example, the flooding-tolerant *Senecio aquaticus* accumulated the highest amount, in particular fructans with a degree of polymerization up to 35 compared with 10 under control conditions. Nearly 70% of the soluble carbohydrates were fructans, compared with 30% under aerated conditions.

In all species tested the starch fraction marginally increased or remained constant.

Fructans were found to accumulate as a response to oxygen deficiency in both flooding-tolerant and intolerant species but with higher absolute values and ratios between fructan and starch in the flooding-tolerant species. At 24 h after the onset of hypoxic treatment, the sugar content rose, in spite of the diminished photosynthetic rates. The ability to accumulate fructans seems to vary in plants coping with oxygen shortage but could be interpreted as advantageous in comparison to starch synthesis in terms of (i) the location of the fructan metabolism in the vacuoles, with no negative feedback in the photosynthetic apparatus and (ii) the possibility of storage of sucrose in the form of fructose polymers without the intermediate stages and energy-using processes which are indispensable for starch synthesis. These results might indicate that fructan metabolism could play a role in the tolerance of oxygen deficiency.

Key words: Fructan, oxygen deficiency, flooding, *Senecio aquaticus*/*S. jacobaea*, *Myosotis palustris*/*M. arvensis*.

INTRODUCTION

Oxygen deficiency during flooding is a major limitation to plant growth and productivity. Maintenance of carbohydrate metabolism is a fundamental prerequisite for survival during a period of oxygen shortage. Substrate consumption associated with glycolysis dominated earlier studies on flooding stress and continues to be one of the most studied aspects of oxygen shortage research (Drew, 1990; Kennedy, Rumpho & Fox, 1992; Perata & Alpi, 1993; Crawford & Brändle, 1996). None the less, the question of availability of extensive energy resources and the control of substrate metabolism under oxygen deprivation needs to be clarified. Although there are no doubts as to the occurrence of alcoholic fermentation in plant tissues under anoxia, there is some debate about the extent and regulation of fermentative metabolism. In contrast to an accelerated fermentation resulting in starvation of carbohydrates (Vantoai *et al.*, 1995) many plants accumulate sugars (Setter *et al.*, 1987) and fructans (Albrecht *et al.*, 1993) at low oxygen concentration. Kubin (1992) presented data that show a relationship between flooding tolerance and the ability of *Iris* species to store fructan as the main storage carbohydrate. In this paper the following questions are addressed.

* To whom correspondence should be addressed.

Table 1. *Dry weight (d. wt) of shoots and roots of* Myosotis *and* Senecio *species with different flooding tolerances*; M. arvensis, S. jacobaea–*flooding sensitive*, M. palustris, S. aquaticus–*flooding tolerant*

	D. wt (mg)						D. wt of Hypoxia (%) Control (100 %)	
	Control			Hypoxia				
	Shoot	Root	S/R-ratio	Shoot	Root	S/R-ratio	Shoot	Root
M. arvensis	77·9	22·2	3·5	60·1	14·0	4·3	77·2	63·1
M. palustris	73·1	20·6	3·5	69·1	19·6	3·5	94·6	95·0
S. jacobaea	76·5	29·5	2·6	58·0	18·9	3·1	75·8	64·1
S. aquaticus	137·3	44·6	3·1	111·3	34·9	3·2	81·1	78·1

Plants were grown under both hypoxic and control conditions. Values are means of 10 measurements.

(a) Is fructan synthesis confined to the shoots or does it also occur in oxygen deficient roots?
(b) Do flooding-tolerant and intolerant species differ in fructan metabolism under restricted oxygen supply?
(c) Could fructan accumulation be of physiological importance in tolerance to flooding?

For our investigations we have chosen plants with different flooding tolerances and the ability to accumulate fructans and starch. We took respectively two species from the genera *Senecio* and *Myosotis* to provide the model for the ecophysiological study in carbohydrate metabolism and flooding tolerance. We compared reactions of related plant species naturally grown either on flooding-prone sites (*Senecio aquaticus* Hill., *Myosotis palustris* (L.) Lehm. Rchb.) or in habitats with only a low risk of oxygen shortage (*Senecio jacobaea* L., *Myosotis arvensis* (L.) Hill.). We investigated the concentrations and pattern of carbohydrates following growth in the aerated, oxygen-sufficient (control) and in nitrogen-flushed, oxygen-deficient nutrient solution (hypoxia).

MATERIALS AND METHODS

Plant growth

Seeds of *Myosotis arvensis*, *M. palustris*, *Senecio jacobaea* and *S. aquaticus* were grown on a vermiculite substrate. At 7 wk they were transferred to aerated hydroponics (35 % Knops-solution). After 1 wk the root medium was continuously flushed with either air (C–control, oxygen concentration > 8·3 mg l^{-1}) or pure nitrogen (H–hypoxia, oxygen concentration < 0·05 mg l^{-1}). The shoots were cultivated in air under light/dark cycles (16 h, 22 °C 180 μmol s^{-1} m^{-2}; 8 h, 17 °C).

Samples were taken on day 9 of each treatment.

Assay of carbohydrates

Shoot and root samples (always 1 g f. wt) were boiled once for 15 min in 10 ml of 20 % ethanol and three times for 15 min in 10 ml of water. The extracts were pooled, purified with anion-exchange resin (Amberlite IRA-68) and cation-exchange resin (Amberlite IRC-50) then by C-18 cartridges (Sep-Pak, Waters-Millipore) and dried under vacuum (Albrecht *et al.*, 1993). The residues were dissolved in 1 ml water. The water-soluble carbohydrates (wsC) were analysed by different HPLC systems. The carbohydrates were separated using a cation-exchange column (Sugar-Pak, Waters-Millipore) and eluted with water at a flow rate of 0·45 ml min^{-1}. Sugars were detected with a differential refractometer (Waters 410) and quantified using fructose, glucose, sucrose and inulin (from Jerusalem Artichokes – Sigma) as standards. Fructan oligomers were quantified individually using HPLC (Heinze & Praznik, 1991) with an rp-column (Spherisorb ODS II, Knauer-Berlin) eluted with a gradient (1–10 % methanol in 60 min) at 0·8 mol min^{-1} and detected by Varex ELSD (Evaporative light scattering detector). Additionally fructans were quantified on a Dionex-System using the following parameters: detector, Fa. Salus-Braumapharm; flow rate, 0·8 ml min^{-1}; column, PA-100; eluent, 100 mM sodium hydroxy – 600 mM sodium hydroxide/500 mM sodium acetate in 90 min. Fructan standards with DP 3, DP 4, and DP 5 were prepared from kestose syrup (FA. Bio-Rad, 1 m × A2·5 cm internal diameter).

Starch contents were measured on residues from the water-soluble carbohydrate extraction or on replicate tissue samples. Either the iodine precipitation method of Lustinec, Hadacova & Prochazka (1983) or the anthrone method as modified by Brändle (1985) were used.

Gas-exchange measurements

Photosynthetic CO_2-exchange rates were measured, 2 h into the daily light period using an open infra-red gas-analyser (model LCA 4, Hoddesdon). For these measurements single leaves from five different plants were selected and placed in the cuvette. This was repeated twice 1 h before the carbohydrates were extracted.

Table 2. *Content of carbohydrates (mg* g^{-1} *f. wt) in* Myosotis *and* Senecio *species with different flooding tolerances following hypoxic treatment (Hypoxia, nitrogen flushed nutrient solution; Control, aerated nutrient solution)*

Sugars	*M. arvensis*		*M. palustris*		*S. jacobaea*		*S. aquaticus*	
mg g^{-1} f. wt	Control	Hypoxia	Control	Hypoxia	Control	Hypoxia	Control	Hypoxia
(*a*) Shoots								
Fructose	0·7 (0·31)	1·08 (0·19)	0·86 (0·33)	1·52 (0·35)	1·17 (0·26)	3·49 (0·44)	1·22 (0·28)	2·86 (0·64)
Glucose	0·74 (0·19)	1·24 (0·24)	0·76 (0·29)	1·76 (0·23)	1·22 (0·18)	3·67 (0·57)	2·37 (0·43)	3·91 (0·67)
Sucrose	1·19 (0·21)	1·57 (0·25)	1·32 (0·27)	2·99 (0·13)	3·85 (0·33)	8·24 (0·62)	5·95 (0·54)	9·88 (0·94)
F+G+S	2·63	3·89	2·94	6·27	6·24	15·42	9·54	16·65
Fructan	1·95 (0·51)	4·06 (0·57)	2·44 (0·23)	5·78 (0·16)	3·62 (0·45)	13·52 (1·56)	5·04 (0·60)	30·88 (6.09)
wsC (F+G+S+Fr)	4·58	7·95	5·38	12·05	9·86	28·92	14·58	47·53
Starch	3·26 (0·24)	4·87 (0·07)	3·6 (0·01)	3·12 (0·39)	3·19 (0·5)	4·37 (0·6)	5·09 (0·04)	10·07 (2·43)
Starch+wsC	7·84	12·82	8·98	15·71	13·05	33·29	19·67	57·6
(*b*) Roots								
Fructose	0·41 (0·10)	0·65 (0·15)	0·46 (0·13)	0·62 (0·1)	1·16 (0·57)	2·98 (0·52)	0·88 (0·04)	1·07 (0·23)
Glucose	0·30 (0·03)	1·24 (0·27)	0·46 (0·09)	0·86 (0·2)	2·72 (0·59)	3·74 (0·45)	1·21 (0·26)	1·52 (0·28)
Sucrose	0·31 (0·03)	0·86 (0·10)	0·85 (0·12)	1·56 (0·11)	4·48 (0·63)	7·89 (1·02)	4·58 (1·36)	3·19 (0·63)
F+G+S	1·02	2·75	1·77	3·04	8·36	14·61	6·67	5·78
Fructan	1·51 (0·08)	2·78 (0·22)	1·41 (0·08)	2·91 (1·19)	4·21 (0·63)	10·14 (2·18)	4·4 (0·69)	24·93 (3·62)
wsC (F+G+S+Fr)	2·53	5·53	3·18	5·95	12·57	24·75	11·07	30·71
Starch	1·29 (0·23)	1·17 (0·17)	1·48 (0·07)	1·73 (0·05)	1·48 (0·2)	2·18 (0·50)	2·19 (0·38)	2·84 (0·34)
Starch+wsC	3·82	6·80	4·66	7·68	14·05	26·93	13·26	33·55

Values are the means of five independent measurements (standard deviation in parentheses).

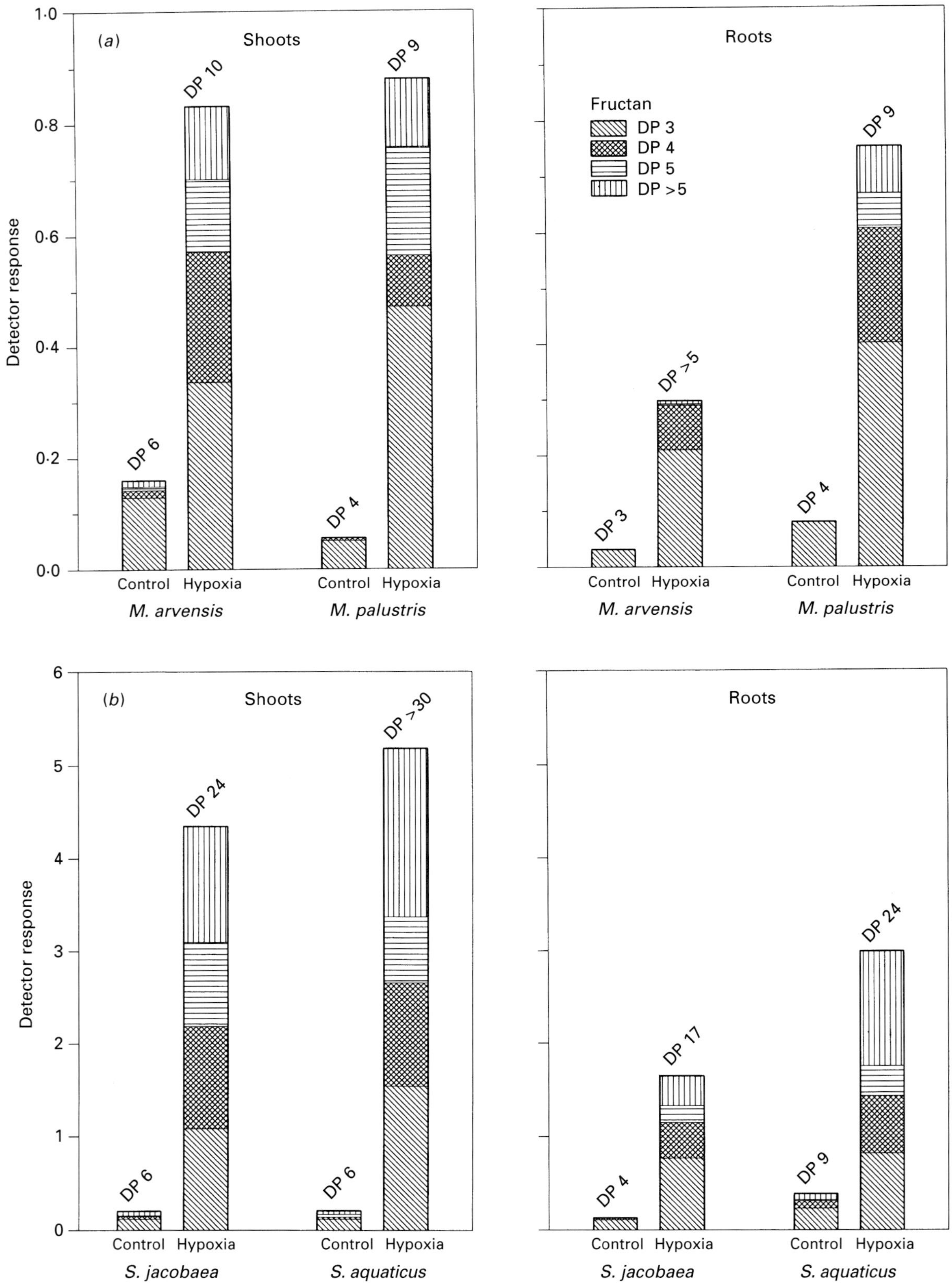

Figure 1. Content of fructans in the shoots and roots of (*a*) *Myosotis* and (*b*) *Senecio* species following hypoxic treatment. The fructans were separated accordingly to their DP and detected by an evaporative light-scattering detector.

RESULTS

The species differed significantly in their biomass (d. wt) increase with the onset of the hypoxic treatment. Roots suffered immediate growth reductions resulting from oxygen deficiency and were much more affected than the shoots (Table 1). The species most affected by oxygen deficiency to its roots was *M. arvensis*, followed by *S. jacobaea*, neither of which is flooded in its natural habitat.

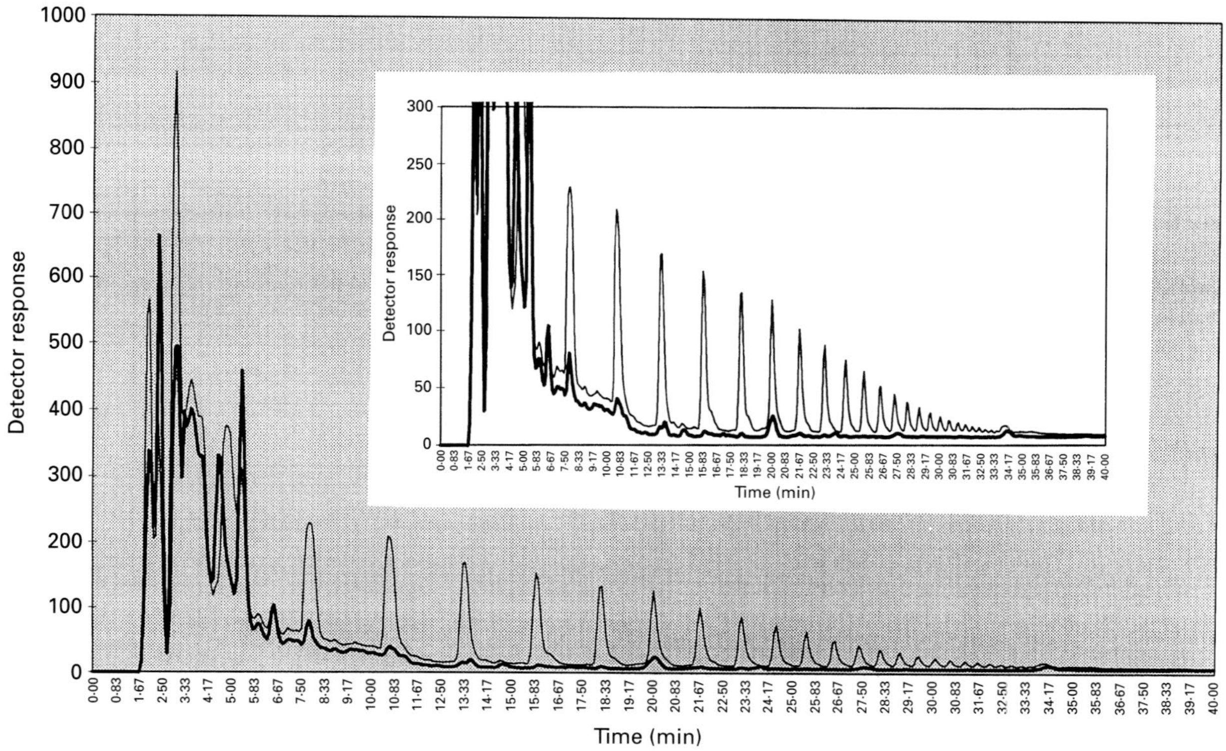

Figure 2. Chromatogram of fructans from the shoots of *Senecio aquaticus* either following 9 d of nitrogen-flushing of the nutrient solution (H, ——) or cultivation in aerated nutrient solution (C, ——).

Flooding tolerance of each species was reflected by growth under hypoxia; *M. palustris* > *S. aquaticus* > *S. jacobaea* > *M. arvensis*.

Remarkably, root and shoot retardation in the tolerant species were similar, with only slight increases in the shoot:root ratios, in contrast to significant increases in the intolerant species *S. jacobaea* and especially in *M. arvensis* (Table 1).

Following nitrogen-flushing of the nutrient solution for 9 d, there was an increase in water-soluble carbohydrates (wsC) of shoots, compared with those of the aerated control specimens (Table 2*a*). The increases were fourfold in *S. aquaticus*, threefold in *S. jacobaea*, and at least twice as much was found in the *Myosotis* species. Similar remarkable increases also occurred in the roots (Table 2*b*).

Whilst the amounts of glucose, fructose and sucrose changed only slightly, and peaked at a maximum approximately double that of the aerated control specimens, those of fructans were increased twofold to tenfold (Fig. 1). Most of the changes in carbohydrate content were due to fructans. In the species studied, fructans differed strongly both quantitatively and qualitatively in their response to oxygen deficiency. The shoots of the flooding-tolerant species *S. aquaticus* (Fig. 2) and *M. palustris* formed more fructans when grown under oxygen deficiency than did their flooding-intolerant relatives *S. jacobaea* and *M. arvensis*. Following oxygen shortage in the root environment, the fructans became the main pool of wsC in all species (Table 2*a*, *b*). This is also true if we take the starch fraction into account. In contrast to only a slight increase in starch, a significant increase in fructan content was induced by restricted oxygen supply. Only in the most flooding-sensitive species (*M. arvensis*) were the amounts of fructans and starch not significantly different, because of only a slight increase in fructans. In the other species, the fructan pool under hypoxia clearly exceeded the starch pool. An appreciable starch increase (a doubling) was measured only in the shoots of *S. aquaticus*.

The genera differed strongly in the content and size of fructans. *Myosotis* species in particular had lower concentrations of fructans than did *Senecio* species (Fig. 1). Genus-specific differences were also observed in the degree of polymerization (DP). Shoots of the *Myosotis* species reached a DP of 10 (*M. arvensis*) and 9 (*M. palustris*) under oxygen shortage, compared with a maximum DP of 6 (*M. arvensis*) and 4 (*M. palustris*) in the control specimens. The DP in *S. jacobaea* increased from 6 to 24 and in *S. aquaticus* from 6 to more than 30 after the roots were subjected to hypoxia for 9 d. The oligomers of DP 3 and 4 in *Myosotis* species seem to be distinctive, and the structure of the representatives of the *Boraginales* seems not to be so 'symmetrical' as those of the *Asterales* (*Senecio*). Structural details require further research.

A substantial increase in fructan content occurred in all four species in spite of diminished photosynthetic rates under restricted oxygen supply (Fig. 3). Immediately after the onset of oxygen shortage in the nutrient solution, the photosynthetic rates de-

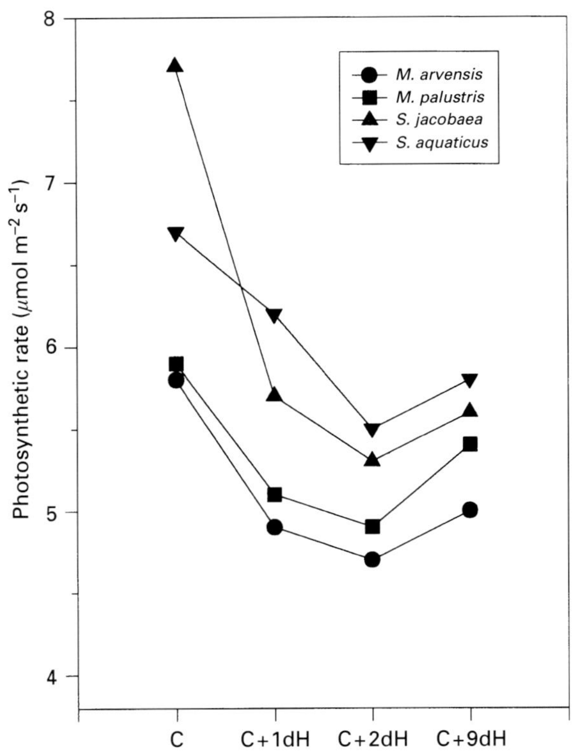

Figure 3. Photosynthetic rates of *Senecio* and *Myosotis* species following 1, 2 or 9 d of hypoxia. Each data point represents the means of at least four independent experiments. (SD < 15 %).

creased with a more marked drop in the flooding-sensitive species *M. arvensis* and *S. jacobaea*. After 2 d, photosynthetic rates began to increase. After 9 d of oxygen shortage, rates were nearly 90 % of the flooding-tolerant counterparts.

DISCUSSION

The accumulation of fructans in shoots and roots during oxygen shortage indicates a rate of carbon fixation that exceeds the rate of utilization (Albrecht *et al.*, 1993; Schlüter, Albrecht & Wiedenroth, 1996). Under hypoxia, both processes are restricted. Immediately after the onset of hypoxic treatment there is a more significant drop in carbon fixation in the flooding-sensitive species (*M. arvensis*, *S. jacobaea*) than in that of their tolerant counterparts (*M. palustris*, *S. aquaticus*). The average value for the decrease in the photosynthetic rate is nearly 25 % in the tolerant and 35 % in the intolerant species. Following 9 d of hypoxia to the roots, the average value for the photosynthetic rate again increased to about 80 % of the control, without a significant difference between flooding-tolerant and -intolerant species. One may conclude that the accumulation of sugars was mainly due to differences in carbon partitioning and utilization and was not caused by significant differences in carbon fixation. Low ambient temperatures also alter the balance between carbon assimilation and utilization and result in a pronounced increase in fructan content (Pollock, 1984; Bancal & Gaudillère, 1989; Wagner & Wiemken, 1989; Tognetti *et al.*, 1990). However, it is not known whether fructans are directly involved in the mechanism of cold adaptation.

After 1 d of hypoxic treatment the reducing sugars and fructans increase although photosynthetic rates are reduced. This response could be caused by a less sensitive photosynthetic rate and a very sensitive growth rate in plants exposed to restricted oxygen supply to the root environment. Growth is a good indicator of hypoxic stress (Thomson *et al.*, 1990); apparently roots stop elongation shortly after transfer from aerated to nitrogen-flushed nutrient solution (Wiedenroth & Erdmann, 1985).

The roots acclimatised to hypoxia over 9 d and resumed elongation. However, flooding-tolerant and flooding-intolerant species differed strongly in growth rates. Roots suffered directly from oxygen shortage and were more sensitive than the shoots. This resulted in an increase in the shoot to root ratio (S/R). Sensitive species had a larger increase in S/R. The S/R-ratio of flooding-intolerant *M. arvensis* rose from 3·5 to 4·3. In *S. jacobaea* it was less affected, and hardly changed in *S. aquaticus* and *M. palustris*. Thus, accumulation of fructans is not easily explained by retarded growth in species with slightly altered biomass production because they accumulated more soluble sugars. Reduced growth seems to be imposed by strongly restricted respiration due to inadequate supplies of oxygen. nevertheless, sufficient energy must be available to enable plants to survive. It seems reasonable to suppose that part of the energy supply comes from the induction of glycolytic and fermentative enzymes, since there is no other known mechanism by which ATP can be generated. Recently Andrews *et al.* (1994) showed that the induction of transcripts for the glycolytic and ethanolic fermentative pathways in roots during hypoxia differed strongly in rapidity and magnitude. Message levels of glycolytic enzymes (aldolase and enolase) were < twofold whilst mRNAs for pyruvate decarboxylase (fermentative enzyme) increased about 15-fold during the first 12 h of hypoxia, then declined to initial amounts after 18–24 h. These findings agree with the results of Albrecht & Wiedenroth (1994), whose measurement of carbon dioxide release by roots under nitrogen compared with aerated conditions showed that oxygen shortage did not cause a substantial acceleration in CO_2 production, suggesting relatively low glycolytic activity and substrate consumption. Glycolysis is highly controlled by the activity of phosphofructokinase (PFK), whose activity in *S. aquaticus* was not increased under hypoxic conditions (Smith & ap Rees, 1979).

The proposed pathway of sucrose metabolism

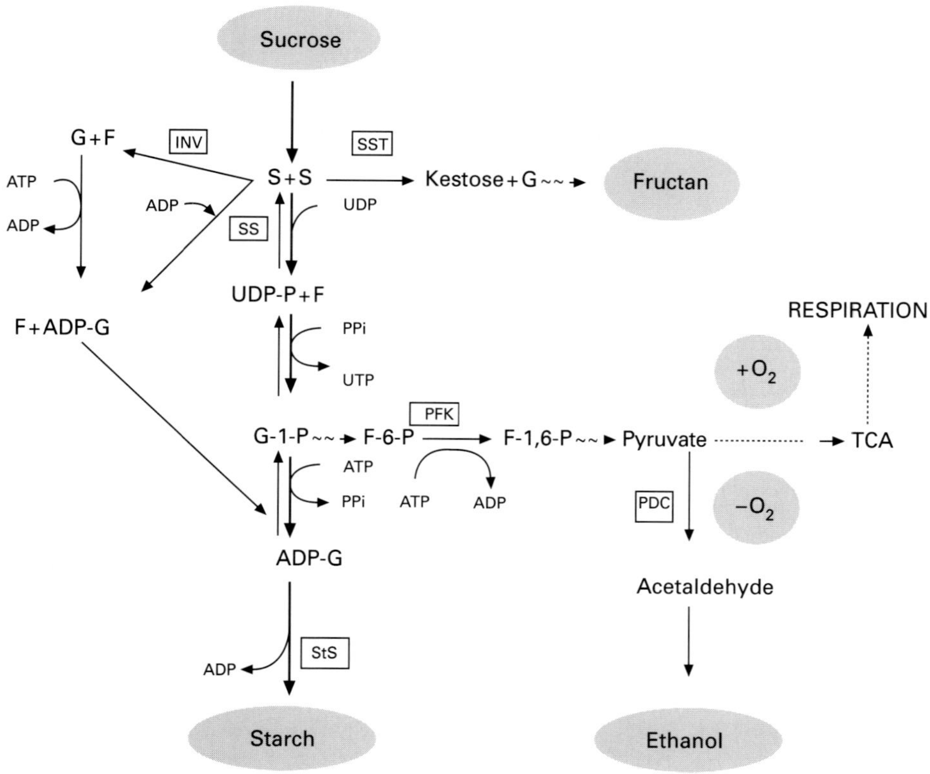

Figure 4. Proposed pathway of sucrose in catabolic (respiration, fermentation) and anabolic (synthesis of fructans and starch) metabolism with varying oxygen availability.

under oxygen deprivation is given in Figure 4. The energy-rich sucrose, synthesized in shoots, is transported to the oxygen-deprived tissues of the roots. Sucrose degradation occurs via invertase and sucrose synthase (SS). The results of several authors (Chourey *et al.*, 1991; Ricard *et al.*, 1991) suggest that under anoxia/hypoxia the SS pathway plays a major role. This is supported by a decreased activity of alkaline invertase activity under anoxia and a strong increase of SS activity (Guglieminetti *et al.*, 1995). Moreover, the flow of carbon from sucrose to hexose monophosphates through the SS pathway results in a considerable ATP saving over the invertase pathway. Afterwards, the UDP-glucose is converted to glu-1-P, which is a branching point between catabolic and anabolic carbon pathways. It has been shown that the carbon flux channelled to the glycolytic pathway is at most slightly accelerated (Albrecht & Wiedenroth, 1994; Andrews *et al.*, 1994). Pyruvate, the end product of glycolysis, can hardly be used in respiration if the tissue is deprived of oxygen. To counteract the decreased ATP production in the absence of oxygen, pyruvate is partially channelled to ethanolic fermentation.

Sucrose which is not used for catabolic degradation in the absence of oxygen might be consumed as storage carbohydrates. Fructans were accumulated in response to oxygen deficiency in both flooding-tolerant and flooding-intolerant species but with higher absolute values and ratios between fructan to starch in the flooding-tolerant species. The ability to accumulate fructans seems to differ between plants that cope with oxygen deficiency but could be an advantage in comparison to starch synthesis in terms of (i) lack of negative feedback on the photosynthetic rates and (ii) the storage of sucrose as fructose polymers in the vacuoles without the intermediate stages and energy conversion processes required for starch synthesis. For the contemporary flora, there is no single description of distribution of fructan-rich species. The fructan flora is not favoured by wet nor by cold temperatures (Hendry, 1993). However it appears that plants with the ability to form fructans are important components of ecosystems that experience frequent environmental changes. The intermediate storage of sucrose in the form of fructans seems to be of importance in overcoming temporary stresses as imposed by hypoxia. The stored fructans can be used immediately after re-aeration, leading to a greater re-growth of the roots previously suffering from hypoxia (Albrecht *et al.*, 1993).

The accumulation of stored carbohydrates under oxygen deficiency is on the one hand considered as a metabolic dysfunction in carbohydrate utilization, but on the other hand the roots are able to maintain important metabolic functions. The roots continue to operate as a sink by using the imported sucrose to form fructans and do not use it in a wasteful and harmful way (Pasteur effect). We conclude, therefore, that fructan metabolism might play a role in tolerance of oxygen deficiency.

ACKNOWLEDGEMENTS

We wish to thank Birgit Haß for technical assistance and Roman Sommerfeld for skilful help during the gas-exchange measurements.

REFERENCES

Albrecht G, Wiedenroth EM. 1994. Is long-term hypoxia met by the Pasteur effect in roots of wheat seedlings? *Proceedings of the Royal Society of Edinburgh* **102b**: 407–412.

Albrecht G, Kammerer S, Praznik, W, Wiedenroth EM. 1993. Fructan content of wheat seedlings (*Triticum aestivum* L.) under hypoxia and following re-aeration. *New Phytologist* **123**: 471–476.

Andrews, DL, MacAlpine DM, Johnson JR, Kelley PM, Cobb BG, Drew MC. 1994. Differential induction of mRNAs for glycolytic and ethanolic fermentative pathways by hypoxia and anoxia in maize seedlings. *Plant Physiology* **106**: 1575–1582.

Bancal P, Gaudillère JP. 1989. Rate of accumulation of fructan oligomers in wheat seedlings (*Triticum aestivum* L.) during the early stages of chilling treatment. *New Phytologist* **112**: 459–463.

Brändle R. 1985. Kohlenhydratgehalte und Vitalität isoloierter Rhizome von *Phragmites australis*, *Schoenoplectus lacustris* und *Typha latifolia* nach mehrwöchigem O_2-Mangelstreß. *Flora* **177**: 317–321.

Chourey PS, Taliercio EW, Kane EJ. 1991. Tissue-specific expression and anaerobically induced posttranscriptional modulation of sucrose synthase genes in *Sorghum bicolor* M. *Plant Physiology* **96**: 485–490.

Crawford RMM, Brändle R. 1996. Oxygen deprivation stress in a changing environment. *Journal of Experimental Botany* **47**: 145–159.

Drew MC. 1990. Sensing soil oxygen. *Plant, Cell and Environment* **13**: 681–693.

Guglielminetti L, Perata P, Alpi A. 1995. Effect of anoxia on carbohydrate metabolism in rice seedlings. *Plant Physiology* **108**: 735–741.

Hendry GAF. 1993. Evolutionary origins and natural function of fructans – a climatical, biogeographical and mechanistic appraisal. *New Phytologist* **123**: 3–14.

Heinze B, Praznik, W. 1991. Separation and purification of inulin oligomers and polymers by reversed-phase high performance liquid chromatography. *Journal of Applied Polymer Science* **48**: 207–225.

Kubin P. 1992. Fructans: the role in adaptation to hypoxia. In: Pollock CJ, ed. *Second International Symposium on Fructans, Aberystwyth, UK*, 1992, 2–13.

Kennedy RA, Rumpho ME, Fox TC. 1992. Anaerobic metabolism in plants. *Plant Physiology* **100**: 1–6.

Lustinec J, Hadacova V, Prochazka Z. 1983. Quantitative determination of starch, amylose, amylopectin in plant tissue using glassfiber paper. *Analytical Biochemistry* **132**: 265–271.

Perata P, Alpi A. 1993. Plant responses to anaerobiosis. *Plant Science* **93**: 1–17.

Pollock CJ. 1984. Sucrose accumulation and the initiation of fructan biosynthesis in leaves of *Lolium temulentum* L. *New Phytologist* **96**: 527–534.

Ricard B, Rivoal J, Spiteri A, Pradet A. 1991. Anaerobic stress induces the transcription of sucrose synthase in rice. *Plant Physiology* **95**: 669–674.

Schlüter U, Albrecht G, Wiedenroth EM. 1996. Content of water-soluble carbohydrates under oxygen deprivation in plants with different flooding tolerance. *Folia Geobotanica Phytotax* **31**: 57–64.

Setter TL, Waters I, Atwell BJ, Kupkanchanakul T, Greenway H. 1987. Carbohydrate status of terrestrial plants during flooding. In: Crawford RMM, ed. *Plant Life in Aquatic and Amphibious Habitats*, British Ecological Society. Oxford: Blackwell Scientific Publishers.

Smith AM, ap Rees T. 1979. Pathways of carbohydrate fermentation in the roots of marsh plants. *Planta* **146**: 327–334.

Thomson CJ, Armstrong W, Waters I, Greenway H. 1990. Aerenchyma formation and associated oxygen movement in seminal and nodal roots of wheat. *Plant Cell and Environment* **13**, 395–403.

Tognetti JA, Salerno GI, Crespi MD, Pontis HG. Sucrose and fructan metabolism of different wheat cultivators at chilling temperatures. *Physiologia Plantarum* **78**: 554–559.

Vantoai TT, Saglio P, Ricard B, Pradet A. 1995. Developmental regulation of anoxic stress tolerance in maize. *Plant, Cell and Environment* **18**: 937–942.

Wagner W, Wiemken A. 1989. Fructan metabolism in expanded primary leaves (*Hordeum vulgare* L. cv. Gerbel): change upon ageing and spatial organization along the leaf blade. *Journal of Plant Physiology* **134**: 237–242.

Wiedenroth EM, Erdmann B. 1985. Morphological changes in wheat seedlings (*Triticum aestivum* L.) following root anaerobiosis and partial pruning of the root system. *Annals of Botany* **56**: 307–316.

New Phytol. (1997), **136**, 145–152

Fructan to nitrogen ratio as an indicator of nutrient stress in wheat crops

BY V. B. MCGRATH*, A. B. BLAKENEY AND G. D. BATTEN

NSW Agriculture, Yanco Agricultural Institute, Yanco, NSW 2703, Australia

(*Received 24 July 1996*)

SUMMARY

Wheat stems serve as a store for fructans to buffer the plant against nutritional and environmental influences. It has been suggested that fructan storage influences yield stability and tolerance of environmental factors. Near infra-red spectroscopy (NIR) analysis provides a rapid and accurate assessment of the fructan content of the wheat stem, as well as allowing detection of growth-limiting nutrient stresses, and so is proving to be a useful technique for making crop management decisions. Commercial laboratories using NIR analysis have been tissue-testing crops in the eastern Australian wheat belt since 1993. In healthy, normally developing crops not under stress there is a predictable relationship between nitrogen and fructan. Investigation of the nitrogen and fructan concentrations in commercial crops has confirmed an inverse relationship between these two constituents. The function: Fructan (%) $= a + b\text{N}\,\% + c\text{N}\,\%^2$ accounted for up to 81% of variation in tissue fructan concentration. In commercial tissue-testing this relationship is used to detect crops under stresses other than nitrogen deficiency. If the fructan concentration deviates by more than 4%, cereal growers are advised that their crop might be subject to other stresses which might reduce its response to applied nitrogen.

Key words: Fructan, wheat, near infra-red spectroscopy, non-structural carbohydrate, nitrogen.

INTRODUCTION

Fructan-rich cereals grown in temperate zones of the world have great economic importance for their human value and use in animal production. Wheat, barley, oats and rye are the dominant cereal crops outside the tropics (Hendry & Wallace, 1993). The fructan component is the major carbohydrate in vegetative tissue of wheat and accumulates in the stems, as well as leaf shoots, shoot apices, and growing zones of the leaves and roots. Higher concentrations of fructans normally occur in young tillers, during the tillering growth stage of a wheat crop (Suzuki, 1993). Fructan reserves might be utilized by the plant to overcome interference with regular photosynthesis (Geiger & Servaites, 1991) or be used by the plant during ear development (Wardlaw, 1990).

Non-structural carbohydrates stored in the stem can be mobilized easily and contribute to the growth of grain (Borrel *et al.*, 1989). They can also accumulate in, or be depleted from, tissue when growth is impeded by an environmental stress such as low temperature or moisture deficiency, disease or nutritional influences including the deficiencies of essential elements (Batten *et al.*, 1993). In order to optimize crop production, potential stresses affecting plant growth should be identified and corrected *in situ* if possible.

The determination of non-structural carbohydrates is usually a slow and expensive process, and so has not been done on a routine basis. Near infra-red spectroscopy (NIR) allows rapid, accurate and inexpensive routine analysis of total nitrogen (Batten *et al.*, 1991) and fructan (Batten & Blakeney, 1992*a*) content of samples of the shoots of wheat crops. We have developed an NIR calibration for wheat tissue fructan using an acid thiobarbituric acid colorimetric method as the reference method. The original calibration set included 20 spring wheat cvs, with a range of fructan concentrations from 2·4% to 28%; R^2 (correlation coefficient) of 0·99; SEC (standard error of calibration) of 0·92 and SEP (standard error of prediction) of 0·98 (Blakeney, Batten & McGrath, 1992). Whole plants are cut off at ground level by the farmer or agronomist and microwave-dried (Batten *et al.*, 1995) to ensure stability of the sample during transport to the laboratory.

From the NIR analysis of several hundred samples from commercial wheat crops in 1992 the following function was derived:

$$\text{Fructan } (\%) = a + b\text{N}\,\% + c\text{N}\,\%^2 \; (R^2 = 0{\cdot}81).$$

* To whom correspondence should be addressed.

The determination of N and fructan simultaneously by NIR has shown that an inverse relationship exists between these two constituents (Batten *et al*., 1993). This paper reports the relation between fructan and N for wheat grown over several seasons, for several cultivars and for other cereals such as oats and barley.

Fructans are widely distributed in all parts of the wheat plant, but the amount that occurs in a particular tissue will depend on the stage of development. Commercial testing of Australian wheat crops at the tillering to stem elongation stage has shown that a wide range of fructan concentration exists in wheat tissue and the upper limit reaches 36%. The distribution of fructan and N in a wheat variety at booting will be discussed in this paper as well as the changing concentration of fructan over the growing season of the same wheat variety.

Suzuki (1993) states that fructan reserves can make a major contribution to grain yield under stressed conditions. This paper reports how by utilizing the NIR tissue testing for the identification, and possible correction of stresses affecting growth increases in grain yield can be realized (McGrath *et al*., 1995). NIR testing does not reveal the nature of the stress, but the laboratory advises the grower to check for disease or nutritional causes. In the near future we expect to be able to measure other elements, such as P and S, simultaneously with N and fructan using NIR spectrophotometry. The value of measuring the fructan to N ratio will be enhanced with the correction of deficiencies of these two elements.

MATERIALS AND METHODS

Industry-funded research at Yanco Agricultural Institute led to a commercial wheat tissue-testing service being available to wheat growers in New South Wales (NSW) in 1993. Near-infra-red (NIR) spectroscopy is utilized to measure fructan and N content of shoot samples from wheat crops. The samples are dried by a microwave technique, and sent to a laboratory for grinding and analysis (Fig. 1). This testing will be available in all wheat-growing states of Australia in 1996.

Calibration of the NIR spectrophotometer by the researchers at Yanco Agricultural Institute involves relating the absorbance of energy in the wavelength range 1100–2500 nm to the known composition of a set of samples. Samples with known concentrations of fructan, uniformly spread across the range available, were packed into 50-mm diameter NIR cells and scanned using a 19 filter NIR instrument (Batten *et al*., 1993). Calibrations were developed using the log 1/reflectance (1/R) values. The advanced instrument design and complex software permit precision equal to that of wet chemistry methods. Thus, the chemical composition of unknown samples can be predicted accurately (SEP = 0·98 for fructan and 0·09 for N, Table 1) using NIR analysis. The use of a scanning NIR instrument for the past 4 yr and the application of partial least squares regression has refined the calibrations further.

The chemical determination of values for the NIR calibrations was obtained by the Kjeldahl (Bradstreet, 1965) method for N and the thiobarbituric acid method for fructan (Percheron, 1962; Blakeney & Mutton, 1980; Pontis, 1990). This method is specific for ketohexoses, and the total apparent fructose determined is used as a measure of fructan content. Monosaccharides and sucrose can be determined and deducted (Blakeney & Mutton, 1980) or destroyed (Pontis, 1966). The ballmilled plant sample (0·1 g) is extracted in diammonium phosphate solution (20 ml, 50 mM), mixed and heated for 20 min with two further vortex mixings, then filtered through Whatman No. 40 filter paper. An aliquot (100 μl) is heated with hydrochloric acid (1 ml, conc.) and 2-thiobarbituric acid (Tokyo Kosai G.R., 1 ml, 0·02 M) in a boiling water bath for 6 min. The fructan is hydrolysed and the fructose formed is dehydrated to give 5-hydroxymethylfurfural (HMF) (Fuchs, 1993). Thiobarbituric acid reacts with HMF to give an intense yellow colour which can be measured spectrophotometrically at 430 nm. We use both fructose and inulin standards. To achieve complete recovery of fructan by the ammonium phosphate solution, samples must be ballmilled to rupture the cell walls. We have attained higher and more reproducible extraction with ballmilled wheat tissue than with samples ground to pass a 0·5 mm screen. For the NIR analysis, samples are ground to pass a 0·5 mm screen.

Robust calibrations have been developed to measure N and fructan. The calibration set included samples from a number of years, growing environments and varieties (Table 1). These calibrations are based on whole-shoot samples with a fructan concentration ranging from 2·4 to 28% fructan.

The testing of a plant sample for N content makes the assumption that N is the factor limiting plant growth. To aid in testing the validity of this assumption we have developed the concept of the nutrient energy index (NEIDX). NIR-determined tissue N is used to predict the expected fructan concentration, which is then subtracted from the fructan value determined by the NIR. The mathematical absolute value of this subtraction is defined as the NEIDX, so that plants with either excess or deficient fructan for their N content are identified. Plants where growth is limited by disease, environmental and/or nutritional stresses, will have a high absolute NEIDX and will be accumulating or depleting their fructan reservoir more than a healthy plant. Crops showing irregularities are reported to the cereal grower as being affected by some factor

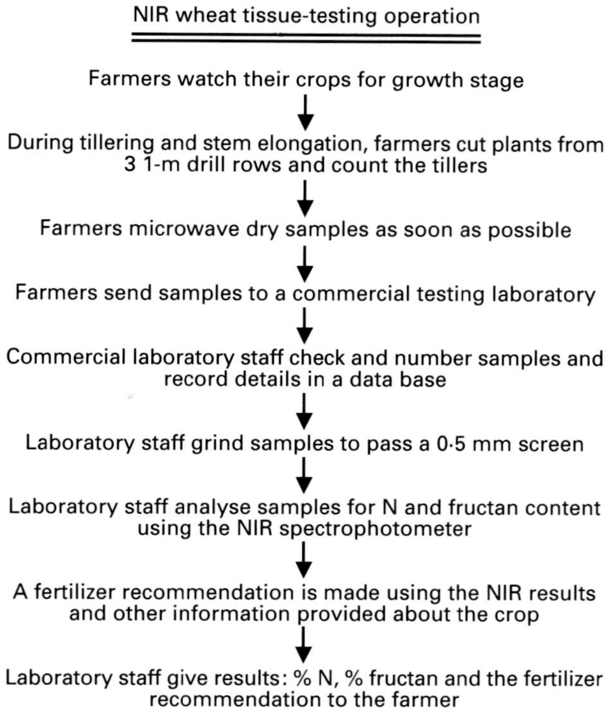

Figure 1. The NIR (near infra-red) wheat tissue-testing operation; from crop sampling to NIR analysis and final report including fertilizer recommendation and fructan:nitrogen ratio indicator to the wheat grower.

other than, or in addition to, N. Growth inhibition without a decrease in photosynthesis due to stress causes an accumulation of fructans. Primary inhibition of photosynthesis causes depletion of fructans, and accumulation of fructans is the trait associated with deficiencies in nutrients other than N. Where NEIDX is shown to be in the expected range and N is deficient, growers can apply top-dressing fertilizer with a good chance of increasing yield.

All grower and crop details are stored in an interactive database which is used to collate data relevant to each sample, calculate averages and generate a comprehensive report on each crop. All the data collected on crops are transferred to the master database at our laboratory.

To test the uniformity of distribution of fructan and N in our tissue-testing samples we used two cultivars of wheat plants which were dissected into (i) 5-cm segments of the stem, (ii) individual leaves and (iii) roots. These were analysed chemically for fructan, N and P.

The concentration of fructan in whole-plant samples was also measured at key growth stages from early tillering to physiological maturity. These wheat plants were grown in adjacent plots which had pre-designed low, moderate and high soil N status.

RESULTS AND DISCUSSION

Fructan : nitrogen relationship

Over 5000 wheat samples have been analysed for N and fructan by the NIR method in the past 3 yr. Studies of wheat crops grown in eastern Australia have revealed that generally there is an inverse relationship between wheat tissue fructan and tissue nitrogen content. Every year of commercial testing has shown this relationship, where the function Fructan (%) $= a + b\text{N}\,\% + c\text{N}\,\%^2$ (Batten *et al.*, 1993) accounts for up to 81 % of the variation in fructan content. Analysis of acid digests of some samples with a high NEIDX has revealed S-deficiency to be the stress. This is similar to findings in rice crops where a high starch:nitrogen index was linked to S-deficiency (Batten *et al.*, 1993). Samples collected during frosts, drought or from crops affected by waterlogging also have a high NEIDX. By studying the fructan:nitrogen ratio or NEIDX, conclusions can be drawn about environment–genotype interactions. This paper extends previous

Table 1. *Statistics for the fructan and nitrogen calibrations for wheat shoots using a filter NIR instrument*

	Nitrogen	Fructan
No. of samples	108	108
Range of values	0·94–4·94 %	2·4–28 %
R^2	0·995	0·99
SEC	0·114	0·915
SEP	0·092	0·98
No. varieties	20	20
F-ratio	2371	1285
Filter wavelength	2180, 2100, 1680	2310, 1818, 1778, 2100

R^2 = correlation coefficient; SEC = standard error of calibration, i.e. a measure of calibration fit calculated by determining the standard deviation of the difference between the instrument calculation and the reference method value on the calibration sample set; SEP = standard error of prediction, i.e. a measure of prediction calculated by determining the standard deviation of the difference between the instrument calculation and the reference method value on the validation sample set.

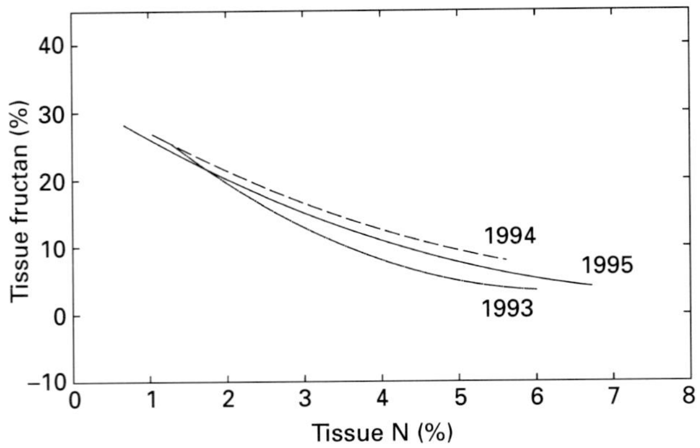

Figure 2. Comparison of the fructan : nitrogen relationship in wheat shoots from different seasons (1993, 1994 and 1995). Equation of the curve is: $y = a + bx + cx^2$ where y = %Fructan(NIR), x = %N(NIR), a = 39·2 (1993), 33·6 (1994) and 33·1 (1995); $b = -11·4$ (1993), $-7·0$ (1994) and $-7·3$ (1995); and c = 0·9 (1993), 0·4 (1994) and 0·4 (1995).

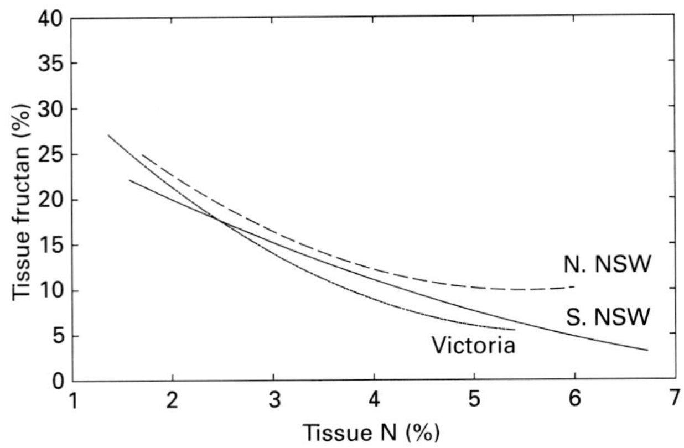

Figure 3. Comparison of the fructan : nitrogen relationship in wheat shoots from different environments (northern New South Wales, southern New South Wales and Victoria). Equation of the curve is: $y = a + bx + cx^2$ where y = %Fructan(NIR), x = %N(NIR), a = 32·9, $b = -11·5$ and c = 0·6 (northern New South Wales); a = 31·4, $b = -6·4$ and c = 0·7 (southern New South Wales); and a = 42·6, $b = -12·8$ and c = 0·7 (Victoria).

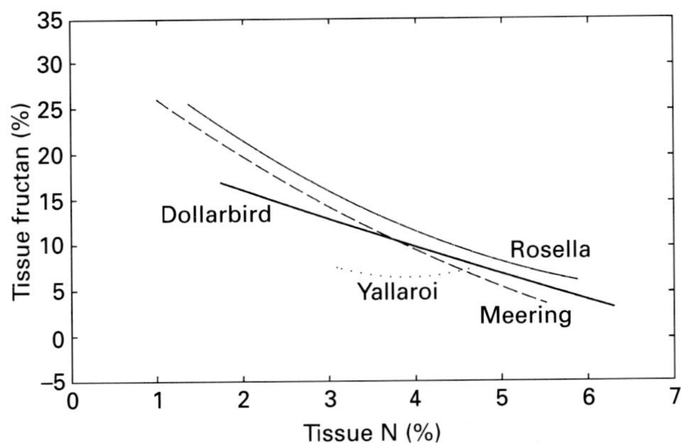

Figure 4. Comparison of the fructan : nitrogen relationship in wheat shoots of four wheat varieties ('Rosella', 'Dollarbird', 'Meering' and 'Yallaroi') in 1995. Equation of the curve is: $y = a + bx + cx^2$ where y = %Fructan(NIR), x = %N(NIR), a = 32·9, $b = -7·3$, and c = 0·9 ('Meering'); a = 35·7, $b = -8·2$ and c = 0·6 ('Rosella'); a = 31·8, $b = -13·0$ and c = 0·2 ('Dollarbird'); and a = 22·8, $b = -3·5$ and c = 0·7 ('Yallaroi').

studies on the accumulation and distribution of fructan within wheat plants and the relationship with N in cereal crops.

Seasonal variation

Figure 2 compares the regression curves for crops tested in three consecutive years. The 1994 season was a drought year, and the lack of subsoil moisture, combined with severe frosting, contributed to the relatively higher fructan concentrations observed. Fructans have been considered to play a role as osmotic regulators (Darbyshire & Henry, 1981; Pontis & del Campillo, 1985) and in chill and freeze tolerance (Eagles, 1967; Edelman & Jefford, 1968; Gunn & Walton, 1985; Pontis, 1989). The 1995 season was generally a better season, but frosts and waterlogging in some districts could be responsible for the upward movement of the curve, relative to that of 1993, which was quite favourable for winter crops, producing satisfactory yields and grain quality.

Differences between wheat-growing environments within the Australian wheat belt can be detected in the samples received at laboratories in northern NSW, southern NSW and Victoria (Fig. 3). The increase in mean temperatures and decrease in rainfall as one moves north through the wheat belt parallels the general increase in fructan to tissue N of crops tested.

Genotype and species variation

Significant genotypic variation in fructan accumulation exists in wheat (Dubois, Winzeler & Nösberger, 1990). A comparison of four wheat varieties revealed a noticeable difference between: 'Rosella', a NSW Australian Standard White grade (ASW) wheat; 'Meering', a Victorian ASW grade wheat; 'Dollarbird', a Hard to Prime Hard variety (Fig. 4). 'Yallaroi', a durum species, was the most divergent, but there were fewer samples, covering a narrower range of N content. Plant breeders can use NIR tissue-testing to assist in the selection of genotypes with high yield and high grain protein (Noaman, Taylor & Martin, 1990; O'Brien *et al.*, 1992; Rostami & O'Brien, 1993). Sampling the trials for tissue N might contribute to improving grain yield and protein content, as a high availability of N is essential for optimum expression of both traits. Varietal variation has also been observed in the starch : nitrogen ratio in rice crop testing. The higher-yielding, medium-grain cv. Amaroo had a higher starch content at a given N status than the taller, long-grain cv. Pelde (Batten *et al.*, 1993).

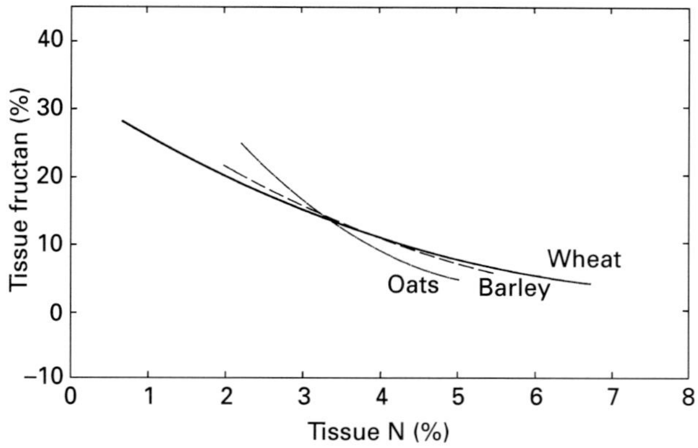

Figure 5. Comparison of the fructan:nitrogen relationship in shoots of different cereals (wheat, barley and oats) in 1995. Equation of the curve is: $y = a+bx+cx^2$ where y = %Fructan(NIR), x = %N(NIR), $a = 32{\cdot}9$, $b = -7{\cdot}2$, and $c = 0{\cdot}7$ (wheat); $a = 36{\cdot}5$, $b = -8{\cdot}5$ and $c = 0{\cdot}8$ (barley); $a = 59{\cdot}5$, $b = -19{\cdot}3$ and $c = 0{\cdot}8$ (oats).

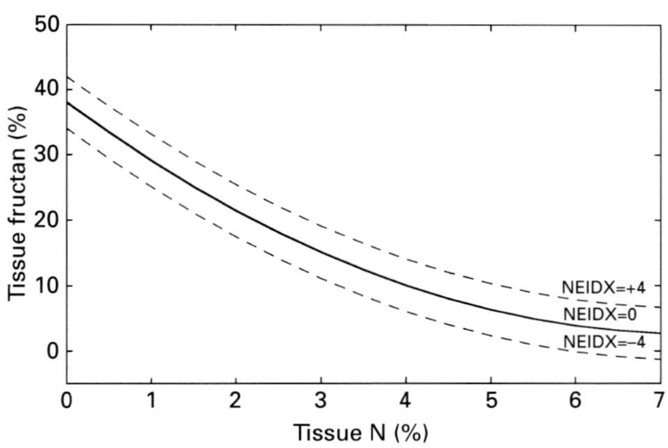

Figure 6. Relationship between N and fructan in wheat shoots. NEIDX = nutrient energy index, i.e. the absolute value of the subtraction of the expected fructan concentration (calculated by the NIR nitrogen value) from the fructan concentration measured by the NIR. The equation of this curve is Fructan(%) $= a+b\text{N}\,\% + c\text{N}\,\%^2$ where $a = 38{\cdot}1$, $b = -9{\cdot}6$ and $c = 0{\cdot}65$. The a, b and c values are the average of the coefficient values for 3 yr.

NIR calibrations have also been developed at Yanco Agricultural Institute to measure the N and fructan content of barley and oats. We are able to compare the fructan:nitrogen relationship of these two cereals to that of wheat in the same year (Fig. 5). Barley has a curve very similar to that found in wheat tissue-samples. Wheat, barley and rye accumulate large amounts of fructan during cold hardening (Meier & Reid, 1982). Oat is related to wheat and barley taxonomically and agronomically, but contains fructans with different structures (Livingston, Chatterton & Harrison, 1993), and this might explain why oat displays a very different fructan:nitrogen curve from those of wheat and barley.

Relevance to cereal growers

For the purpose of reporting results of crop analysis to growers (Fig. 1) we have developed a standard curve relating fructan and nitrogen by averaging the

Table 2. *Comparison of nutritional status of crops tissue-tested by NIR in 1995 (growth stage: z21–32)*

Status of crops	No. of crops
No growth-limiting factors (i.e. rec.fert. = 0; $\lvert$neidx$\rvert$ < 4)	293
Nitrogen is not limiting growth, but other factors might be (i.e. rec.fert. = 0; $\lvert$neidx$\rvert$ ⩾ 4)	120
Nitrogen is the growth-limiting factor (i.e. rec.fert. > 0; $\lvert$neidx$\rvert$ < 4)	444
Nitrogen and other factors are limiting growth (i.e. rec.fert. > 0; $\lvert$neidx$\rvert$ ⩾ 4)	285

Categorized according to recommended fertilizer rate (rec.fert.) based on nitrogen analysis only, and nutrient energy index (NEIDX), based on the fructan:nitrogen ratio (z = Zadok scale).

a, b and c constants in the equation Fructan (%) $= a+b\text{N}\,\% + c\text{N}\,\%^2$ for three years' data (Fig. 6). Crops which have $> \pm 4\,\%$ fructan than predicted using this equation from the NIR-N analysis are identified. The cereal grower is alerted on a grower report form. Of the 1142 samples analysed last season, one third were affected by a stress other than, or as well as, N-deficiency (Table 2). Correction of stresses, such as S-deficiency, has been possible as a result of the growers having the fructan:nitrogen ratio available as a management tool.

Distribution of fructan in wheat at booting

There is a mixture of soluble carbohydrates in the leaves, stems, apices (Pollock & Cairns, 1991) and roots of wheat plants at the growth stage we focused on. Fructans are the main component of these carbohydrates, the stems storing more than the leaves or roots (Fig. 7). The two wheat varieties stored more fructan in the stem segment: the basal stem containing about three times more fructan than the leaf with the highest fructan content. The stem is the major store of carbohydrate in the wheat crop, and fructans tend to accumulate there when the rate of production in the photosynthetic tissue exceeds the demand for them by other sinks (Borrell *et al.*, 1989). It has been previously reported (Smith, 1967; Pollock & Jones, 1979) that the lower stem internodes store more fructan than the upper, and our data confirm this distribution (Fig. 7), with fructan accounting for almost half the dry weight of the stem in the lowest 5 cm.

The roots of the cv. Wyuna were found to contain a high (11·5 %) concentration of fructan. McGrath (1988) stated that roots of perennial and Italian ryegrass generally had 5–10 % fructan. It has been suggested (Austin *et al.*, 1977) that this assimilate might be used as an energy source in reducing nitrate to nitrite to ammonia in the roots. Root growth which is dependent on this available assimilate

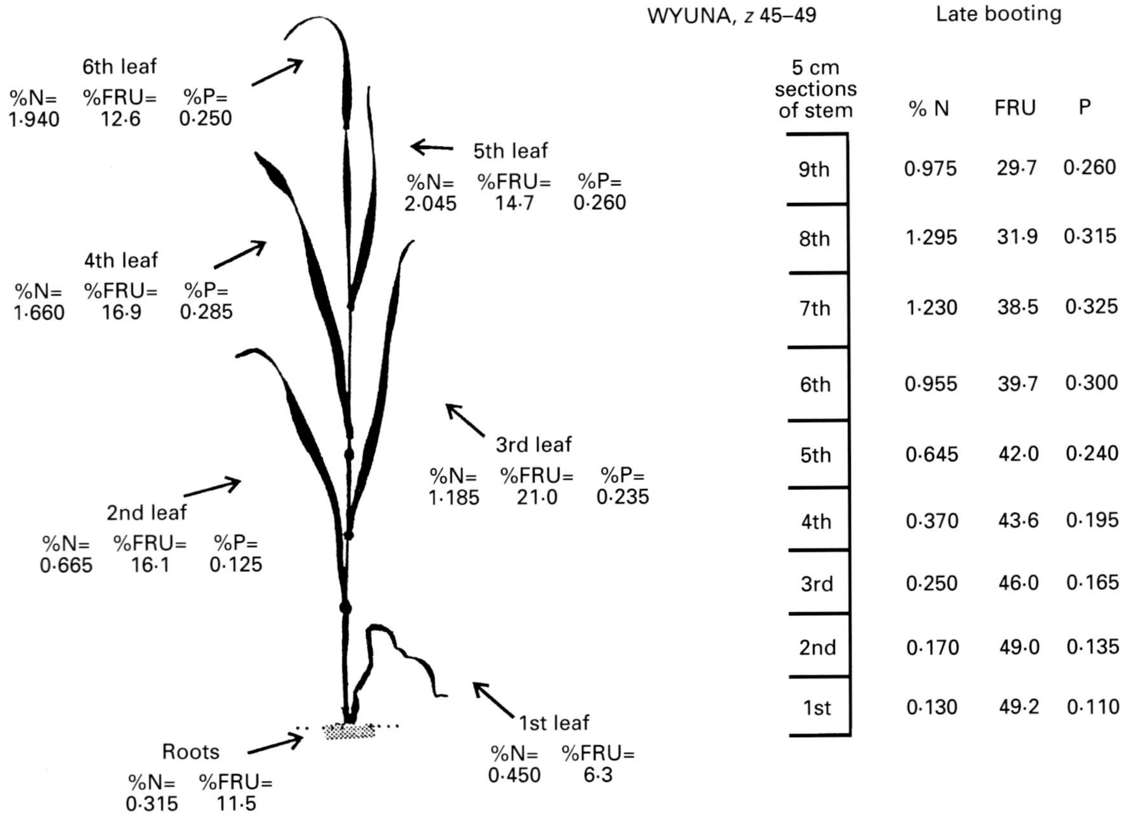

Figure 7. Dissection of wheat plant (cv. Wyuna) at booting stage into leaf, root and 5-cm stem segments. Analysis of different segments for fructan, N and P content.

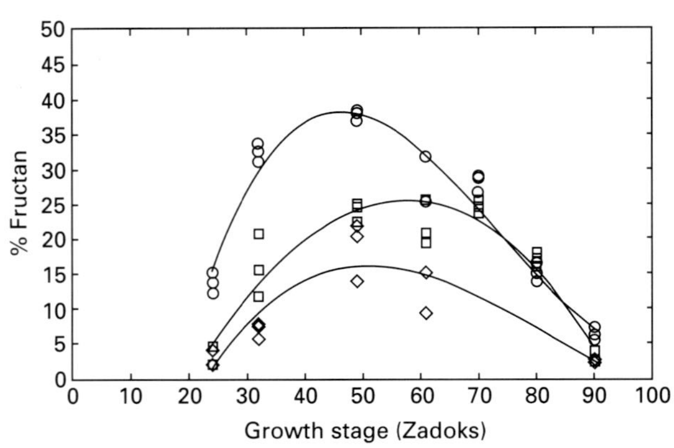

Figure 8. Fructan content in wheat shoot samples from three fertility sites (○, low; □, medium; and ◇, high) at different growth stages.

performs the essential task of 'foraging' for nitrate in the soil (Austin *et al.*, 1977). Our routine tissue-testing unfortunately cannot provide data on this component of the fructan store.

Accumulation of fructan at various growth stages

Important variations in amounts of fructan occur during the life cycle (Meier & Reid, 1982). Accumulation of fructans in whole-plant tissue was examined at growth stages, from early tillering to physiological maturity of the variety 'Wyuna' growing on sites of low, medium and high fertility (Fig. 8). On the range of sites of pre-designed fertility, the fructan content peaked around the late booting stage until the emergence of the inflorescence. The significant comparison to be made is that up until grain development stage, the fructan content of the total plant from a low-fertility site is always higher than that from a medium-fertility site, which in turn is higher than that from a high-fertility site. This supports earlier studies (Rawson & Evans, 1971; Evans & Wardlaw, 1976) which show that in a less than optimum environment, fructan accumulates and can be remobilized to sustain the plant in grain-filling.

Pollock & Jones (1979) reported a similar pattern of fructan accumulation in the growth cycle of forage grasses. In their study, fructan content was low in the early stages of rapid growth, rising to a peak in the stages of tiller and reproductive growth and declining when the potential for leaf elongation increased.

Conclusion

The application of NIR spectroscopy has become an integral part of mid-season crop-management programmes in Australia. The technique involves samples being quickly and accurately assessed for many and varied constituents, and now has expanded into commercial testing. The tissue analysis report of NIR values for N and fructan concentration and appropriate fertilizer recommendations allows the cereal grower very rapid evaluation of the growing crop and the opportunity to remedy deficiencies impeding crop growth.

The detection of growth-limiting stresses other than or in addition to N, using the NEIDX concept, is improving the reliability of fertilizer recommendations which have been based on data from N-rate trials which relate critical tissue nitrogen concentrations at defined growth stages and maximum yield, and thus the amount of nitrogen fertilizer to be applied to reach that yield. By considering the fructan concentration in addition to that of N, the likely effectiveness of applied N can be gauged. The accumulation (or depletion) of fructan relative to that expected at a particular N concentration indicates a growth-limiting stress. Further studies are required to define the impact of stresses, such as temperature, water deficiency, disease, lodging or mineral deficiency, in commercial crops on the fructan:N ratio.

We have also demonstrated the feasibility of developing calibrations to identify plant age (Blakeney & Batten, 1992) and to determine other nutrients such as P and S (Batten & Blakeney, 1992*b*). Work on providing these and other nutrient calibrations for inclusion in the wheat, rice, barley and oats tissue-testing service continues. The use of combinations of these elements with fructan and N might provide additional crop-management information in much the same way as NEIDX.

ACKNOWLEDGEMENTS

We thank the Grains Research and Development Corporation through the Australian Grain Industries Centre for NIR for financial support. We thank the staff of the collaborating tissue-testing laboratories and the cereal producers who participated in the tests for their cooperation in providing samples for this study.

REFERENCES

Austin RB, Ford MA, Edrich JA, Blackwell RD. 1977. The nitrogen economy of winter wheat. *Journal of Agricultural Science, Cambridge* **88**: 159–167.

Batten GD, Blakeney AB. 1992*a*. Using NIR to monitor crop energy reserves. In: Murray I, Cowe IA, eds. *Making Light Work: Advances in Near Infrared Spectroscopy*. Weinheim: VCH, 303–308.

Batten GD, Blakeney AB. 1992*b*. The determination of sulphur in plants using NIR. In: Hildrum KI, Isaksson T, Naes T, Tandberg A, eds. *Near Infra-red Spectroscopy: Bridging the Gap between Data Analysis and NIR Applications*. New York: Ellis Horwood, 185–190.

Batten GD, Blakeney AB, Glennie-Holmes MG, Henry RJ, McCaffery AC, Bacon PE, Heenan DP. 1991. Rapid determination of shoot nitrogen status in rice using near infrared reflectance spectroscopy. *Journal of the Science of Food and Agriculture* **54**: 191–197.

Batten GD, Blakeney AB, McGrath VB, Ciavarella S. 1993. Non-structural carbohydrate: analysis by near infrared reflectance spectroscopy and its importance as an indicator of plant growth. *Plant and Soil* **155/156**: 243–246.

Batten GD, McGrath VB, Boland OW, Blakeney AB, Ciavarella S. 1995. Spectral stability of microwave-dried rice and wheat shoots. In: Batten GD, Flinn PC, Welsh LA, Blakeney AB, eds. *Leaping Ahead with Near Infrared Spectroscopy*. North Melbourne, Australia: Royal Australian Chemical Institute, 252–255.

Blakeney AB, Batten GD. 1992. Using NIR to determine plant age. In: Murray I, Cowe IA, eds. *Making Light Work: Advances in Near Infrared Spectroscopy*. Weinheim: VCH, 298–302.

Blakeney AB, Batten GD, McGrath VB. 1992. Fructan analysis of winter cereals using NIR. *Proceedings of NIR Users Group Meeting*, RACI 9th National Convention, Monash University, Melbourne, Royal Australian Chemical Institute, Parkville, Australia, 12.

Blakeney AB, Mutton LL. 1980. A simple colorimetric method for the determination of sugars in fruit and vegetables. *Journal of the Science of Food and Agriculture* **31**: 889–897.

Bolland MDA, Bowden JW, Brennan RF, Gartrell JW, Mason MG, Porter WM. 1991. Nutrition. In: Perry M, Hillman B, eds. *The Wheat Book*. Bulletin 4196. Perth, Australia: Department of Agriculture, 53–77.

Borrell AK, Incoll LD, Simpson RJ, Dalling MJ. 1989. Partitioning of dry matter and the deposition and use of stem reserves in a semi-dwarf wheat crop. *Annals of Botany* **63**: 527–539.

Bradstreet RB. 1965. *The Kjeldahl method for organic nitrogen*. New York: Academic Press.

Darbyshire B, Henry RJ. 1981. Difference in fructan content and synthesis in some *Allium* species. *New Phytologist* **87**: 249–256.

Dubois D, Winzeler M, Nösberger J. 1990. Fructan accumulation and sucrose:sucrose fructosyltransferase activity in stems of spring wheat genotypes. *Crop Science* **30**: 315–319.

Eagles CF. 1967. Variation in the soluble carbohydrate content of climatic races of *Dactylis glomerata* L. at different temperatures. *Annals of Botany* **31**: 645–651.

Edelman J, Jefford TG. 1968. The mechanism of fructosan metabolism in higher plants as exemplified in *Helianthus tuberosus* L. *New Phytologist* **67**: 517–531.

Evans LT, Wardlaw IF. 1976. Aspects of the comparative physiology of grain in cereals. *Advances in Agronomy* **28**: 301–359.

Fuchs A. 1993. Production and utilization of inulin. Part II. Utilization of inulin. in: Suzuki M, Chatterton NJ, eds. *Science and Technology of Fructans*. Boca Raton, FL: CRC Press, 319–352.

Geiger DR, Servaites JC. 1991. Carbon allocation and response to stress. In: Mooney HA, Winner WE, Pell EJ, eds. *Response of Plants to Multiple Stresses*. San Diego, CA: Academic Press, 103–127.

Gunn TC, Walton DWH. 1985. Storage carbohydrate production and overwintering strategy in a winter-green tussock grass on South Georgia (Sub Antarctica). *Polar Biology* **4**: 237–242.

Hendry GA, Wallace RK. 1993. The origin, distribution, and evolutionary significance of fructans. In: Suzuki M, Chatterton NJ, eds. *Science and Technology of Fructans*. Boca Raton, FL, USA: CRC Press, 119–139.

Livingston DP, Chatterton NJ, Harrison PA. 1993. Structure and quantity of fructan oligomers in oat (*Avena* spp.). *New Phytologist* **123**: 725–734.

McGrath D. 1988. Seasonal variation in the water-soluble carbohydrates of perennial and Italian ryegrass under cutting conditions. *Irish Journal of Agricultural Research* **27**: 131–139.

McGrath VB, Blakeney AB, Batten GD, Boland OW. 1995. Tissue analysis improves quality and yield of wheat. In: Williams YA, Wrigley CW, eds. *Proceedings of the 45th Australian Cereal Chemistry Conference*. North Melbourne, Australia: Royal Australian Chemical Institute, 538–541.

Meier H, Reid JSG. 1982. Reserve polysaccharides other than starch in higher plants. In: Loewus FA, Tanner W, eds. *Encyclopedia of Plant Physiology, New Series, vol. 13A. Intracellular Carbohydrates*. Berlin: Springer-Verlag, 418–471.

Noaman MM, Taylor GA, Martin JM. 1990. Indirect selection for grain protein and grain yield in winter wheat. *Euphytica* **47**: 121–130.

O'Brien L, Fisher J, Ballantyne B, Brennan JP. 1992. Using NIR determined shoot nitrogen content to predict grain yield and protein content in wheat. In: Humphrey-Taylor VJ, ed. *Proceedings of the 42nd Australian Cereal Chemistry Conference*, Cereal Chemical Division, Royal Australian Chemical Institute, Parkville, Australia, 232–234.

Percheron F. 1962. Colorimetric measures of fructose and fructofuranosides by thiobarbituric acid. *Comptes Rendus* **255**: 2521–2523.

Pollock CJ, Cairns AJ. 1991. Fructan metabolism in grasses and cereals. *Annual Review of Plant Physiology and Plant Molecular Biology* **42**: 77–101.

Pollock CJ, Jones T. 1979. Seasonal patterns of fructan and metabolism in forage grasses. *New Phytologist* **83**: 9–15.

Pontis HG. 1966. Observations on the *de novo* synthesis of fructosans *in vivo*. *Archives of Biochemistry and Biophysics* **116**: 416–423.

Pontis HG. 1989. Fructans and cold stress. *Journal of Plant Physiology* **134**: 148–150.

Pontis HG. 1990. Fructans. In: Dey PM, Harborne JB, eds. *Methods in Plant Biochemistry*, Vol. 2, *Carbohydrates*. New York: Academic Press, 353–369.

Pontis HG, del Campillo E. 1985. Fructans. In: Dey PM, Dixon RA, eds. *Biochemistry of Storage Carbohydrates in Green Plants*. New York: Academic Press, 205–227.

Rawson HM, Evans LT. 1971. The contribution of stem reserves to grain development in a range of wheat cultivars of different height. *Australian Journal of Agricultural Research* **22**: 851–863.

Rostami MA, O'Brien L. 1993. Focused plant improvement: towards responsible and sustainable agriculture. In: Imrie BC, Hacker JB, eds. *Proceedings of the 10th Australian Plant Breeding Conference*, Vol. 2. Organising Committee, Australian Convention and Travel Service: Canberra, 109–110.

Smith D. 1967. Carbohydrates in grasses. II. Sugar and fructosan composition of the stem bases of bromegrass and Timothy at several growth stages and at different plant parts at anthesis. *Crop Science* **7**: 62–67.

Suzuki, M. 1993. Fructans in crop production and preservation. In: Suzuki M, Chatterton NJ, eds. *Science and Technology of Fructans*. Boca Raton, FL, USA: CRC Press, 227–255.

Wardlaw IF. 1990. The control of carbon partitioning in plants. *New Phytologist* **116**: 341–381.

New Phytol. (1997), **136**, 153–161

Growth and fructan content of plants of *Vernonia herbacea* (Asteraceae) regenerated from rhizophores

By M. ANGELA MACHADO DE CARVALHO*, LILIAN B. P. ZAIDAN AND SONIA M. C. DIETRICH

Seção de Fisiologia e Bioquímica de Plantas, Instituto de Botânica, Caixa Postal 4005, CEP 01061–970 São Paulo – SP, Brazil

(Received 24 July 1996)

SUMMARY

The growth and fructan contents of plants of *Vernonia herbacea* (Vell.) Rusby regenerated from fragments of the underground organ (rhizophore) were analysed during two developmental cycles. When cultivated away from their natural ecosystem, these plants went through the same phenological phases as those naturally grown in the cerrado. The first aerial shoots started developing from initial fragments before the new underground system, the former acting as the main metabolic sink for the fructan present in the initial fragments. The new rhizophores started to develop by the sixth month, as photosynthates were transformed into fructans. Growth parameters measured reached higher values in the second growth cycle, exhibiting growth curves synchronized with those obtained for fructan contents in the new rhizophores. Fructo-oligosaccharides were available mostly during the period of fast growth of aerial organs whilst the fructo-polysaccharides were reduced. Fructan variation indicates that these compounds are metabolized during development. Additionally, the rapid growth, intense production of rhizophores and high fructan content make *Vernonia herbacea* a promising source of fructan for commercial use.

Key words: Asteraceae, *Vernonia herbacea*, fructan, underground organs, cerrado vegetation.

INTRODUCTION

The cerrado vegetation covers 22% of the Brazilian land area, and has been characterized by Eiten (1972) as a floristically and physiognomically diverse savanna vegetation. Mantovani & Martins (1988) carried out a phenological survey of the vascular plants in a cerrado area and concluded that the phenological behaviour of the vegetation represents a series of adaptive strategies to overcome thermal, hydric and nutritional stresses.

Vernonia herbacea (Vell.) Rusby (Asteraceae) is a perennial herb native to the cerrado. It shows a well defined seasonal developmental pattern which involves a sprouting phase in spring, followed by vegetative growth and flower development. Vegetative growth lasts until early autumn; stem growth is then arrested, leaves start to senesce, aerial organs abscise and the plants enter a dormant state until the end of winter. In early spring, which is coincident with rainfall, higher temperatures and longer days, new buds from the rhizophores sprout, starting a new growth cycle (Carvalho & Dietrich, 1993).

* To whom correspondence should be addressed.

The underground perennating organ of *V. herbacea*, denominated rhizophore for the genus by Menezes, Muller & Sajo (1979), is considered both a reserve structure accumulating fructans as a major polysaccharide (Carvalho & Dietrich, 1993), and an organ of vegetative propagation.

Fructans from *V. herbacea* exhibit an homologous series of oligosaccharides similar to that found in tubers of *Helianthus tuberosus* and could therefore be fructans of the inulin type. Indeed, fructan from *V. herbacea* was shown to be as suitable for kidney clearance determination as inulin from *Dahlia*, commercialized by Sigma (Dias-Tagliacozzo, Dietrich & Mello-Aires, 1996). Fructan contents of *V. herbacea* can reach as much as 90% of the rhizophore dry mass (Carvalho & Dietrich, 1993). Since such characteristics could place this species among the important inulin producers, the determination of crop yield in this plant is highly important.

Additionally, growth analysis is a useful tool for elucidating the responses of plants to environmental conditions and to cultural treatments, including yields of underground organs (Sobulo, 1972).

Shoot growth has frequently been positively

correlated with tuber development, as shown for yams (Sobulo, 1972) and for corms of *Xanthosoma sagittifolium* (Graziano, Dietrich & Figueiredo-Ribeiro, 1992), and inversely when plants are subjected to different environmental factors (Sobulo, 1972).

The aim of the present work was primarily to study growth parameters of *V. herbacea* plants during two growth cycles at different phenological stages in relation to the storage and mobilization of fructans from the underground organs and, secondly, to provide data for the establishment of the best harvest period for rhizophores.

MATERIALS AND METHODS

Plant material

Adult plants of *Vernonia herbacea* (Vell.) Rusby were collected from a conserved cerrado area in Moji-Guaçu, SP, Brazil (22° 18′ S, 47° 11′ W) at early dormancy, according to the phenological phases described in Carvalho & Dietrich (1993). The plants were brought to the laboratory where the rhizophores were isolated, washed thoroughly, blotted dry and cut into fragments approx. 3-cm long.

For obtaining new plants, these initial fragments were planted in boxes *c.* 2 cm below the soil surface and kept humid in a glasshouse at ambient temperature and daylength. The planting of fragments was considered the start of the analysis. After sprouting, the plants were transferred to individual pots outdoors, and watered regularly throughout the experiment.

Growth and carbohydrate analyses were performed on samples of 20 plants collected every 45 d during the first 3 months then approx. every 3 months for a total of 26 months. The aerial organs were first measured 3 months after the fragments were planted.

For carbohydrate analysis, rhizophores from the plants were pooled and four samples of each of the initial and of the newly formed rhizophores weighing between 4 and 10 g fresh mass were used separately.

Growth analysis

Each of the 20 plants sampled was separated into aerial organs (stem and leaves) and rhizophores (initial and newly formed, separately). Average fresh and dry mass were obtained for these portions, as well as length of the highest stem, leaf number and total leaf area per plant. Leaf area was calculated by using a linear regression curve ($y = 0{\cdot}2737 + 0{\cdot}7158x$), where x is the product of leaf length vs. leaf width, with a correlation coefficient of 0·9658, determined specifically for this species (Carvalho, 1991).

Dry mass was determined in samples by oven-drying at 80 °C to constant weight. Net assimilation rate (NAR) was calculated according to Williams (1946). Leaf area ratio (LAR) and relative growth rate (RGR) were determined as described in Hunt (1978) whilst the harvest index was calculated according to Medeiros, Pereira & Miranda (1990).

Carbohydrate analysis

Quantitative determinations. Fructo-oligosaccharides together with free fructose and sucrose and fructo-polysaccharides were extracted from rhizophores as reported by Carvalho & Dietrich (1993). Free and combined fructose were measured in both fractions by the resorcinol-HCl reagent (McRary & Slattery, 1945) and reducing sugars in the fructo-oligosaccharide fraction, according to Somogyi (1945), using fructose as standard.

Thin-layer chromatography. The fructo-oligosaccharides were analysed by ascending thin layer chromatography (TLC). Concentrated extracts were deionized through an ion-exchange column consisting of half anionic (Dowex 1 × 8–200 – chloride form) and half cationic resin (Dowex 50 × 8–100 – hydrogen form), as described by Pollock & Jones (1979). TLC of fructo-oligosaccharides was carried out according to Kanaya, Chiba & Shimomura (1978). Carbohydrates were visualized by spraying with the ketose-specific urea-phosphoric acid (Wise *et al.*, 1955). Oligosaccharides of the inulin series from tubers of *H. tuberosus* were used as reference material.

Anion-exchange chromatography. The fructo-oligosaccharides were also analysed on a 4 × 250 mm CarboPac PA-1 anion-exchange column using a Dionex DX-300 gradient chromatography system (HPAEC/PAD). The gradient was established according to Shiomi (1993), by mixing eluant A (150 mM NaOH) and eluant B (500 mM sodium acetate in 150 mM NaOH) as follows: 0–1 min, 25 mM; 1–2 min, 25–50 mM; 2–14 min, 50–500 mM; 14–22 min, 500 mM; 22–30 min, 25 mM. The flow rate through the column was 1·0 ml min^{-1}. The applied PAD potentials for E1 (300 ms), E2 (120 ms) and E3 (300 ms) were 0·04, 0·60 and −0·80 V, respectively, and the output range was 1000 nA.

Gel-permeation chromatography. Fructo-polysaccharides were analysed by gel-permeation chromatography (GPC), using polyacrylamide gel (Bio-Gel P-10) as described by Carvalho & Dietrich (1993). Apparent molecular mass was estimated according to Andrews (1965). Free and combined fructose in the fractions were measured by the anthrone reaction (Jermyn, 1956).

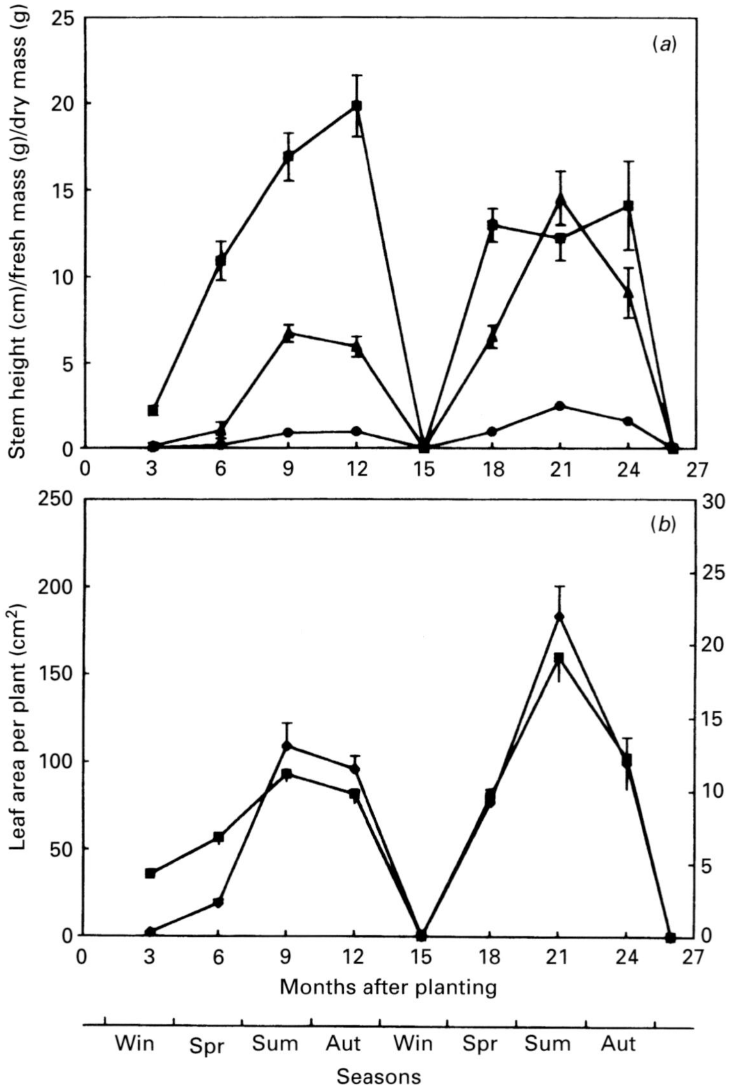

Figure 1. Growth of aerial organs from plants of *Vernonia herbacea* regenerated from fragments of rhizophores. (*a*) Stem height (■), fresh mass (▲), dry mass (●). (*b*) Leaf number (■), leaf area (◆). Bars represent SE and, where not shown, were smaller than the symbol.

RESULTS

Growth analysis

Sprouting of aerial shoots began approx. 50 d after planting the fragments. During the 26 months of observation, plants went through two growth cycles, showing vigorous vegetative growth in the first cycle followed by senescence of the aerial organs after 12–15 months (end of autumn/beginning of winter), leading to the dormant stage. Flowering did not occur in the first cycle.

Newly sprouted shoots were seen by month 16 (end of winter), and were followed by flowering (16–18 months). Most growth parameters analysed reached their highest values when plants were in the vegetative phase (9 and 12 months in the first cycle and 21 months in the second), which coincided with the end of summer (Fig. 1).

In the second growth cycle, plants usually had more than one aerial shoot. This led to greater leaf number, leaf area, and fresh and dry masses (Fig. 1*a*, *b*). In both cycles, senescence was accompanied by a decrease in these parameters.

Dry mass of rhizophores increased during the 26-month-period exclusively as the result of the development and growth of newly formed rhizophores (Fig. 2). Growth of rhizophores was interrupted between the beginning of dormancy (12 months) and flowering (18 months). The harvest index decreased (Fig. 2) as aerial organs showed a period of rapid growth (Fig. 1), compared with low rhizophore mass gain (Fig. 2). An increase in the harvest index was observed as growth of the rhizophore became more conspicuous, by month 26. Maximum values occurred at the dormant phases when shoots were absent.

Rapid aerial growth during the summer (Fig. 1*a*, *b*) was followed by marked increases in rhizophore mass, especially in the second growth cycle (Fig. 2). At this point, leaf area (Fig. 1*b*) and the NAR reached the highest values (Fig. 3*a*). A

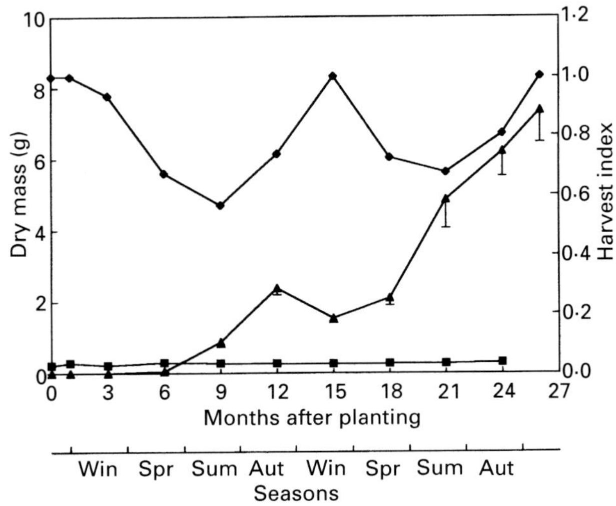

Figure 2. Time-course of harvest index (◆) and dry mass of rhizophores during growth of plants of *Vernonia herbacea* regenerated from fragments of rhizophores. Initial fragments (■) and new rhizophores (▲). Bars represent SE and where not shown, were smaller than the symbol.

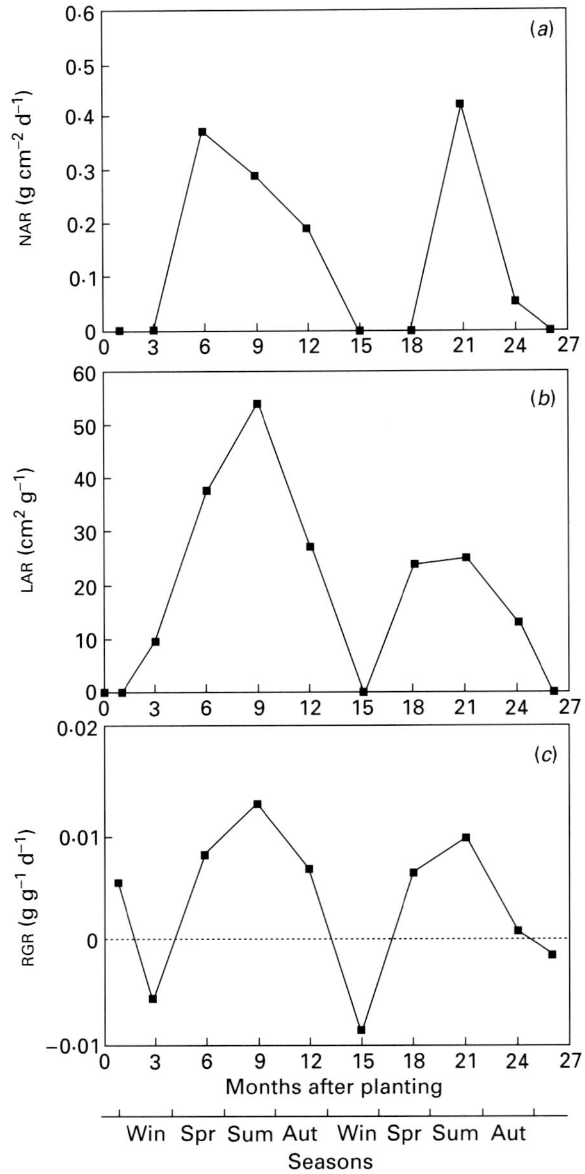

Figure 3. Time-course of net assimilation rate (NAR) (*a*), leaf area ratio (LAR) (*b*) and relative growth rate (RGR) (*c*) of plants of *Vernonia herbacea* regenerated from rhizophores.

decrease in the relation between aerial parts (synthesizing system) and total dry mass during plant growth can be seen from values obtained for LAR, when the slopes of the first and second growth cycles are compared (Fig. 3*b*). The RGR generally increased towards the end of summer (9 and 21 months) and declined during the months preceding dormancy or immediately after planting (Fig. 3*c*).

Carbohydrate analysis

Fructo-polysaccharides present in the initial fragments decreased from 302·39 to 6·34 mg g^{-1} dry mass during the 24 months of analysis, being impossible to measure in month 26 owing to tissue deterioration (Fig. 4*a*).

Fructo-oligosaccharides increased in the initial fragment between 1–9 months, a period when growth rate of plants is high. This was followed by a decrease in month 12 concurrent with early dormancy and a gradual increase up to month 18, while plants went through dormancy, sprouting and flowering. At the end of the second developmental cycle (24 months), the amount of fructo-oligosaccharides was low, and similar to that at the end of the first growth cycle (Fig. 4*a*). Reducing sugars were present in this fraction in very low concentrations, increasing slightly in month 3 (Fig. 4*a*), when all fragments had already sprouted.

Contents of fructo-polysaccharides in the newly formed rhizophores increased from 93 mg g^{-1} dry mass in month 6 to 472 mg g^{-1} dry mass by the month 12 (Fig. 4*b*), accompanying the growth and development of the underground organs. In month 15 (dormancy), amounts of these compounds were still high, declining during sprouting and flowering (18–21 months). High amounts were subsequently restored. Contents of fructo-oligosaccharides were inversely related to those of fructo-polysaccharides, i.e. at early stages of development, rhizophores had high amounts of fructo-oligosaccharides (Fig. 4*b*) which declined as polysaccharides increased towards dormancy of the plants. High levels of oligosaccharides were detected again in month 21, concurrent with intense growth of aerial organs (Fig. 1 *a*, *b*). Small amounts of reducing sugars were found in these tissues throughout most of the experiment.

Analysis of the fructo-oligosaccharide fractions by TLC revealed that all components of the homologous series of fructo-oligosaccharides based on the trisaccharide 1-kestose were present in the initial fragments and new rhizophores throughout the development of the plants. Variations in proportions of components were observed, especially in the

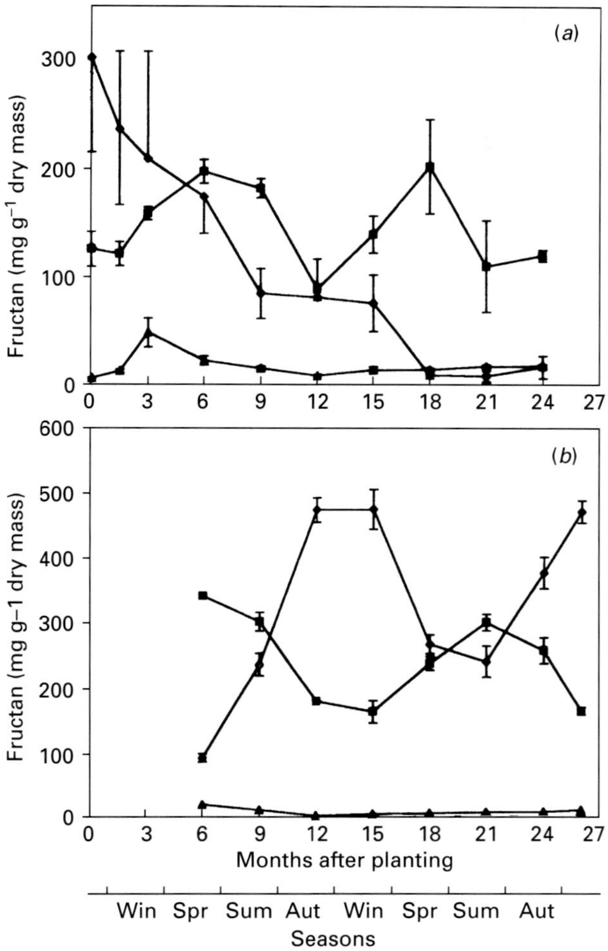

Figure 4. Time-course of fructo-polysaccharides (◆), fructo-oligosaccharides (■) and reducing sugars (▲) in initial fragments (*a*) and new rhizophores (*b*) of plants of *Vernonia herbacea* regenerated from rhizophores. Bars represent SE and, where not shown, were smaller than the symbol.

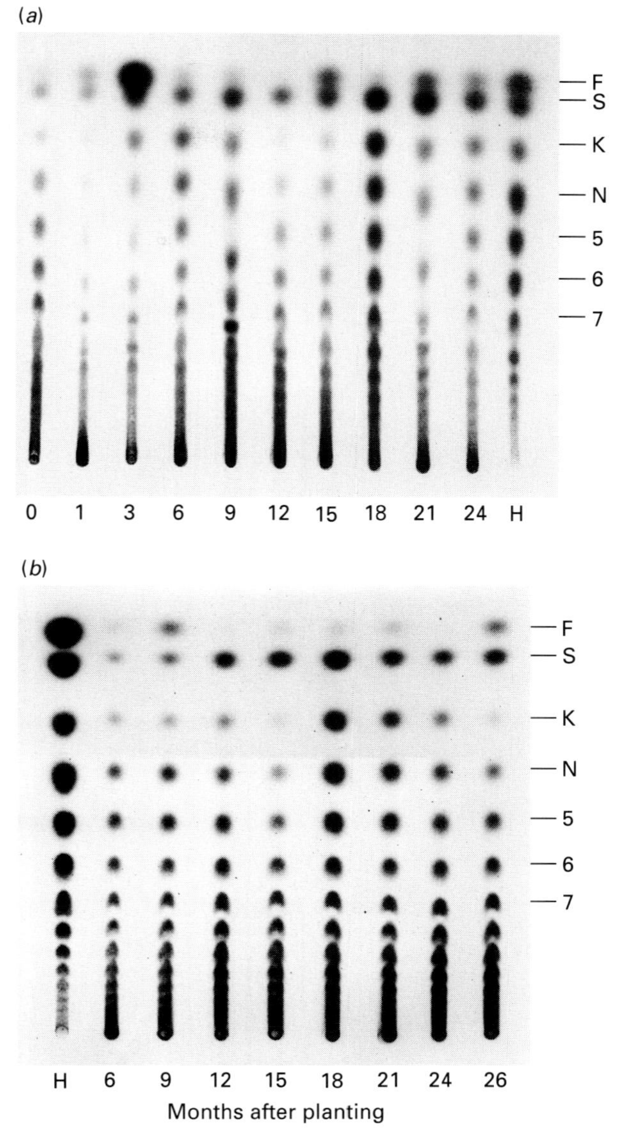

Figure 5. TLC of fructo-oligosaccharides from *Vernonia herbacea* regenerated from rhizophores. Each lane contained approx. 100 μg fructose equivalents. (*a*) Initial fragments. (*b*) New rhizophores. F, fructose; S, sucrose; K, 1-kestose; N, nystose; H, fructo-oligosaccharides from *Helianthus tuberosus*. Numbers 5–7 mark the mobilities of inulin oligosaccharides of DP 5–7.

initial fragments (Fig. 5). Free fructose was most abundant in month 3, in the initial rhizophores, whereas 1-kestose was particularly low in both months 15 and 26, in newly formed rhizophores.

The chromatographic profile of fructo-oligosaccharides was confirmed through PAD/AEC of selected samples (Fig. 6). The increase in the proportion of fructose in month 3, shown in the initial fragments in Figure 5 was confirmed. Generally, considering the size pattern of peaks, fructose predominated over glucose and sucrose predominated over 1-kestose, whilst the opposite was found in the sixth month in the newly formed rhizophores.

Intermediary peaks following 1-kestose, nystose and the pentasaccharide of the homologous series, respectively, were detected in tissues of initial fragments in months 3 and 15 (Fig. 6*b*, *d*).

The mean molecular mass of fructo-polysaccharides from the initial fragment was generally higher than that found in new rhizophores (Table 1). In the former it varied from 5·1 kDa, in the first month of the experiment, to 7·2 kDa. In the latter, it increased gradually from the sixth month until dormancy, repeating this pattern in the second growth cycle. These variations paralleled those found in amounts of fructo-polysaccharides (Fig. 4*b*).

DISCUSSION

The average annual temperature in the cerrado is 20–26 °C and extreme minimum temperatures vary between 14 °C and −4 °C, coinciding with the driest months (July or August), when the average rainfall is 10–30 mm (Eiten, 1972). Therefore, herbaceous plants in the cerrado are subjected to long dry periods in the winter, when dormancy occurs (Mantovani & Martins, 1988). Plants originated

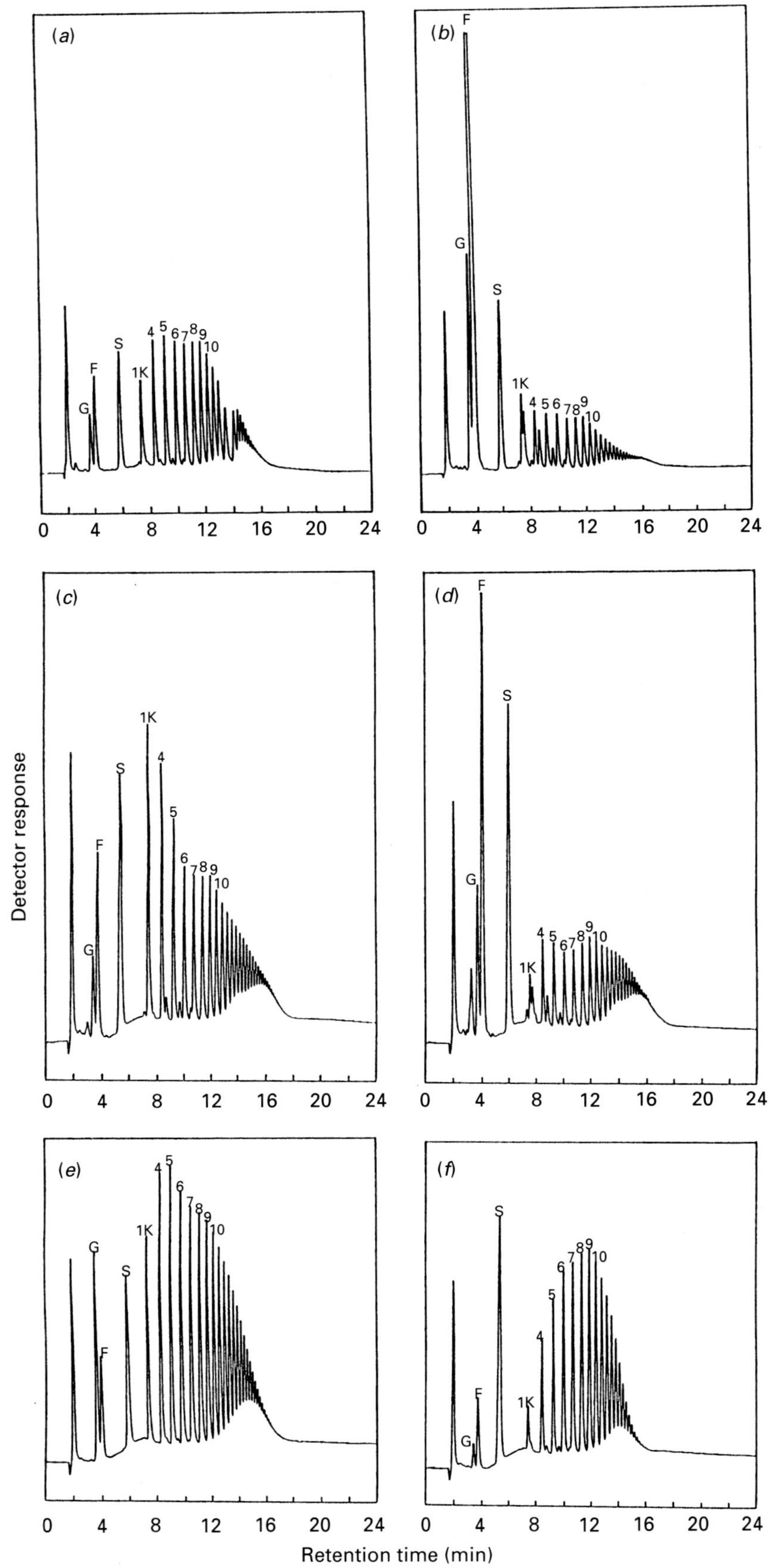

Figure 6. HPAEC/PAD profiles of fructosaccharides from rhizophores of *Vernonia herbacea*. Initial fragments: (*a*) time 0; (*b*) 3 months; (*c*) 6 months; (*d*) 15 months. New rhizophores: (*e*) 6 months; (*f*) 15 months. G, glucose; F, fructose; S, sucrose; K, 1-kestose. Numbers 4–10 identify peaks of inulin oligosaccharides of DP 4–10.

Table 1. *Molecular mass and degree of polymerization of fructo-polysaccharides from initial fragments and new rhizophores of plants of* Vernonia herbacea *regenerated from rhizophores*

	Initial fragments		New rhizophores	
Months after planting	Molecular mass (kDa)	DP*	Molecular mass (kDa)	DP
0	6·6	41	—	—
1	5·1	31	—	—
3	5·8	36	—	—
6	7·2	45	4·4	27
9	5·7	35	4·6	29
12	5·7	35	4·6	29
15	5·9	36	5·9	36
18	—	—	5·7	35
21	—	—	4·8	30
24	5·5	34	4·6	29
26	—	—	5·2	32

* DP, degree of polymerization.

from fragments of rhizophores of *V. herbacea*, cultivated away from the cerrado and irrigated regularly, went through the same phenological phases in a synchronized way as plants growing naturally in the cerrado (Carvalho & Dietrich, 1993). This was also observed in other species from the cerrado bearing underground reserve organs containing fructans (Isejima & Figueiredo-Ribeiro, 1991) or other storage carbohydrates (Figueiredo-Ribeiro & Dietrich, 1981; Sá e Carvalho & Dietrich, 1996) and indicates that irrigation did not affect the phenological behaviour of these species. Other growth conditions such as temperature and photoperiod, were similar to th cerrado environment.

In the present work, in which the regeneration of new plants from fragments of rhizophore was studied, shoots started to develop before the underground system, the former acting as the main metabolic sink for the fructans present in the initial fragments of rhizophores. In month 6, the new underground system of the plants started to grow, acting as a sink for the excess of photosynthates, which were there transformed into storage fructans. These compounds are therefore responsible for most of the increase in dry mass of the underground organ, which, like the aerial parts, presented growth periods which were interrupted during dormancy and early sprouting.

In the second growth cycle, growth values registered for the aerial organs were higher, except for the length of the highest stem, probably because in the second cycle more than one stem was usually present on the plants and also because of the increase in available supply of energy from the metabolism of stored fructans.

Lower values of LAR in the second cycle indicate an increase in the underground system in relation to the aerial organs or leaf area, whilst a shorter period of high NAR in the second cycle reflects a higher leaf area increment. In fact, it reflects high photosynthetic activity at the end of the summer, when growth parameters are high and fructo-oligosaccharides are most available in the tissues.

Harvest index values opposed RGR, indicating that periods of intensive growth are not favourable for harvesting root crops. Data presented here do not demonstrate whether or not there is significant loss of underground dry mass after it reaches maximum values in the first cycle, but if harvesting is the aim it must be pointed out that there is a much higher increase in underground dry mass in the second growth cycle than in the first. Additionally it seems that the dormant phase of the plants is the most favourable for harvesting.

When comparing the growth of aerial organs and rhizophores, the following points must be considered.

(1) There are growth cycles for both systems interrupted by the dormant period of the plants, the second cycle showing a more prominent growth.

(2) The maximum values for the growth curves of the rhizophore occur with some delay in relation to that of aerial parts, showing that growth of the latter leads to more storage.

(3) Both curves are synchronized in some way with those obtained for the contents of fructans in the new rhizophores.

(4) Fructan reserves increase with growth of the rhizophore, although only a small part is mobilized during sprouting of the aerial organs in each new growth cycle.

The storage strategy in *V. herbacea* might be related to extra demand for carbohydrates during extreme drought, occasional occurrence of near-freezing temperatures and spontaneous or provoked fire reported for cerrado areas (Eiten, 1972; Coutinho, 1990).

The coincident variation in growth of aerial organs and in fructo-oligosaccharides of the new rhizophores shows that these compounds are especially available during the periods of most intense growth. Variations in fructo-polysaccharides occur in the opposite direction and seem to be related to accumulation by the new rhizophore. As in plants naturally growing in the cerrado (Carvalho & Dietrich, 1993), the large amounts of high DP fructans found in early dormancy and dormancy decrease between sprouting and flowering, when shoots and flowers act as strong metabolic sinks for the reserve carbohydrate.

Some deductions about the activity of enzymes of fructan metabolism can be made from the data on fructan variation. The increases in amount and mean DP of fructo-polysaccharides between 6 and 15 months and between 21 and 26 months, indicate the expression of FFT-like activity during these periods; this is reinforced by the low amounts of 1-kestose in the new rhizophores in months 15 and 26 (Fig. 5). Increase in glucose and 1-kestose in newly formed rhizophores (6 months), evidenced by HPAEC analyses, is suggestive of SST activity in this phase.

Reducing sugars are mainly represented by glucose and fructose in new rhizophores. According to Frehner, Keller & Wiemken (1984), fructose yielded during mobilization of fructans in the vacuoles is promptly transformed into sucrose in the cytoplasm and either translocated to sink organs, like shoots and inflorescences, or utilized by tissues of the source organs as metabolites, which seems also to be the case in *V. herbacea*.

Nevertheless, free fructose is clearly more abundant in the third month in the initial rhizophore, when reducing sugars show a slight increase, corresponding to the period of intense sprouting of the fragments and drastic reduction of fructo-polysaccharides, suggesting high exohydrolase activity, even beyond the requirements of the sprouting tissues.

Ernst, Chatterton & Harrison (1996) detected a new fructan series in reserve underground organs of *Taraxacum officinalis*, *H. tuberosus* and *Cichorium intybus*, which was called inulo-*n*-ose, according to the nomenclature of Lewis (1993). This series was thought to originate as breakdown products of the inulin series. In *V. herbacea*, the presence of compounds with retention times intermediate to those of the fructan series, in initial fragments in months 3 and 15 of the experiment, suggests that elements of the inulo-*n*-ose series might be present in rhizophores of *V. herbacea* during mobilization of fructans. In this case, the three peaks with retention times following 1-kestose, nystose and the pentasaccharide could correspond to inulobiose, inulotriose and inulotetraose, respectively.

An inverse relation between contents of oligosaccharides and polysaccharides, not to mention the fructan variation itself, indicates that these compounds are metabolized during the development of the plant, as demonstrated in the classical works of Bacon & Loxley (1952) and Jefford & Edelman (1963), for tubers of *H. tuberosus*.

Rapid growth and easy environmental adaptation, intense production of rhizophore mass and the high level of fructans accumulated in these organs make *V. herbacea* an interesting material for studies on fructan metabolism and adaptive strategies of plants to unfavourable environmental conditions, as well as a promising source of fructo-oligosaccharides and fructo-polysaccharides for commercial exploitation.

ACKNOWLEDGEMENTS

Thanks are due to Fundação de Amparo à Pesquisa do Estado de São Paulo (FAPESP) for financial support (Proc. 91/3588-8) and to Conselho Nacional de Desenvolvimento Científico e Tecnológico (CNPq) for a grant to M. A. M. Carvalho (Proc. 300667/87-6). The authors also thank Dr R. C. L. Figueiredo-Ribeiro for a critical review of the manuscript.

REFERENCES

Andrews P. 1965. The gel filtration behaviour of proteins related to their molecular weights over a wide range. *Biochemical Journal* **96**: 595–606.

Bacon JSD, Loxley R. 1952. Seasonal changes in the carbohydrates of the Jerusalem artichoke tuber. *Biochemical Journal* **51**: 208–213.

Carvalho MAM. 1991. *Variações no conteúdo e na composição de frutanos em rizóforos de Vernonia herbacea (Vell.) Rusby*. Ph.D. thesis, Universidade Estadual de Campinas, Brasil.

Carvalho MAM, Dietrich SMC. 1993. Variation in fructan content in the underground organs of *Vernonia herbacea* (Vell.) Rusby at different phenological phases. *New Phytologist* **123**: 735–740.

Coutinho LM, 1990. Fire in the tropical biota – ecosystem processes and global challenges. In: Goldammer JG, ed. *Ecological Studies vol. 84*. Berlin: Springer-Verlag, 82–105.

Dias-Tagliacozzo GM, Dietrich SM, Mello-Aires MM. 1996. Glomerular filtration rate measurement using inulin from *Vernonia herbacea*, a Brazilian native species. *Brazilian Journal of Medical and Biological Research* **29**: 1393–1396.

Eiten G. 1972. The cerrado vegetation of Brazil. *The Botanical Review* **38**: 201–341.

Ernst M, Chatterton NJ, Harrison PA. 1996. Purification and characterization of a new fructan series from species of Asteraceae. *New Phytologist* **132**: 63–66.

Frehner M, Keller F, Wienken A. 1984. Localization of fructan metabolism in the vacuoles isolated from protoplasts of Jerusalem artichoke tubers (*Helianthus tuberosus* L.). *Journal of Plant Physiology* **116**: 197–208.

Figueiredo-Ribeiro RCL, Dietrich SMC. 1981. Variações estacionais nos compostos de reserva e no metabolismo do xilopódio de *Ocimum nudicaule* Benth. var. *anisifolia* Giul. (Labiatae). *Revista Brasileira de Botânica* **4**: 73–82.

Graziano TT, Figueiredo-Ribeiro RCL, Dietrich SMC. 1992. Characterization of starch of the underground system of *Xanthosoma sagittifolium* (L.) Schott (Araceae) during plant development. *Revista Brasileira de Fisiologia Vegetal* **4**: 7–10.

Hunt, R. 1978. *Plant growth analysis*. London: Edward Arnold.

Isejima EM, Figueiredo-Ribeiro RCL. 1991. Partição de matéria seca durante o desenvolvimento de *Viguiera discolor*. *Revista Brasileira de Botânica* **14**: 107–114.

Jefford TG, Edelman J. 1963. The metabolism of fructose polymers in plants. 2. Effect of temperature on the carbohydrate changes and morphology of stored tubers of *Helianthus tuberosus* L. *Journal of Experimental Botany* **14**: 56–62.

Jermyn MA, 1956. A new method for the determination of ketohexoses in presence of aldohexoses. *Nature* **177**: 38–39.

Kanaya KI, Chiba S, Shimomura T. 1978. Thin-layer chromatography of linear oligosaccharides. *Agricultural Biological Chemistry* **42**: 1947–1948.

Lewis DH. 1993. Nomenclature and diagrammatic representation of oligomeric fructans – a paper for discussion. *New Phytologist* **124**: 583–594.

Mantovani W, Martins FR. 1988. Variações fenológicas das espécies de cerrado da Reserva Biológica de Moji-Guaçu, Estado de São Paulo. *Revista Brasileira de Botânica* **11**: 101–112.

McRary, WL, Slattery MC. 1945. The colorimetric determination of fructosan in plant material. *Journal of Biological Chemistry* **157**: 161–167.

Medeiros JG, Pereira W, Miranda JEC. 1990. Análise de crescimento em duas cultivares de batata-doce (*Ipomoea batatas* (L.) Lam.). *Revista Brasileira de Fisiologia Vegetal* **2**: 23–29.

Menezes NL, Muller C, Sajo MG. 1979. Um novo e peculiar tipo de sistema subterrâneo em espécies de *Vernonia* da Serra do Cipó (Minas Gerais, Brasil). *Boletim de Botânica, Universidade de São Paulo* **7**: 33–38.

Pollock CJ, Jones T. 1979. Seasonal pattern of fructan and metabolism in forage grasses. *New Phytologist* **83**: 8–15.

Sá e Carvalho CG, Dietrich SMC. 1996. Carbohydrates in tuberous roots of *Cochlospermum regium* in different stages of development. *Revista brasileira de Botânica* **19**: 127–131.

Shiomi N. 1993. Structure of fructopolysaccharide (asparagosin) from roots of asparagus (*Asparagus officinalis* L.). *New Phytologist* **123**: 263–270.

Sobulo RA. 1972. Studies on white yam (*Dioscorea rotundata*) I. Growth analysis. *Experimental Agriculture* **8**: 99–106.

Somogyi M. 1945. A new reagent for the determination of sugars. *Journal of Biological Chemistry* **160**: 61–63.

Williams RF. 1946. The physiology of plant growth with special reference to the concept of net assimilation rate. *Annals of Botany* **10**: 41–72.

Wise CS, Dimler RJ, Davis HA, Rist CE. 1955. Determination of easily hydrolyzable fructose units in dextran preparation. *Analytical Chemistry* **27**: 33–36.

Author Index

Subject Index